# Plasmodesmata

# Annual Plant Reviews

A series for researchers and postgraduates in the plant sciences. Each volume in this series focuses on a theme of topical importance, and emphasis is placed on rapid publication.



*Titles in the series:*

1. **Arabidopsis** Edited by M. Anderson and J.A. Roberts
2. **Biochemistry of Plant Secondary Metabolism** Edited by M. Wink
3. **Functions of Plant Secondary Metabolites and their Exploitation in Biotechnology** Edited by M. Wink
4. **Molecular Plant Pathology** Edited by M. Dickinson and J. Beynon
5. **Vacuolar Compartments** Edited by D.G. Robinson and J.C. Rogers
6. **Plant Reproduction** Edited by S.D. O'Neill and J.A. Roberts
7. **Protein–Protein Interactions in Plant Biology** Edited by M.T. McManus, W.A. Laing and A.C. Allan
8. **The Plant Cell Wall** Edited by J.K.C. Rose
9. **The Golgi Apparatus and the Plant Secretory Pathway** Edited by D.G. Robinson
10. **The Plant Cytoskeleton in Cell Differentiation and Development** Edited by P.J. Hussey
11. **Plant–Pathogen Interactions** Edited by N.J. Talbot
12. **Polarity in Plants** Edited by K. Lindsey
13. **Plastids** Edited by S.G. Møller
14. **Plant Pigments and their Manipulation** Edited by K.M. Davies
15. **Membrane Transport in Plants** Edited by M.R. Blatt
16. **Intercellular Communication in Plants** Edited by A.J. Fleming
17. **Plant Architecture and its Manipulation** Edited by C. Turnbull
18. **Plasmodesmata** Edited by K.J. Oparka

# Plasmodesmata

Edited by

KARL J. OPARKA, FRSE
Scottish Crop Research Institute
Dundee
UK

Blackwell
Publishing

Editorial Offices:
Blackwell Publishing Ltd, 9600 Garsington Road, Oxford OX4 2DQ, UK
Tel: +44 (0)1865 776868
Blackwell Publishing Professional, 2121 State Avenue, Ames, Iowa 50014-8300, USA
Tel: +1 515 292 0140
Blackwell Publishing Asia, 550 Swanston Street, Carlton, Victoria 3053, Australia
Tel: +61 (0)3 8359 1011



First published 2005 by Blackwell Publishing Ltd

Library of Congress Cataloging-in-Publication Data
Plasmodesmata / edited by Karl J. Oparka.
p. cm.
Includes bibliographical references and index.
ISBN 1-4051-2554-3 (hardback: alk. paper)
1. Plasmodesmata. I. Oparka, K.J. II. Title.
QK725.P594 2005
571.6′2–dc22

ISBN-10 1-4051-2554-3
ISBN-13 978-14051-2554-3

A catalogue record for this title is available from the British Library

Set in 10/12 pt Times
by Newgen Imaging Systems (P) Ltd, Chennai, India
Printed and bound in India
by Replika Press Pvt. Ltd

The publisher's policy is to use permanent paper from mills that operate a sustainable forestry policy, and which has been manufactured from pulp processed using acid-free and elementary chlorine-free practices. Furthermore, the publisher ensures that the text paper and cover board used have met acceptable environmental accreditation standards.

For further information on Blackwell Publishing, visit our website:
www.blackwellpublishing.com

# Contents


**Contributors** xiii

**Preface** xv

**1 Plasmodesmal structure and development** 1
ALISON G. ROBERTS
1.1 Introduction 1
1.2 Structure of plasmodesmata 1
1.2.1 Formation of plasmodesmata 1
1.2.2 General structure 2
1.2.3 Historical notes on plasmodesmatal research 5
1.2.4 The advent of electron microscopy 6
1.2.5 Intercellular transport 8
1.3 Additional components of plasmodesmata 10
1.3.1 The cytoskeleton and cytoskeletal-associated proteins 10
1.3.2 Callose 14
1.3.3 Additional components of plasmodesmata 16
1.4 Developmental changes to plasmodesmata 18
1.4.1 Branched plasmodesmata 18
1.4.2 Loss, reduction or occlusion of plasmodesmata 20
1.4.3 Formation of secondary plasmodesmata 21
1.4.4 The future 23
Acknowledgements 23
References 23

**2 Evolution of plasmodesmata** 33
JOHN A. RAVEN
2.1 Introduction 33
2.2 The distribution of plasmodesmata among extant photosynthetic organisms 33
2.3 The phylogeny of photosynthetic organisms and its relation to the occurrence of plasmodesmata 37
2.4 Functional aspects of the distribution and evolution of plasmodesmata 39
2.4.1 Background 39
2.4.2 Cyanobacteria 40
2.4.3 Chlorophyta 41
2.4.4 Heterokontophyta 42
2.4.5 Conclusions 43

2.5 Functioning of complex photosynthetic organisms which lack plasmodesmata 43
2.5.1 Introduction 43
2.5.2 Multicellular algae lacking plasmodesmata or analogues of plasmodesmata 43
2.5.3 Rhodophyta and pit plugs: (trans)mission impossible? 44
2.5.4 Morphologically complex acellular macroalgae 46
2.5.5 Symbioses of (mainly) unicellular algae with fungi and metazoan 48
2.6 Conclusions 49
Note 50
References 50

**3 Plasmodesmata: protein transport signals and receptors 53**
FRIEDRICH KRAGLER
3.1 Introduction 53
3.2 Components of the PD transport pathway 53
3.2.1 Principles of signal-mediated protein transport 53
3.2.2 Putative PD pathway components 56
3.2.3 NCAPP1 – the initial PD receptor 58
3.2.4 Potential role of the cytoskeleton in the PD transport pathway 59
3.2.5 HSP70-related proteins and the PD transport pathway 60
3.2.6 Potential role of ISE1 – a mutant plant with impaired PD function 61
3.2.7 A PD pathway model 62
3.3 Identifying PD transport signal(s) 63
3.3.1 Definition of a targeting signal 63
3.3.2 KNOTTED1 – existence of a PD-targeting and SEL increase signal 63
3.3.3 Phloem Thioredoxin h – charged amino acids as PD motifs 65
3.3.4 Proteolytic processing facilitates transport of CmPP36 65
3.3.5 A short PD-targeting motif in phloem HSC70 65
3.3.6 The elusive nature of PD-targeting signals 66
3.4 Conclusions and future prospects 67
Acknowledgements 67
References 68

**4 Comparative structures of specialised monocotyledonous leaf blade plasmodesmata 73**
C.E.J. BOTHA, R.H.M. CROSS and L. LIU
4.1 Introduction 73
4.2 Maturity-related changes in plasmodesmatal structure 74
4.3 The plasmodesmatal cell wall interfaces in monocot leaves 75

4.4 Plasmodesmata crossing the suberin lamella – constrictions and asymmetry 77
4.5 Regulation at the neck – structural considerations 85
4.6 Changes in wall structure and plasmodesmatal form – secondary modification 86
References 87

**5 Plasmodesmata and plant morphogenesis 90**
KEN KOBAYASHI, INSOON KIM, EUNA CHO and PATRICIA ZAMBRYSKI
5.1 Introduction 90
5.2 Role of PD and symplasmic domain boundaries during plant development 90
5.3 Generic changes in structure and function of PD during different developmental processes 92
5.3.1 PD type and frequency 93
5.3.2 PD are dynamic 94
5.3.3 PD are passageways for nontargeted and targeted transport 95
5.4 Changes in PD conductivity during vegetative development 97
5.5 Changes in PD conductivity during floral induction 98
5.6 Changes in PD conductivity during embryogenesis 100
5.7 Importance of intercellular communication: roles of transported macromolecules in plant development 105
5.8 Concluding remarks 107
Acknowledgements 109
References 109

**6 Transcription factor movement through plasmodesmata 113**
DAVID JACKSON
6.1 Introduction 113
6.2 Plant TFs 114
6.3 Evidence for cell-to-cell trafficking of TFs 114
6.4 Plant TFs that traffic cell to cell 115
6.4.1 Homeobox proteins – KNOTTED1 and homologs 115
6.4.2 MADS box proteins 118
6.4.3 FLORICAULA and LEAFY 120
6.4.4 Shortroot 122
6.4.5 Caprice 125
6.4.6 Other candidate cell-to-cell trafficking TFs 126
6.5 Long-distance signaling by TFs, TF mRNAs and miRNAs 127
6.6 Mechanisms of TF movement 128
6.7 Perspectives 130

Acknowledgements 130
References 130

**7 Role of plasmodesmata in solute loading and unloading 135**
ALEXANDER SCHULZ
7.1 Introduction 135
7.2 Role of plasmodesmata in collection phloem 136
7.2.1 Pre-phloem transport 137
7.2.2 The two modes of phloem loading 138
7.2.2.1 Symplasmic loading 140
7.2.2.2 Apoplasmic loading 141
7.2.2.3 The evolution of minor vein phloem loading 141
7.2.2.4 The default mode of phloem loading 143
7.2.2.5 Symmetrical and asymmetrical branching of plasmodesmata in the loading pathway 144
7.3 Role of PPUs between SE and CCs 145
7.4 Role of plasmodesmata in the transport phloem 147
7.5 Role of plasmodesmata in release phloem 148
7.5.1 General considerations 149
7.5.2 Unloading in permanent sinks 150
7.5.3 Unloading in temporary sinks 152
7.5.4 Unloading in storage sinks 153
7.6 Concluding remarks 155
References 156

**8 Plasmodesmata and the phloem: conduits for local and long-distance signaling 162**
ROBERT L. GILBERTSON, MARIA R. ROJAS and WILLIAM J. LUCAS
8.1 Introduction 162
8.2 Phloem is well suited for long-distance trafficking of macromolecules 162
8.2.1 Sieve elements are the building blocks of a network of dynamic living tubes for macromolecular trafficking 162
8.2.2 Specialized companion cells maintain living enucleate SEs 164
8.2.3 SEs and CCs are interconnected by specialized PD 164
8.3 Macromolecular trafficking in the phloem 165
8.3.1 Source to sink: there is a strong directional movement in the phloem translocation stream 165
8.3.2 Do all long-distance signals in the phloem move in a source-to-sink direction? 166

8.3.3 P-proteins and the wound response 166
8.3.4 A large and diverse population of proteins can be detected in SEs 166
8.3.5 Synthesis and trafficking of macromolecules across the SE–CC PD 167
8.3.6 Green fluorescent protein as a reporter for both trafficking in the phloem and SE–CC SEL 167
8.3.7 Phloem proteins can increase the SEL of mesophyll PD 169
8.3.8 What is the function of the nonstructural phloem proteins? 170
8.3.9 Phloem proteins as molecular chaperones 170
8.4 Nucleic acids in the phloem and their role in long-distance signaling 171
8.4.1 Why do most plant viruses have RNA genomes? 171
8.4.2 Going with the flow: phloem-mediated delivery of viral reporters 171
8.4.3 Endogenous mRNAs in the phloem: possible long-distance developmental signals 172
8.4.4 Phloem translocation of small RNAs 173
8.4.5 Are small RNAs the long-distance signal involved in PTGS? 174
8.4.6 Do miRNAs serve as long-distance developmental signals? 175
8.5 Developmental regulation influences the number and properties of PD: impacts on long-distance trafficking in the phloem 175
8.6 Signaling and systemic defense responses 176
8.6.1 Signaling in the wound response: the challenges of identifying the long-distance signal 177
8.6.2 The systemin receptor: unexpected homolog of the brassinosteroid hormone receptor 177
8.6.3 Is JA the long-distance signal involved in the systemic wound response? 179
8.6.4 A JA biosynthesis pathway in the phloem is poised for a rapid wounding response 179
8.6.5 Prosystemin in phloem parenchyma cells: compartmentalization of different signals? 180
8.6.6 Systemic acquired resistance: similar challenges in the identification of the long-distance signal 180
8.6.7 A need to better understand trafficking of lipid-based signals 181
8.6.8 Cross-talk in systemic signaling systems 181
8.7 Prospectives 182
References 183

**9 Movement of viruses to and through plasmodesmata** **188**
RICHARD S. NELSON
9.1 Introduction 188
9.2 Virus cell-to-cell movement 189
9.2.1 Potexviruses and Hordeiviruses 190
9.2.2 Tobamoviruses 193
9.2.3 Comoviruses, Nepoviruses and Tospoviruses 201
9.2.4 Other viruses 204
9.2.5 Summary 205
Note 206
Acknowledgements 206
References 206

**10 Systemic RNA silencing** **212**
MANFRED HEINLEIN
10.1 Introduction 212
10.2 The RNA silencing pathway 213
10.3 Virus-induced RNA silencing 218
10.4 Systemic silencing 219
10.5 Nature of the systemic silencing signal 223
10.6 Pathway of the silencing signal in plants – plasmodesmata 226
10.7 Concluding remarks 228
Acknowledgements 230
References 230

**11 Techniques for imaging intercellular transport** **241**
KARL OPARKA and PETRA BOEVINK
11.1 Introduction 241
11.2 Microinjection 241
11.2.1 Pressure injection 242
11.2.2 Iontophoretic injection 243
11.2.3 Iontophoretic versus pressure injection 243
11.2.4 Some milestones in microinjection 243
11.2.5 Dye loading 245
11.2.6 Caged probes 245
11.3 Fluorescent reporter proteins 246
11.3.1 GFP 246
11.4 Red fluorescent proteins 247
11.4.1 DsRed 247
11.4.2 mRFP 247
11.5 Biolistic bombardment 248
11.6 Viral vectors 250
11.7 Tissue or cell-specific promoters 250
11.8 Photoactivatable GFP 251
11.8.1 Other photoactivatable proteins 252

11.9 Fluorescence recovery after photobleaching 253
11.10 FlAsH and ReAsH 254
11.11 Ablation techniques 257
11.11.1 Laser ablation 257
11.11.2 Diphtheria toxin 257
11.12 Quantum dots 258
11.13 Summary 258
References 259

**12 Electrical signalling via plasmodesmata 263**
AART J.E. VAN BEL and KATRIN EHLERS
12.1 Introduction 263
12.2 Electrical coupling as an initial probe for symplasmic cell coupling 263
12.3 Physical essentials of electrical coupling measurements 264
12.4 Membrane potential mapping as a tool for recognition of symplasmic domain borders 265
12.5 Interrelations between current injection and plasmodesmatal permeability 266
12.6 Potential functions of electrical cell–cell (un)coupling 267
12.7 Relationships between electropotential waves, plasmodesmatal conductance and physiological activities 269
12.8 Role of plasmodesmata in the propagation of electropotential waves and long-distance signalling 270
12.9 Are integrated electrical and chemical long-distance signals mediated by plasmodesmata during systemic induced resistance? 272
12.10 Concluding remarks 274
References 274

**13 Regulation of plasmodesmal conductance 279**
TERENA L. HOLDAWAY-CLARKE
13.1 Introduction 279
13.2 Monitoring plasmodesmal conductance 279
13.2.1 Terminology 279
13.2.2 Monitoring plasmodesmata as diffusion conduits 280
13.2.3 Transport through the desmotubule 280
13.2.4 Dye and GFP-fusion studies 281
13.2.5 Inferences from electron microscopy 281
13.2.6 Measuring plasmodesmal conductance/electrical resistance 282
13.2.7 Time resolution 283
13.2.8 Difficulties and some solutions 283

13.2.9 Coupling ratio 283
13.2.10 Patch-clamping? 284
13.3 Regulation of plasmodesmal conductance 285
13.3.1 Algal plasmodesmata 285
13.3.2 Plasmodesmata in higher plants 286
13.3.3 Factors decreasing conductance 287
13.3.3.1 Divalent cations/$[Ca^{2+}]_{cyt}$ 287
13.3.3.2 Plasmolysis and pressure differentials 289
13.3.4 Factors increasing plasmodesmal conductance 289
13.3.4.1 Deplasmolysis 289
13.3.4.2 Metabolic inhibition 289
13.3.5 Other possibilities 290
13.3.5.1 Cytoplasmic pH 290
13.3.5.2 Voltage gating 290
13.4 Mechanisms 290
13.4.1 Myosin 291
13.4.2 Centrin and $Ca^{2+}/pH_{cyt}$ regulation of plasmodesmata 291
13.4.3 How do plasmodesmata dilate? 292
13.5 Conclusion 292
References 293
Appendix 298
13.6 Solving the electric circuit model of a sister-cell pair 298
13.6.1 Model for cytoplasmic location of current-injecting electrode in a sister-cell pair 298
13.6.1.1 Model for vacuolar location of electrode tip in a sister-cell pair 299
13.6.2 Solution of the circuit model for a single-ended file 300
**Index** **303**

# Contributors

**Dr Petra Boevink** Unit of Cell–Cell Communication, Scottish Crop Research Institute, Invergowrie, Dundee DD2 5DA, UK

**Professor C.E.J. Botha** Botany Department, Rhodes University, PO Box 94, Grahamstown 6140, South Africa

**Dr Euna Cho** Department of Plant and Microbial Biology, 111 Koshland Hall, University of California, Berkeley, CA 94720-102, USA

**Dr R.H.M. Cross** Electron Microscopy Unit, Rhodes University, PO Box 94, Grahamstown 6140, South Africa

**Dr Katrin Ehlers** Institute of General Botany, Justus-Liebig University, Senckenbergstrasse 17, D-35390 Giessen, Germany

**Professor Robert L. Gilbertson** Department of Plant Pathology, University of California-Davis, 1 Shields Ave, Davis, CA 95616, USA

**Professor Manfred Heinlein** Institute of Botany, University of Basel, Hebelstrasse 1, 4056 Basel, Switzerland

**Dr Terena Holdaway-Clarke** 32 Carcoola Crescent, Normanhurst, New South Wales 2076, Australia

**Professor David Jackson** Cold Spring Harbor Laboratory, Cold Spring Harbor, NY 11724, USA

**Dr Insoon Kim** Department of Plant and Microbial Biology, 111 Koshland Hall, University of California, Berkeley, CA 94720-3102, USA

**Dr Ken Kobayashi** Department of Plant and Microbial Biology, 111 Koshland Hall, University of California, Berkeley, CA 94720-3102, USA

**Dr Friedrich Kragler** Vienna Biocenter, Dr Bohr Gasse 9/5, A-1030 Vienna, Austria

**Dr William J. Lucas** Section of Plant Biology, Division of Biological Sciences, University of California-Davis, 1 Shields Ave, Davis, CA 95616, USA

**Dr L. Liu** Linyi University, Linyi 276005, People's Republic of China

**Dr Richard S. Nelson** Samuel Roberts Noble Foundation, 2510 Sam Noble Pky, Ardmore OK 73401, USA

**Professor Karl Oparka** Unit of Cell–Cell Communication, Scottish Crop Research Institute, Invergowrie, Dundee DD2 5DA, UK

**Professor John A. Raven** Division of Environmental and Applied Biology, University of Dundee, Dundee DD1 5EH, UK

**Dr Alison G. Roberts** Unit of Cell–Cell Communication, Scottish Crop Research Institute, Invergowrie, Dundee DD2 5DA, UK

**Dr Maria R. Rojas** Department of Plant Pathology, University of California-Davis, 1 Shields Ave, Davis, CA 95616, USA

**Professor Alexander Schulz** Department of Plant Biology, The Royal Veterinary and Agricultural University, Frederiksberg C, Copenhagen, Denmark

**Professor Aart J.E. van Bel** Institute of General Botany, Justus-Liebig University, Senckenbergstrasse 17, D-35390 Giessen, Germany
**Professor Patricia Zambryski** Department of Plant and Microbial Biology, 111 Koshland Hall, University of California, Berkeley, CA 94720-3102, USA

# Preface

Since their discovery over 100 years ago, plasmodesmata have been the focus of intense investigation. Plasmodesmata are unique to plants and form an intercellular continuum for the transport of solutes, signals and ribonucleoprotein complexes. It is now clear that plasmodesmata formation and regulation are central to a diverse range of plant functions that include developmental programming, host–pathogen interactions and systemic RNA signaling.

This book provides a state-of-the-art overview of the diverse forms and functions of plasmodesmata. It covers the structure and evolution of plasmodesmata, their role in plant development and solute transport, and their central function in systemic signaling via the phloem. It includes critical evaluations of current methods used to study intercellular transport via plasmodesmata. The material is suitable both for researchers in the field of plant cell biology and for postgraduates and advanced graduates in the plant sciences generally. It is comprehensively referenced throughout.

I am grateful to my colleagues who have contributed to this book for providing thought-provoking treatments of their subject areas. It is some years since a book was devoted to the subject of plasmodesmata, and I hope that this volume of the *Annual Plant Reviews* series will serve as an up-to-date source of reference on this fascinating research area.

Finally, I am indebted to Morven Pearson and Philip Smith for their help and patience with the editing of this volume.

Karl Oparka

# 1 Plasmodesmal structure and development

Alison G. Roberts

## 1.1 Introduction

This chapter provides an overview of the structure and development of plasmodesmata that will act as an introduction to the other chapters of this book. Although we now have a reasonable understanding of the major structural components of plasmodesmata, relatively little is known about the modes of regulation of the components, or of the detailed protein constituents that make up, and likely control, plasmodesmatal functions. There have been many excellent reviews published recently in this area, in addition to the original research papers, and so throughout this chapter, the reader will be directed to the primary literature for additional information.

Plasmodesmata are sub-cellular structures that are unique to plants. This is because plant cells also contain, or rather are contained by, the cell wall. The cell wall acts as a barrier between the cell contents and the external environment, providing protection from environmental influences and pathogens, and it also provides rigidity and mechanical strength to the plant body. However, this wall also acts as a barrier between cells. Plasmodesmata have evolved to allow the movement of information molecules between cells thereby enabling communication between them, and permitting coordinated growth, development and response to stimuli (Ding, 1998; Hashimoto and Inze, 2003). Plasmodesmata are basically pores in the plant cell wall that connect the cytoplasm and membrane systems of one cell with its neighbour. For many years, plasmodesmata were thought to be little more than static pores – membrane-lined channels that were functionally comparable to animal gap junctions. However, recent advances have shown them to be highly fluid, dynamic structures that not only influence the flux of molecules and metabolites between cells, but also play a key role in controlling intercellular trafficking of proteins and RNA, and in so doing can act to control plant development (Lucas *et al.*, 1993; Blackman and Overall, 2001; Ehlers and Kollmann, 2001; Roberts and Oparka, 2003).

## 1.2 Structure of plasmodesmata

### *1.2.1 Formation of plasmodesmata*

All tissues in a growing plant are derived from meristematic tissue. Within the apical meristem, cell initials undergo division and differentiation to form leaves, stems and lateral meristems. A similar process occurs in the root apical meristem to give rise to the root system. As each cell divides, newly synthesised cell-wall material is laid down between daughter cells, and strands of cortical endoplasmic reticulum (ER) become trapped across the developing wall. At these points, cell plate vesicles are unable to fuse and wall material cannot be deposited, leading to the entrapment of ER between blocks of cell-wall material (Porter and Machado, 1960; Hepler, 1982; Staehelin and Hepler, 1996). The plasma membrane surrounds the new wall material, forming a sleeve that also surrounds the trapped ER. The ER becomes appressed, presumably due to the forces exerted by the developing phragmoplast, and forms a condensed, membranous tube, called the desmotubule, that runs through the centre of the plasma-membrane-lined pore (Jones, 1976). The plasmodesmata that form during the division of a cell are called primary plasmodesmata (Jones, 1976). Those that form *de novo*, in existing cell walls, are called secondary plasmodesmata (Lucas *et al*., 1993).

The number of primary plasmodesmata that form in a cell wall can be used to predict the extent of cell expansion (Overall, 1999), but how this is regulated is not understood. However, the number of primary plasmodesmata cannot predict the final number of plasmodesmata since this can be altered considerably during development; plasmodesmatal frequency can range from 0.1 to 10 per $\mu m^2$ of cell wall (Fisher, 2000). An artificial increase in plasmodesmatal density has been achieved by inhibiting cellulose biosynthesis (Vaughn *et al*., 1996). This resulted in an increase in the amount of callose deposition, which may, in turn, have an effect on the positioning or frequency of plasmodesmata.

### *1.2.2 General structure*

Plasmodesmata have not been isolated or biochemically analysed and so structural models have been based largely on data from transmission electron micrographs (Robards, 1971; Robards and Lucas, 1990; Beebe and Turgeon, 1991; Ding *et al*., 1992b; Botha *et al*., 1993; Overall and Blackman, 1996; Waigmann *et al*., 1997). Plasmodesmata are found in many forms in different tissues and at different stages of development. Based on ultrastructure, they are generally described as being either simple or branched. Some additional modifications are found in some tissues, primarily in the phloem, and these will be discussed in the following chapters. However, for the general description of plasmodesmal structure, simple plasmodesmata will be described, and only angiosperms will be considered. The detailed structure of monocotyledonous plasmodesmata will be discussed fully in Chapter 4. A diagrammatic representation of the structure of a simple plasmodesma that includes all sub-structural components that have been confirmed to date is shown in Figure 1.1.

A simple plasmodesma consists of a linear, cylindrical pore that traverses the cell wall. The pore is lined by the plasma membrane which is continuous from one cell to the next. The ER is also continuous between cells, but is modified to form an appressed

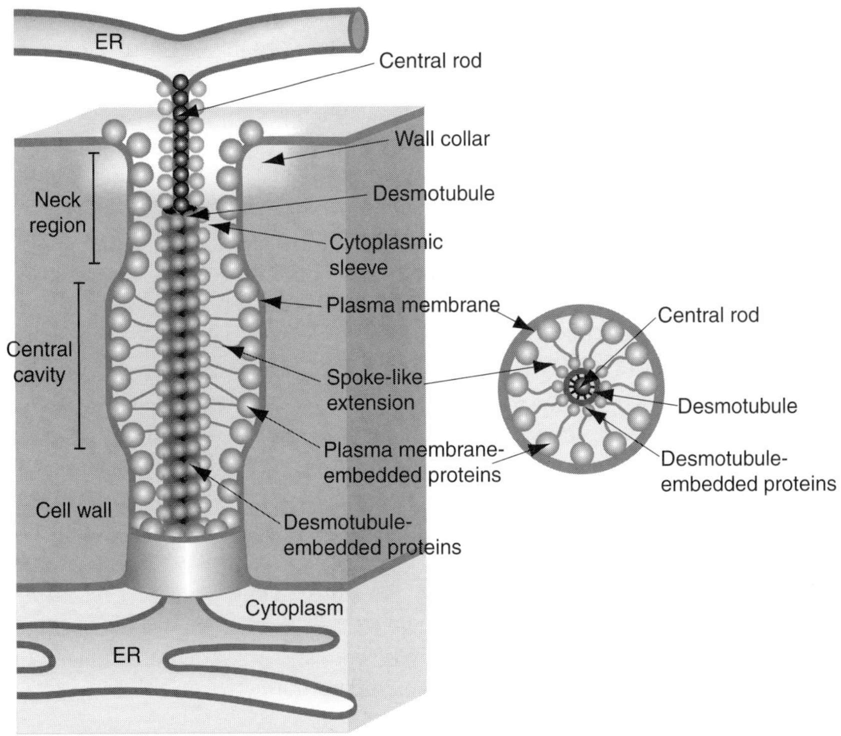


**Figure 1.1** Diagrammatic representation of the structure of a simple plasmodesma including all substructural components that have been confirmed to date. A longitudinal section through the pore is shown to the left, while a transverse section through the central cavity region is shown to the right. Based primarily on the model in Ding *et al.* (1992b) and modified from Roberts and Oparka (2003).

tubule known as the desmotubule, and frequently appears in micrographs as a solid central rod. In this way, the plasma membrane and ER both provide membrane continuums throughout the plant (Robards and Lucas, 1990; Denecke, 2001). In the model proposed by Overall and Blackman (1996), a cable of actin runs through the centre of the pore, closely associated with the desmotubule. Filling the space between the ER and the plasma membrane is cytoplasm. This region is known as the cytoplasmic annulus or sleeve (Esau and Thorsch, 1985), and it is through this fluid continuum that the bulk of molecules can be exchanged between cells (Lucas and Wolf, 1993; Epel, 1994; Kragler *et al.*, 1998). It should be noted, however, that cytoplasmic streaming apparently has no direct effect on the movement of molecules and does not provide the driving force for intercellular transport (Tucker, 1987). Plasmodesmata themselves appear to ultimately control the rate of cytoplasmic exchange.

At both ends of the plasmodesma, just inside the cell wall, lies the neck region. At this point, the plasmodesmal pore is frequently occluded, especially in simple plasmodesmata, bringing the plasma membrane into close proximity with the desmotubule. The cell wall follows the contour of the plasma membrane, producing a collar effect in the neck region, and callose deposits are commonly, but not always, found at the neck of plasmodesmata. Whether this state (with a restricted region at both ends of the pore) is natural or an artefact remains to be proven since application of an inhibitor of callose synthesis cause plasmodesmata in *Allium cepa* to become funnel-shaped rather than constricted at the neck (Radford *et al.*, 1998). Although callose may be used by the plant to control the plasmodesmal aperture, it is very commonly deposited as a result of wounding or chemical fixation and so its presence at the neck may be artefactual as a result of tissue preparation. Electron micrographs have also shown an electron-dense ring of proteinaceous material surrounding the neck of the plasmodesmata that has been described as a wall collar or sphincter. This is consistently visible if tannic acid is applied during fixation (Fisher, 2000), and may function to alter the size exclusion limit (SEL; Olesen, 1979; Olesen and Robards, 1990; Beebe and Turgeon, 1991; Badelt *et al.*, 1994; Turner *et al.*, 1994).

In addition to the wall collar constrictions, the cytoplasmic sleeve is also partially occluded with globular subunits in the neck region. The subunits are thought to be protein complexes and surround the desmotubule (Ding *et al.*, 1992a; Overall and Blackman, 1996). They appear to exist in a helical arrangement, at a pitch of 20–30° relative to the desmotubule axis, and are thought to functionally divide the cytoplasmic sleeve into a number of spiralling micro channels through which cytosolic exchange takes place (Zee, 1969; Robards, 1976; Olesen, 1979; Overall *et al.*, 1982; Wolf *et al.*, 1989; Olesen and Robards, 1990; Robards and Lucas, 1990; Ding *et al.*, 1991; Robinson-Beers and Evert, 1991; Lucas and Wolf, 1993; Lucas *et al.*, 1993; Overall and Blackman, 1996; Waigmann *et al.*, 1997). These proteins are sometimes seen to be in close proximity to the plasma membrane, and at other times the cytoplasmic sleeve is larger and they appear to be connected to the plasma membrane by projecting spokes. The spokes may be composed of myosin and linked to the actin cable at the centre of the pore, and could form an important regulatory structure that can alter the SEL of plasmodesmata (see Chapter 13). It is unknown whether these spokes are present throughout the entire length of a plasmodesma. At the very centre of the plasmodesma, when viewed in transverse section, the appressed lipid membrane of the desmotubule also appears to have a series of proteins embedded within it, and this is referred to as the central rod (Ding *et al.*, 1992b). Filamentous structures that link the proteins of the central rod to the wall of the desmotubule have also been seen (Ding *et al.*, 1992b). While the structure of most plasmodesmata conforms to this model, a range of substructural variations have been found within tissues of the same plant and between plant species (Robinson-Beers and Evert, 1991; Waigmann *et al.*, 1997). In addition to these major structural components of plasmodesmata, many other molecules have been found to be associated with them and these will be discussed in later sections of this chapter, and in greater detail in later chapters of this book.

### *1.2.3 Historical notes on plasmodesmatal research*

In 1879, Tangl described 'open communications' between protoplasts of endosperm cells (see Figure 1.2). At that point, many people claimed to have seen them earlier, but only Tangl realised the implications of what we now call symplasmic continuity. In the subsequent three decades, there was a huge output of papers describing cellular communications in a range of plants and tissues, postulating the origin, development and function of these connections (Carr, 1976). Strasburger first called the connections 'plasmodesmata' in 1901. The reason for this flurry of activity was that many researchers were attempting to expand on the cell theory, as proposed by Schleiden and Schwann in 1804 (see Baker, 1948), which stated that each cell acted autonomously. This did not explain results obtained in the latter half of the nineteenth century regarding growth, tropisms and solute transport throughout the whole plant, and hence mechanisms of possible intercellular transport were an important concept at the time. In 1873, Heitzmann suggested that during cell division, protoplasmic strands (the term used at that time for plasmodesmata) were laid down and allowed subsequent protoplast continuity (Carr, 1976). The plasma membrane was postulated as early as 1855 (Nägeli and Cramer, 1855), and was proposed to line plasmodesmata by Strasburger (1901). At this time, it was also acknowledged that plasmodesmata would assert some control, or 'filtration' to the movement of substances between cells (see Carr, 1976).

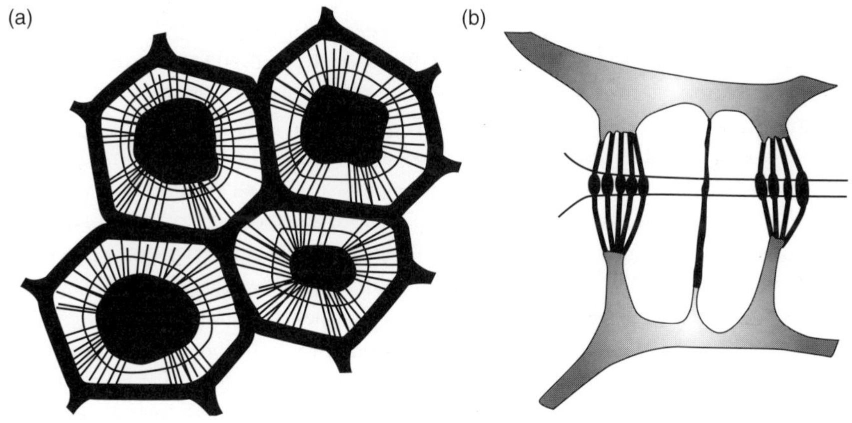


**Figure 1.2** (a) The first documented example of plasmodesmata. Reproduction of one of Tangl's drawings from 1879 showing 'Offene Communicationen', or open communications, in endosperm tissue of *Strychnos nux-vomica*. Thick section showing 'protoplasmic processes in the connecting canals ... stained with iodine-potassium iodide'. Reproduced with permission from Intercellular Communication in Plants: Studies on Plasmodesmata (1976). Edited by B.E.S. Gunning and A.W. Robards, Springer-Verlag, Berlin. (b) A representation of plasmodesmal structure from the early twentieth century. Protoplasmic connections between two cells of the mesophyll of *Viscum album* (after Keinitz-Gerloff). Reproduced with permission from Physiological Plant Anatomy (Translated from the 4th edn; 1914) by Haberlandt, MacMillan Press, London.

Given this early foresight, one might assume that plasmodesmatal research has continued apace since the 1920s. By 1920, not only were many of the major hypotheses in place, but it was known that plasmodesmata were not distributed evenly across all walls, that they were absent from some walls and that secondary plasmodesmata must arise following cytokinesis. Also, some early counts of plasmodesmatal frequency had been attempted. However, by 1920, the scope for additional information was limited by the available methodologies; further clarification of plasmodesmal structure and distribution was not possible using light microscopy. For the next three decades, research in the area was largely abandoned and many of the earlier advances were forgotten (Carr, 1976). Key publications in related fields, such as Münch's symplasmic concept, and the possibility that plasmodesmata could realistically be involved in short-distance transport (Münch, 1930) continued, but plant physiologists were predominantly studying ion uptake at this time. In the 1950s, the invention of the electron microscope opened up new avenues for plasmodesmatal research. As early as 1956, a plasmodesma is shown in a diagram of plant cell ultrastructure, although the plasma membrane is absent (Mercer, 1956), and the existence of the plasma membrane within plasmodesmata remained in question by some authors until about 1965 (Carr, 1976). Over the 125 years since their discovery, our understanding of plasmodesmal structure increased sporadically; in fact, some substructures were discovered several times before they were generally accepted.

### *1.2.4 The advent of electron microscopy*

Electron microscopy kindled a renewed interest in plasmodesmata, and another flurry of research papers followed. The ultrastructural demonstration of cytoplasmic continuity came in 1957 (Buvat, 1957; Strugger, 1957) with the use of osmium-tetroxide-stained cells. Early electron micrographs also showed that in plants the cortical ER was closely associated with plasmodesmata, but for the next 20 years, despite many different hypotheses being proposed, the relationship of ER to plasmodesmata was unresolved. In 1960, the continuity of ER was first proposed by Whaley *et al.* (1960) using potassium permanganate fixation to improve the preservation of membrane structure in their ultrastructural studies, and in the same year, plasmodesmatal formation was firmly linked to cell-plate formation (Porter and Machado, 1960). In 1963, Katherine Esau – a pioneer of plant ultrastructural studies – still avoided the view that the ER necessarily traversed the cell wall: '... *Endoplasmic reticulum is typically connected to the plasmodesmata; in fact, many investigators think that tubules of ER are structural elements of plasmodesmata. ...*', but she was not alone; a number of publications at the time were not prepared to accept direct ER continuity. It was not until 1966 that the ER was observed to be appressed into a tubule, revealing the inner opaque layer of the ER membrane to be as a central rod (López-Sáez *et al.*, 1966).

The term 'desmotubule' was first used in 1968 by Robards (1968b), who suggested that it had a structure similar to a microtubule and was abutted to, but not continuous with, the ER (Figure 1.3a). However, by 1971, further studies had

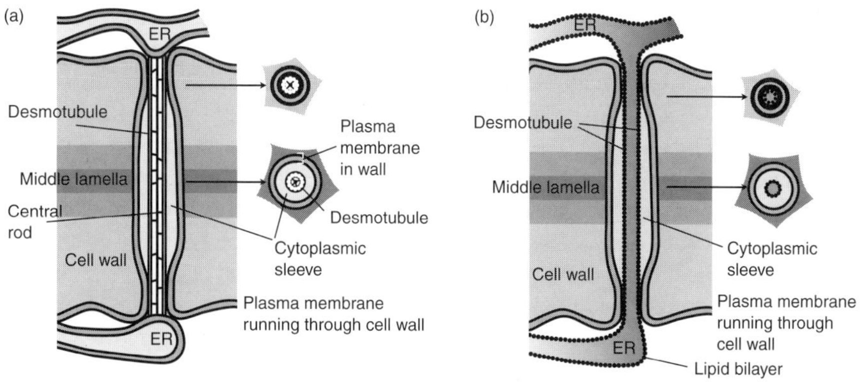


**Figure 1.3** (a) Diagram reproduced from Robards (1968b) showing the structure of a simple plasmodesma as understood at the time. A longitudinal section through the pore is shown to the left, while transverse sections through the neck and central cavity regions are shown to the right. Redrawn with permission from Intercellular Communication in Plants: Studies on Plasmodesmata (1976). Edited by B.E.S. Gunning and A.W. Robards, Springer-Verlag, Berlin. The figure legend reads 'The diagram closely reflects the image as *seen* in the electron microscope (this model was constructed from observations on plasmodesmata in the walls of xylem ray cells of *Salix fragilis*, but it is very similar to the situation in other plants – e.g. barley root plasmodesmata)'. At this time, the desmotubule was thought to be discontinuous with the ER although many of the fine substructures associated with the desmotubule are present. (b) Diagram reproduced from Robards (1971) showing the structure of a simple plasmodesma as understood at the time. A longitudinal section through the pore is shown to the left, while transverse sections through the neck and central cavity regions are shown to the right. Re-drawn with permission from Intercellular Communication in Plants: Studies on Plasmodesmata (1976). Edited by B.E.S. Gunning and A.W. Robards, Springer-Verlag, Berlin. By the early 1970s, it was thought that there was open communication between the desmotubule and the ER, and the ER was depicted as an expanded lipoprotein bilayer (although Robards states this is not an essential feature of the model). Additional labelling as in (a).

discredited that suggestion and proposed a model in which there was continuity between the desmotubule and the ER, and the central rod was thought to be an artefact (Robards, 1971; Figure 1.3b). By 1976, when Gunning and Robards edited their book '*Intercellular Communication in Plants: Studies on Plasmodesmata*', there were still two major unresolved features of simple plasmodesmata: the presence or absence of a central rod through the desmotubule, and the constriction of the neck region (Robards, 1976).

The central rod is not a recent discovery, and historically it was thought to be either a trapped spindle fibre (Tangl, 1879; Russow, 1883; Gardiner, 1900, 1907) or a microtubule. In fact, until relatively recently, the similarity of the desmotubule to a cytoplasmic microtubule led many researchers to assume that it was produced when a nuclear or cell-plate microtubule became trapped and embedded in the developing cell-plate (Allen and Bowen, 1966; Robards, 1968a; Juniper and Barlow, 1969). By the 1980s, most researchers in the field concluded that the ER ran through the pore linking the endomembrane system of adjacent cells, and that the desmotubule was comprised of modified ER; but doubt persisted, and plant anatomy books published in the late 1980s were still stating that there was 'little support for the interpretation' that ER ran through the pore (Mauseth, 1988). ER continuity between cells is now widely accepted, and it has recently been suggested that phragmoplastin, a dynamin-like protein that is active during cell-plate formation (Gu and Verma, 1996), may cause the constriction of the ER tubule to produce the desmotubule (Overall, 1999). Even after it was understood to be composed of ER, different structures of the desmotubule were postulated. Gunning and Overall (1983) showed it to have a cylindrical, membranous structure with an internal lumen, while Tilney *et al.* (1991) suggested that it was a solid proteinaceous rod. Current interpretation is that the electron-lucent region of the desmotubule is continuous with the unstained lipid interior of the ER membrane, while the central rod is continuous with the dark surface of the ER lumen. This model suggests that the desmotubule, while appressed, is essentially lipid with a few embedded proteins at the core of the tightly furled lipid bilayer (Fisher, 2000). The ER lumen must effectively be eliminated in this configuration, and any aqueous pathway removed, but this configuration may not be permanent. Indeed, it may be another regulatable aspect of plasmodesmal structure (see Chapter 13).

We may again have reached an impasse where advances in plasmodesmal structural research are concerned. Despite many comprehensive microscopical studies that have elucidated the general structures of plasmodesmata, the finer details may not be obtainable until future breakthroughs in imaging are made. Other modern methods, such as atomic force microscopy, have attempted to look at cell walls and plasmodesmata, but with little success and with insufficient resolution (Kirby *et al.*, 1996). Some structural changes that occur in plasmodesmata have gross effects that are visible using electron microscopy, such as the conversion of simple to branched plasmodesmata during the sink–source transition (Roberts *et al.*, 2001). However, many more subtle changes, such as the up- or down-regulation of the SEL, must also require structural changes, given that the size of the channel through plasmodesmata must alter in some way. Such structural changes can be inferred from experimental data on the movement of molecules but not from direct physical evidence, since the preparation of tissue for electron microscopy requires relatively brutal protocols that tend to inflict inherent swelling or shrinkage of the tissue in some way. Quantifying preparation artefacts is difficult, but it is perhaps unlikely that we would be able to 'see' structural changes brought about by subtle protein–protein interactions that might have major implications for the transport capabilities of plasmodesmata.

### *1.2.5 Intercellular transport*

Three possible routes of transport through plasmodesmata exist: the cytoplasmic sleeve, the ER and the plasma membrane. The majority of transport through plasmodesmata appears to occur via the cytoplasmic sleeve. Low molecular weight dyes that are either passively loaded (Duckett *et al.*, 1994; Roberts *et al.*, 1997) or microinjected (Goodwin, 1983; Erwee *et al.*, 1985; Madore and Lucas, 1986; Oparka *et al.*, 1991) are known to use this pathway, and larger molecules such as GFP also move, apparently passively, via the cytoplasmic connection. Barclay *et al.* (1982)

showed that dye movement was not disrupted when cytoplasmic streaming was inhibited by treatment with cytochalasin B but plasmodesmal disruption by plasmolysis did cause a much-reduced rate of fluorescein movement. This suggests that the movement of small solutes occurs by diffusion, and that movement through plasmodesmata is the rate-limiting step for transport (Barclay *et al.*, 1982; Tucker *et al.*, 1989).

More recently, the ER has been implicated in transport to and through plasmodesmata (Roberts and Oparka, 2003; Oparka, 2004). This route is conceivable for plasmodesmata in all tissues, but may be particularly important in the phloem, where the cortical ER of the companion cell is linked via the desmotubule to the parietal ER of the sieve element – a cell type that contains a limited cytoplasm (Blackman *et al.*, 1998a; Oparka and Santa Cruz, 2000; van Bel and Knoblauch, 2000). The desmotubule has been found in both an appressed and a dilated state (Overall *et al.*, 1982; Robinson-Beers and Evert, 1991), and trafficking of fluorescent probes between cells via the ER has been reported (Cantrill *et al.*, 1999). Lipids can diffuse through the ER (Grabski *et al.*, 1993) and this has led several authors to predict that the ER is likely to be involved in selective trafficking of molecules through the plasmodesmata by 'hitching a ride' on the endomembrane system (Blackman *et al.*, 1998a; Oparka and Santa Cruz, 2000; van Bel and Knoblauch, 2000; Roberts and Oparka, 2003). Interestingly, it has been shown that lipids are unable to diffuse within the plasma membrane from one cell to another, and so it has been postulated that the components of the plasma membrane that run through plasmodesmata may be different from those in the membrane that lines the rest of the cell (Grabski *et al.*, 1993). These combined results could suggest that the ER flows through the plasmodesmata while the plasma membrane is a more stationary structure. If intrinsic ER proteins naturally move between cells in this way, is it possible that other molecules, both endogenous and foreign, could hitch a ride on the ER flow? Even if the internal aperture of the desmotubule is too small to allow the passage of molecules, trafficking along the outer membrane of the ER, inside the cytoplasmic sleeve, could still be a possibility.

Trafficking via the ER could potentially occur in three different ways: as a passive or directional flow within the ER lumen, attached to the inner membrane of the desmotubule or attached to the outer, cytoplasmic face of the desmotubule. If molecules can move along the ER, and the ER is thought of as a membranous 'river', is it possible for the membrane to flow through the desmotubule of plasmodesmata? The ER is closely associated with actin filaments (Quader *et al.*, 1987; Boevink *et al.*, 1998) and the highly motile cortical ER has been postulated to run along actin cables (Quader *et al.*, 1987), possibly using myosin motors. Myosin or other motor proteins may anchor the endomembrane system to the cytoskeleton. If this is the case, molecules temporarily attached to, or inserted into, the ER membrane could potentially be moved through plasmodesmata in this way. In addition, if myosin motors are required for the movement along actin cables, this could give rise to the directionality of movement sometimes observed through plasmodesmata (Ding *et al.*, 1991; Waigmann and Zambryski, 1995), since all myosins move uni-directionally – towards either the pointed (−) or the barbed (+) end of the actin filament (Wells *et al.*, 1999). Recent work in Bernard Epel's lab has shown that GFP fused with integral ER membrane proteins was able to move through plasmodesmata in *Nicotiana benthamiana* while GFP fused with lumenal proteins did not. This situation was altered by the presence of a stably expressed TMV movement protein (MP) which caused a reduction in the movement of integral proteins but an increase in the movement of lumenal proteins. It has been suggested that this may reflect a dilation of the desmotubule by the MP, allowing lumenal proteins to traffic between cells (Guenoune-Gelbart and Epel, 2004).

The plasma membrane provides a third possibility for movement of molecules through plasmodesmata. Many unique proteins have been discovered in and around plasmodesmata (see Table 1.1), which suggests that a vesicle targeting pathway may exist to deliver them to the pore, and plasmodesmata may be specific sites for vesicle recycling (Oparka, 2004). Protein mediators of both vesicle trafficking and membrane fusion such as Rabs and SNAREs have recently been implicated in trafficking to the plasmodesmal pore (Oparka, 2004) and these proteins are likely to act on the plasma membrane. The N-terminal moiety of Rab11 has been found to localise to plasmodesmata when expressed as a cDNA–GFP fusion (Medina-Escobar *et al.*, 2003) and Rab11 has been shown to be involved in polarised mRNA trafficking in *Drosophila* (Dollar *et al.*, 2002). Additionally, transgenic tomato plants expressing an antisense Rab11 showed abnormal development, similar to that seen when homeobox genes are disrupted (Janssen *et al.*, 1998). Possibly, Rab11 is involved in the trafficking of transcription factors through plasmodesmata. Another protein, called KNOLLE, is a t-SNARE belonging to the syntaxin protein family and has been found to interact with a viral MP that interacts with plasmodesmata (Laporte *et al.*, 2003). t-SNARE proteins are located on specific target membranes and coordinate localised fusion between membranes and Golgi-derived vesicles (Blatt *et al.*, 1999). Vesicles carrying cargo packaged by the Golgi could be directed to specific sites on the plasma membrane near plasmodesmata by Rab proteins, where fusion of the vesicle with the membrane occurs by interaction of specific SNARE proteins. If the integral membrane trafficking machinery of plant cells is used to carry endogenous proteins to sites of plasmodesmata, this system could easily be hijacked by exogenous proteins such as viral MPs to allow their targeting to plasmodesmata. Any protein associated with the vesicle membrane will potentially end up attached to the plasma membrane following vesicle fusion (see model in Oparka, 2004). Whether proteins might then move through the plasmodesmal pore on the plasma membrane is currently unknown, but the potential for such a route certainly exists.

## 1.3 Additional components of plasmodesmata

### *1.3.1 The cytoskeleton and cytoskeletal-associated proteins*

Structurally, cytoskeletal proteins are a major component of plasmodesmata. However, only those aspects that relate to plasmodesmal structure, or changes in

**Table 1.1** Cellular components and proteins that have been localised to plasmodesmata

| Component | Function/Comment | References |
|---|---|---|
| Actin | Cytoskeletal component of plasmodesmata. Possibly used to traffic molecules through the pore, and/or to control the SEL | Tucker, 1987; White *et al.*, 1994; Zambryski and Crawford, 2000; Blackman and Overall, 2001 |
| Actin regulating proteins (Arp3 homologues) | Suggestion that plasmodesmata may act as actin-filament organising centres, or that Arp proteins are involved in plasmodesmal regulation | van Gestel *et al.*, 2003 |
| Acid phosphatase | Function unknown, but possibly involved in callose synthesis | Esau and Charvat, 1975 |
| ATPase | Provision of energy at the plasmodesmal pore. Specific function unknown | Cronshaw, 1980; Belitser *et al.*, 1982; Chen *et al.*, 1994 |
| Callose | Used to seal plasmodesmata during wounding or pathogen attack. May also control the SEL | Allison and Shalla, 1974; Drake *et al.*, 1978; Peterson and Rauser, 1979; Hughes and Gunning, 1980; Kauss, 1987; Robards and Lucas, 1990; Wolf *et al.*, 1991; Stone and Clarke,1992; Brown *et al.*, 1997; Jones *et al.*,1998; Radford *et al.*, 1998; Rinne and van der Schoot, 1998; Botha *et al.*, 2000; Iglesias and Meins, 2000; Sivaguru *et al.*, 2000; Bucher *et al.*, 2001; Rinne *et al.*, 2001 |
| Calreticulin | A calcium-sequestering protein that may regulate $Ca^{2+}$ levels and help control the SEL | Baluška *et al.*, 1999, 2001 |
| Centrin | A calcium-binding cytoskeletal protein that may control the SEL | Blackman *et al.*, 1999 |
| Connexin-like proteins | A number of proteins have either been found in cell-wall extracts or inferred by cross-reaction with connexin antibodies. Most convincingly shown by Yahalom *et al.* (1991), but other studies have failed to provide clear localisation data, and one putative connexin later proved to be a protein-kinase-like protein | Meiners and Schindler, 1989; Meiners *et al.*, 1991; Yahalom *et al.*, 1991; Hunte *et al.*, 1992, 1993; Schulz *et al.*, 1992; Mushegian and Koonin, 1993 |
| Myosin | A motor protein that may be used to traffic molecules through the pore, or may link the desmotubule to the plasma | Blackman and Overall, 1998; Radford and White, 1998; Reichelt *et al.*, 1999; Baluška *et al.*, 2001 |

**Table 1.1** (*Continued*)

| Component | Function/Comment | References |
|---|---|---|
| | membrane. Could be used to control the SEL. A unique plant myosin (myosin VIII) has been localised to, and may be specialised for, plasmodesmata | |
| Pectin | A component of the cell wall that is enriched around plasmodesmata | Roy *et al.*, 1997; Orfila and Knox, 2000 |
| Pectin methylesterase (PME) | An enzyme involved in pectin de-esterification, it interacts with viral movement proteins and has been hypothesised to act as a plasmodesmal receptor protein | Morvan *et al.*, 1998; Dorokhov *et al.*, 1999; Chen *et al.*, 2000; Morozov and Solovyev, 2003; Oparka, 2004 |
| Peroxidase | Function unknown | Schnepf and Sych, 1983 |
| PRms | A pathogenesis-related protein that both localises to, and moves through, plasmodesmata. Function unknown, but may only transiently reside in plasmodesmata | Murillo *et al.*, 1997 |
| Protein kinases | Involved in the phosphorylation of viral movement proteins, and possibly endogenous trafficking proteins. May be used for protein phosphorylation signalling and control of the SEL | Yahalom *et al.*, 1998; Reichelt and Kendrick-Jones, 2000; Waigmann *et al.*, 2000; Lee and Lucas, 2001; Oparka, 2004 |
| RTM1 protein | A protein that restricts the entry of *Tobacco etch virus* in the phloem. May only transiently reside in plasmodesmata | Mahajan *et al.*, 1998 |
| Ubiquitin | May be involved in the turnover of plasmodesmal proteins, or the removal of plasmodesmata | Ehlers *et al.*, 1996 |
| Vesicle trafficking proteins | May be involved in delivery of exogenous structural or cargo proteins to plasmodesmata, and may be hijacked by viral movement protein to gain transport to the pore. May also be involved in transport through the pore on the plasma membrane | Janssen *et al.*, 1998; Laporte *et al.*, 2003; Medina-Escobar *et al.*, 2003; Oparka, 2004 |
| WD40 repeat-containing protein | Function unknown, but likely to be involved in protein–protein interactions, so may be a docking or receptor protein | Medina-Escobar *et al.*, 2003 |

**Table 1.1** (*Continued*)

| Component | Function/Comment | References |
|---|---|---|
| α-amylase | Function unknown | Gubler *et al.*, 1987 |
| 26- and 27-kDa proteins | Functions unknown, but PAP27 cross-reacts with a connexin32 antibody (see above) | Yahalom *et al.*, 1991; Kotlizky *et al.*, 1992 |
| 41-kDa protein | Function unknown | Epel *et al.*, 1996 |
| 45-kDa protein | Function unknown | Blackman *et al.*, 1998 |
| 5′-nucleotidase | Function unknown | Nougarède *et al.*, 1985 |
| Maize root proteins | Antibodies JIM64 and JIM67 raised against two unknown proteins label plasmodesmata. Their functions are unknown | Turner *et al.*, 1994; Waigmann *et al.*, 1997 |

structure, will be dealt within this chapter. Both actin and myosin have been localised along the entire length of plasmodesmata (White *et al.*, 1994; Blackman and Overall, 1998; Radford and White, 1998). This pattern of localisation may allow constriction and relaxation of the entire length of the pore to control intercellular exchange (Zambryski and Crawford, 2000), although other authors suggest that control at the neck regions only would effect the same result. In contrast, having actin and myosin throughout the pore may be evidence that these cytoskeletal elements are involved in transport per se, rather than having a role in regulating the permeability of the plasmodesma (Blackman and Overall, 2001). The latter point is supported by the fact that myosin and actin inhibitors do not influence the movement of small molecules that pass through plasmodesmata by diffusion, indicating that actin and myosin are not involved in controlling the permeability of plasmodesmata at these low SELs (Tucker, 1987). However, depolymerisation of the actin cytoskeleton using cytochalasin has resulted in a widening of the neck region of plasmodesmata in *Nephrolepis exaltata* (White *et al.*, 1994) and has allowed an increase in the SEL of tobacco plasmodesmata from 1 to over 20 kDa (Ding *et al.*, 1996). In contrast, when the actin microfilaments are stabilised by treatment with phalloidin, cell-to-cell transport was inhibited. Additional work using a GFP-mouse talin fusion to label actin has recently shown that the actin microfilaments are static and do not move through plasmodesmata (Crawford and Zambryski, 2000). These results suggest that actin filaments form a scaffold along which molecules move using a myosin-driven mechanism, and that the actin is used to maintain the conformation and structure of the plasmodesmal pore. Baluška *et al.* (2000) speculated that plasmodesmata may act as actin-filament-organising centres and further support for this comes from the localisation of putative Arp3 homologues at plasmodesmata (van Gestel *et al.*, 2003). Interestingly, despite microtubules being implicated in the transport of viral MPs to the pore (Heinlein *et al.*, 1995; Boyko *et al.*, 2000; Mas and Beachy, 2000), and tubulin being found in protein extracts from cell walls containing plasmodesmata, tubulin has not been shown to localise or be enriched at plasmodesmata (Blackman and Overall, 1998).

An alternative view is that the presence of both actin and myosin along the length of the pore, possibly as helically arranged 'spokes' that connect the desmotubule to the plasma membrane (Overall and Blackman, 1996), provides a contractile mechanism that controls the aperture of the cytoplasmic sleeve (Radford and White, 1998; Reichelt *et al.*, 1999). Centrin – a calcium-regulated cytoskeletal protein – has been localised to the neck region of plasmodesmata and could provide regulation of transport through a contractile mechanism in the neck regions only (Blackman *et al.*, 1999).

Myosin VIII, an unconventional myosin found only in plants, has also been localised to plasmodesmata (Reichelt *et al.*, 1999; Chaffey and Barlow, 2002), has been implicated in plasmodesmal regulation (Baluška *et al.*, 2001; Volkmann *et al.*, 2003) and could be calcium-regulated (Knight and Kendrick-Jones, 1993). Of the myosin VIII proteins, seven have been sequenced to date (Baluška *et al.*, 2001), and they all have a characteristic C-terminal structure that includes probable phosphorylation sites for protein kinases as well as four predicted calmodulin-binding motifs (Reichelt and Kendrick-Jones, 2000). Myosin VIII may be attached to the plasma membrane at plasmodesmata by its C-terminal globular region (Baluška *et al.*, 2001; Oparka, 2004) and may anchor or organise actin, or support the cortical ER elements at plasmodesmata (Volkmann *et al.*, 2003). Immunodepletion of myosin VIII resulted in dilation of plasmodesmata, suggesting that myosin-based forces may be involved in maintaining the structural integrity of the pore (Reichelt *et al.*, 1999). It is therefore possible that myosin VIII may alter the structure of plasmodesmata through regulation by calcium or calmodulin, or may act as a calcium-regulated motor to traffic cargo through the plasmodesmal pore along the central actin filaments. Calcium regulation of trafficking may be a common occurrence since both calcium-binding sites and ATPase activity have been localised to plasmodesmata in a number of tissues (Cronshaw, 1980; Belitser *et al.*, 1982; see also Robards and Lucas, 1990; Chen *et al.*, 1994). Calreticulin – a calcium-sequestering protein – has also been localised to plasmodesmata (Baluška *et al.*, 1999, 2001). It is normally found in the lumen of the ER but has calcium-buffering activity that could potentially regulate plasmodesmal transport.

### *1.3.2 Callose*

Deposits of callose [(1→3)-β-glucan] are commonly found at plasmodesmata (see Table 1.1), and are generally thought to be a response to wounding or pathogenesis [see review by Stone and Clarke (1992)]. Whether callose is truly a functional component of plasmodesmata, or merely material that is deposited around them, could be argued since it does not appear to be a true constituent of the plasmodesmal collar (Turner *et al.*, 1994), but nevertheless, callose deposition represents a mechanism to control the movement of molecules through plasmodesmata. The rates of callose deposition are highly variable, ranging from seconds to hours, and it is not generally considered to be a rapid and responsive control mechanism, being used instead to seal off cells during attack or invasion, and as a developmental regulator of symplasmic continuity. Callose plugs at plasmodesmata have been implicated in the maintenance of dormancy by isolating the meristem from symplasmic continuity with surrounding tissues (Rinne and van der Schoot, 1998; Rinne *et al.*, 2001). During wounding, callose is deposited at plasmodesmata (Hughes and Gunning, 1980) but callose deposition and degradation have been shown to be variable both in time and in quantity. Given that callose deposition can occur in seconds in response to chemical fixatives, such rapid deposition may mean that many electron micrographs used for structural analysis of plasmodesmata show artifactual restrictions.

Callose is deposited at plasmodesmata in response to plasmolysis but most disappear 4–6 h after full turgor is regained (Drake *et al.*, 1978). Recovery of electrical coupling in these oat coleoptyle cells followed this time course of callose removal (Drake *et al.*, 1978) taking several hours to resume, while in *Egeria densa*, electrical coupling was re-established within 10 min of deplasmolysis (Erwee and Goodwin, 1984). Additional experiments have shown that callose may be deposited only in response to specific stresses. Callose deposition is known to be induced by micromolar changes in intercellular calcium concentrations (Kauss, 1987). Induction of stress by application of both aluminium and hydrogen peroxide causes an increase in cytoplasmic calcium concentration and yet, callose deposition occurs only in response to aluminium, and not to peroxide application (Robards and Lucas, 1990; Jones *et al.*, 1998). This suggests that callose deposition, and the subsequent blockage of symplasmic transport, may be a mechanism for aluminium toxicity in plants (Sivaguru *et al.*, 2000).

During pathogenesis, callose deposits are used to occlude the plasmodesmal pore in a plant response to reduce the spread of pathogens. Most published reports have concentrated on viral infections, but callose has also been found to block plasmodesmata during infections by the oomycete *Phytophthora sojae*, suggesting that it may be a widespread plant defence response. Callose has been found in pit-fields and walls of plants infected with *Potato virus X* (PVX; Allison and Shalla, 1974). In these plants, callose deposition was extensive within necrotic viral lesions, but could also be detected outside the visible lesion, suggesting that the formation of callose plugs in plasmodesmata could be an early defence strategy against PVX. Many other viruses, including TMV (Wu and Dimitman, 1970; Moore and Stone, 1972; Leisner and Turgeon, 1993; Beffa *et al.*, 1996), *Tomato bushy stunt virus* (Pennazio *et al.*, 1978) and *Maize dwarf mosaic virus* (Choi, 1999) also cause callose deposition at plasmodesmata in infected plants. In addition, plants that are deficient in β-1,3-glucanases (callose-degrading enzymes) show a decreased susceptibility to infection with TMV (Beffa *et al.*, 1996; Iglesias and Meins, 2000) and to PVX and *Cucumber mosaic virus* (Iglesias and Meins, 2000). These plants also showed increased callose deposition in response to viral infection and a reduction in the plasmodesmal SEL, but not an absolute closure of the plasmodesmata. Complementary experiments showed that TMV mutants that overexpressed the β-1,3-glucanase gene showed increased movement and spread on either wild-type or β-1,3-glucanase-deficient plants (Bucher *et al.*, 2001). Similarly, levels of β-1,3-glucanase were increased in leaves infected with a number of other plant viruses (Moore and Stone, 1972). Collectively, these authors have suggested that induction of β-1,3-glucanases by viruses might be used to enhance their cell-to-cell movement and spread, and act as a counter-response to the callose deposition (Moore and Stone, 1972; Beffa *et al.*, 1996; Iglesias and Meins, 2000). Wolf *et al.* (1991) showed that a reduction in the SEL and an inhibition of dye movement in TMV MP transgenics could be alleviated by treatment of the tissue with an inhibitor of callose synthesis, providing evidence that callose deposition is involved in regulating the plasmodesmata SEL. Dye coupling has also been shown to be reduced in cells adjacent to TMV lesions, in areas known to have callose deposits at plasmodesmata (Susi, 2000; see below). Enhanced callose deposition in a β-1,3-glucanase-deficient mutant also caused a reduction in the SEL (Iglesias and Meins, 2000).

Despite the presence of callose deposits, solute transport can still occur through such plasmodesmata. Some callose is always found in sieve plate pores (which develop from plasmodesmata; Esau, 1977; Chaffey and Barlow, 2002) as a result of sieve-pore genesis, but phloem transport across the sieve plates can occur in the presence of large callose deposits (Peterson and Rauser, 1979). However, genetic control of callose deposition can be specifically controlled. In the maize sucrose-export-deficient mutant (*sxd1*), phloem loading is prevented by callose deposits that specifically block plasmodesmata at the interface between bundle sheath and vascular parenchyma cells (Botha *et al.*, 2000). It is interesting to note that although the effects of callose are generally thought to be cruder than other control mechanisms, many publications show that callose can regulate the SEL, rather than simply blocking the plasmodesmal pore.

### *1.3.3 Additional components of plasmodesmata*

Pectin and pectic polysaccharides are enriched in the cell wall around plasmodesmata (Orfila and Knox, 2000), and pectin methylesterase (PME), an enzyme responsible for the de-esterification of pectin, has also been localised to plasmodesmata (Morvan *et al.*, 1998). In addition, PME is known to interact with the MPs of TMV, *Turnip vein clearing virus* (TVCV) and *Cauliflower mosaic virus* (CaMV), and has been suggested to be a plasmodesmal receptor protein (Dorokhov *et al.*, 1999; Chen *et al.*, 2000). It may be that MPs interact with PME at the pore (Morozov and Solovyev, 2003), but it is also possible that the delivery of endogenous PME is hijacked by the virus to allow transport to the pore (Oparka, 2004). Ubiquitin has also been localised to plasmodesmata and may be used to target plasmodesmal proteins for degradation via the ubiquitin pathway, as occurs for cytosolic proteins. High levels of ubiquitin were found at discontinuous plasmodesmata in regenerating protoplasts, while those that established secondary contacts between neighbouring cells did not contain ubiquitin (Ehlers *et al.*, 1996). The accumulation of ubiquitin in these half-plasmodesmata may point to a role for this protein in the degradation and removal of plasmodesmata. Additional endogenous proteins that have been localised to plasmodesmata include peroxidase (Schnepf and Sych, 1983), α-amylase (Gubler *et al.*, 1987), acid phosphatase (Esau and Charvat, 1975), low-esterified pectin (Roy *et al.*, 1997) and 5′-nucleotidase (Nougaréde *et al.*, 1985); however, no specific plasmodesmal functions have been attributed to any of these proteins, but plasmodesmata have generally been found to be sites of high enzymatic activity (Olesen, 1979).

Biochemical purification of plasmodesmal proteins from wall preparations is very difficult due to the minute quantity of protein present in plasmodesmata, and the difficulty in obtaining pure plasmodesmal extracts without wall contaminants (Epel, 1994; Epel *et al.*, 1995). This means that, to date, proteomic analysis of plasmodesmata has had limited success, but attempts may yield data in the future. So far, antibodies raised to such cell-wall extracts have immunolocalised to plasmodesmata in a few cases, but none of the genes encoding the proteins have yet been discovered. To date, a 41-kDa protein has been found in plasmodesmata of maize mesocotyl cells (Epel *et al.*, 1996), and a 45-kDa protein localised to plasmodesmata in the alga *Chara corallina* (Blackman *et al.*, 1998b). Similarly, antibodies raised against two maize root proteins localise to plasmodesmata in trichomes and mesophyll cells of *Nicotiana clevelandii* (Turner *et al.*, 1994; Waigmann *et al.*, 1997). A number of putative connexin-like proteins have either been found in cell-wall extracts or inferred by cross-reaction with connexin antibodies (Meiners and Schindler, 1989; Meiners *et al.*, 1991; Yahalom *et al.*, 1991; Hunte *et al.*, 1992, 1993; Schulz *et al.*, 1992). However, much of this data is unconvincing, with several studies providing unclear localisation data, and one such protein was later revealed to have greater homology to a protein kinase than to a connexin (Mushegian and Koonin, 1993). Cell-wall protein extraction and antibody localisation has a number of disadvantages, and genetic screens are more likely to produce results in the future. However, one of the problems in identifying plasmodesmal genes using whole-plant genetic screens is likely to be that any mutants have a high chance of lethality, making screens at the embryonic stage necessary (Zambryski and Crawford, 2000).

Another problem with localising proteins to plasmodesmata is that our overall understanding of their structure and function is so limited that we are unable to say whether any new protein localised to the pore is a permanent or a temporary 'resident'. Several of the proteins localised to plasmodesmata to date, and especially those linked to plant pathology, are perhaps likely to be temporary residents – there to do a specific job at a specific time. One such protein was discovered using a yeast two-hybrid system in a search for *Arabidopsis* proteins that interact with the CaMV MP – a protein known to be necessary for virus movement through plasmodesmata (Huang *et al.*, 2001). This protein, MPI7, was found to interact with the CaMV MP using both two-hybrid and FRET analyses. MPI7 was widely expressed throughout *Arabidopsis* plants and was localised to punctate spots at the cytoplasmic periphery, which appear to be plasmodesmata. Sequence data suggests that the protein may be akin to a mammalian Rab acceptor protein (Huang *et al.*, 2001), and may therefore be involved in protein transport through plasmodesmata via the endomembrane system. An alternative method to find plasmodesmatal proteins, involving the use of a viral vector system to express and localise random cDNA–GFP fusions, has identified protein fragments that localise to plasmodesmata. A protein containing a WD-40 repeat was among others found in this screen (Medina-Escobar *et al.*, 2003). WD-40 repeats are involved in protein–protein interactions and so it is conceivable that this protein may function at plasmodesmata as a docking or receptor protein. Murillo *et al.* (1997) found that PRms, an 18-kDa maize pathogenesis-related protein that is produced in response to fungal infection, was found in plasmodesmata. This protein can also move independently between cells, since it was detected in both the parenchyma cells of the protoxylem and central pith of maize radicles, while the *PRms* RNA was found only in parenchyma cells of the protoxylem.

Another protein from *Arabidopsis*, RTM1, has been localised to punctate spots in the walls of sieve elements that may be plasmodesmata. This is one of a group of three proteins that restrict the movement of *Tobacco etch virus* (TEV), preventing systemic spread (Mahajan *et al.*, 1998; Chisholm *et al.*, 2000, 2001). Further analyses of these proteins may reveal if they operate by preventing movement of the viral genome through plasmodesmata in the phloem. While it is possible that plants have evolved proteins specifically to deal with pathogen attack, it may be that the same proteins also play a role in regulating plant physiology and development by their action at plasmodesmata. Unfortunately, although our knowledge of plasmodesmatal proteins is constantly expanding, little is currently known about the function of any of the proteins that have been found to localise to plasmodesmata, and few have been incorporated into either structural or functional models (White *et al.*, 1994).

## 1.4 Developmental changes to plasmodesmata

One of the difficulties in providing an overview of plasmodesmal structure is that it varies considerably – not only between plant species and tissues, but also within a given tissue over a developmental period. For instance, plasmodesmal structure has been shown to increase in complexity from immature to mature cells in trichomes, and has been shown to have distinct morphologies between trichome and mesophyll cells (Waigmann *et al.*, 1997). The final section in this chapter aims to briefly describe some of the variations that occur during plant development.

### *1.4.1 Branched plasmodesmata*

Both primary and secondary plasmodesmata are simple in structure at their formation, but during plant development may be converted, post-cytokinetically, to branched structures (Ehlers and Kollmann, 1996, 2001). Simple plasmodesmata may be grouped together in primary pit-fields, or more randomly spread throughout cell walls, and are common in lower plants such as mosses and algae (Franceschi *et al.*, 1994; Cook *et al.*, 1997). The evolutionary origins of plasmodesmata will be dealt with in Chapter 2. In higher plants, simple plasmodesmata have been found to predominate in immature plant tissues, while branched forms are more common in mature tissues (Oparka *et al.*, 1999) and represent a more evolutionarily advanced structure (Lucas *et al.*, 1993). Branching has been postulated to occur in a number of different ways. As the plant cell wall matures, secondary wall material is deposited on the inner face of the primary wall (Fry, 2001), and additional ER tubules may be formed that are continuous with the desmotubule, leading to the production of branched plasmodesmata (Kollmann and Glockmann, 1999). An alternative method of branching can be accomplished during development by neighbouring simple plasmodesmata fusing to create more complex and branched structures. This appears to be achieved via an H-shaped intermediate structure that originates by the formation of a new protoplasmic bridge between neighbouring simple plasmodesmata, across the middle lamella of the wall, leading to the production of a central cavity that links the two plasmodesmata (Figure 1.4) (Lucas *et al.*, 1993; Oparka *et al.*, 1999; Roberts *et al.*, 2001). However, what is not clear is how the H-shaped intermediate is converted to more complex branched forms with many pores on each side. Although two plasmodesmata have apparently been seen to fuse, leading to the H-shaped structure, this has not been observed for multiple, simple plasmodesmata. There must, therefore, be additional mechanisms used to add extra branches, possibly accomplished in a manner similar to the production of new secondary plasmodesmata. However this is achieved, it must represent a highly coordinated programme of structural rearrangements at plasmodesmata. To perforate an intact cell wall, enzymes will be needed; possibly, only pectinases and hemicellulases are required (cellulose microfibrils may be separated mechanically) but otherwise, cellulases will

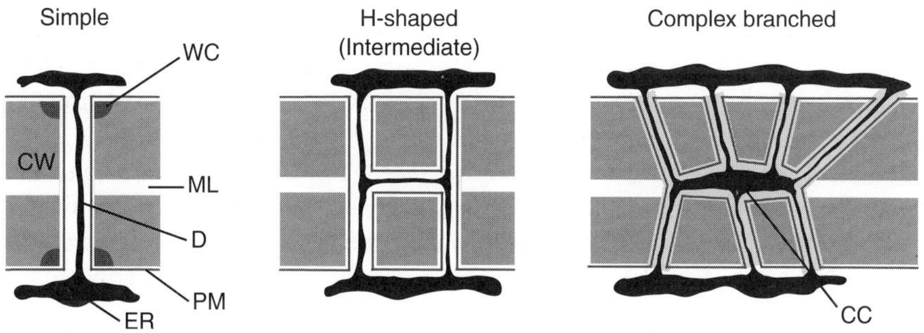


**Figure 1.4** Diagrammatic representation of the structural change from simple to branched plasmodesmata. During development, the simple pore develops into a complex, branched structure with a central cavity aligned along the middle lamella of the cell wall. This transformation occurs via an H-shaped intermediate structure that appears to form by the introduction of a new protoplasmic bridge between neighbouring pairs of simple plasmodesmata. WC, wall collar; ML, middle lamella; D, desmotubule; PM, plasma membrane; ER, endoplasmic reticulum; CW, cell wall; CC, central cavity. Redrawn from Oparka *et al.* (1999).

also be needed (Jones, 1976). These enzymes may be present in the wall (probably in the middle lamella), or the cytoplasm, but must be activated in a highly localised and coordinated manner to allow precise degradation of the wall. Subsequently, vesicles, most likely Golgi-derived but possibly ER-derived, are required to deposit new membrane components, and additional wall material. An additional feature of branched plasmodesmata is that they frequently contain a large central cavity. This is a region of the plasmodesma, corresponding to the middle lamella region of the pre-existing wall, where the lumen of the ER that traverses the pores has expanded, forming a membrane-lined wall cavity. The ER remains appressed, as a desmotubule, in the channels that connect the central cavity to the protoplasts of the neighbouring cells (Lucas *et al.*, 1993). The function of this structure, and the nature of the material that fills the ER in the cavity, is unknown.

### *1.4.2 Loss, reduction or occlusion of plasmodesmata*

While the structure of plasmodesmata can alter during development, their frequency can also change over time or in different tissue types. Plants can seal off or remove plasmodesmata, both temporarily and permanently, possibly by degradation via the ubiquitin pathway (Ehlers *et al.*, 1996). This means that the symplasm is not necessarily a cellular continuum with unlimited communication, but is subdivided into functional domains. These domains, and the limits they place on the movement of signal molecules, are thought to allow coordinated morphogenesis and development in plants (Kragler *et al.*, 1998; Pfluger and Zambryski, 2001; see Chapter 5).

During the sink–source transition in tobacco leaves, the numbers of simple plasmodesmata are dramatically reduced (Roberts *et al.*, 2001). Although some of this loss occurs by the conversion of simple plasmodesmata to branched plasmodesmata (Oparka *et al.*, 1999), many simple plasmodesmata are completely removed during leaf development. Most of these plasmodesmata are lost during the rapid phase of leaf cell expansion when intercellular air spaces are forming in the mesophyll. No trace of these plasmodesmata remains in the mature leaf, and they are literally ripped apart during rapid leaf expansion. It would appear that these simple plasmodesmata are utilised for a brief developmental period to enhance the intercellular communication and rapid unloading of solutes in sink tissues (Roberts *et al.*, 2001). However, once the tissue becomes a source, branched plasmodesmata with a lower SEL are utilised, and appear to give the plant greater control over the movement of molecules in mature tissues.

The development of symplasmic domains in the root epidermis of *Arabidopsis* occurs as the tissue matures (Duckett *et al.*, 1994). Meristematic cells, and those within the elongation zone of the root, are initially coupled but cell–cell communication is down-regulated during development until, by the time of root-hair initiation, epidermal cells are symplasmically isolated. In the *Arabidopsis* shoot, plasmodesmal isolation is used to control communication between cells in different areas of the meristem during the transition from vegetative growth to flowering

(Gisel *et al.*, 1999). By preventing the movement of signal molecules, ions and photoassimilate, plasmodesmal isolation also affects the turgor potential within a given symplasmic domain – a key feature in stomatal function. As guard cells differentiate, cell-wall material is deposited across the pore on the side of the neighbouring epidermal or subsidiary cell wall, effectively truncating all plasmodesmata and symplasmically isolating the guard-cell complex (Wille and Lucas, 1984; Palevitz and Hepler, 1985). Guard-cell isolation has long been understood, but a similar mechanism has also recently been shown to control the development of cotton fibres (Ruan *et al.*, 2001). In these cells, a developmental switch from simple to branched plasmodesmata, concomitant with a transient closure of plasmodesmata, allows increased turgor pressure to drive the rapid elongation required in these specialised cells (Pfluger and Zambryski, 2001; Ruan *et al.*, 2001).

Plasmodesmatal blockages also occur in the vascular tissue. In differentiating xylem tissue of *Sorbus torminalis*, many plasmodesmata are found in the pits that connect immature xylem elements to the surrounding mesophyll cells. However, during the final stages of programmed cell death, the pits become sealed off by the deposition of new cell-wall material across both ends of the plasmodesmatal pores (Lachaud and Maurousset, 1996). Deposition of wall material and/or callose (Russin *et al.*, 1996; Botha *et al.*, 2000) across the plasmodesmatal pores in the phloem can affect phloem loading. The sucrose export deficient (*sxd1*) mutant of maize is unable to load sucrose into minor veins due to a blockage of plasmodesmata at the bundle sheath – vascular parenchyma–cell wall interface (Russin *et al.*, 1996). It has been hypothesised that *sxd1* mutants have a defect in the signalling mechanism between the chloroplasts and nucleus in bundle-sheath cells. This defect in the communication pathway affects the differentiation of bundle-sheath cells during the sink–source transition, leading to deposition of callose across plasmodesmata at the bundle sheath – vascular parenchyma–cell wall interface (Mezitt Provencher *et al.*, 2001). Since the permanent blockage of plasmodesmata may share similar underlying mechanisms, the cloning and study of genes such as *sxd1* may elucidate the molecular events that underlie plasmodesmatal modifications leading to loss of symplasmic continuity.

### *1.4.3 Formation of secondary plasmodesmata*

In addition to their removal or occlusion, new plasmodesmata, with either a simple or branched structure, can be formed across existing cell walls. These are called secondary plasmodesmata and are formed post-cytokinetically by a poorly understood process (Jones, 1976; see Lucas *et al.*, 1993 for a detailed review; Ehlers and Kollmann, 2001). Secondary plasmodesmata cannot be distinguished from primary plasmodesmata by structure alone, since they can also have either a simple or branched morphology (Ding *et al.*, 1992a), although the majority have a branched structure (Lucas *et al.*, 1993). However, at the time of their formation, secondary plasmodesmata, like primary plasmodesmata, have a simple structure that may be modified subsequently. The formation of new plasmodesmata in existing walls must involve the use of enzymes to remove, or break apart, the cell wall in a manner similar to that discussed earlier for the production of branched plasmodesmata. Secondary plasmodesmata can form along any wall in a cell, either increasing the symplasmic continuity between previously connected cells, or allowing the establishment of symplasmic connections between cells that are not cytokinetically related (van der Schoot *et al.*, 1995; Ding and Lucas, 1996; Volk *et al.*, 1996; Itaya *et al.*, 1998; Oparka *et al.*, 1999; van der Schoot and Rinne, 1999). In fact, plasmodesmata can be formed across highly dissimilar boundaries, such as those found in graft unions (Kollmann and Glockmann, 1985, 1991), host–parasite contacts (Dörr, 1987; Dawson *et al.*, 1994; Haupt *et al.*, 2001), plant chimeras (Steinberg and Kollmann, 1994) and during protoplast fusion (Ehlers and Kollmann, 1996). Communication and coordination must be required between adjacent cells to allow the *de novo* formation of plasmodesmata, since the process appears to occur synchronously from both sides of the wall (Ehlers and Kollmann, 2001). How this is achieved is unknown, but various models have been proposed (Jones, 1976; Juniper, 1977; Lucas and Gilbertson, 1994; Ding and Lucas, 1996). The model proposed by Jones (1976) and reviewed by Ehlers and Kollmann (2001), proposes that localised enzymatic thinning of the cell wall occurs to a point where the plasma membrane and ER of both cells can penetrate the digested portion of wall and subsequently fuse, driven by turgor pressure (Lucas *et al.*, 1993). This sequence of events was seen in a detailed study of plasmodesmatal formation in grafted tissue (Kollmann and Glockmann, 1991). They saw that during the period of wall thinning, the ER became attached to the plasma membrane on either side of the thinning wall. Golgi vesicles produced additional membrane material to allow the membrane expansion, and once the wall was removed, membrane fusion occurred. Following the membrane fusion, the ER may branch and new cell-wall material is laid down to complete the plasmodesmal structure. In cases where cell-wall thinning was incomplete and asymmetrical, half plasmodesmata were formed (Ehlers and Kollmann, 2001). Given the central importance of the ER to the plasmodesmal structure, and the branching of the ER in secondary plasmodesmata, it has been suggested that the mechanism that produces secondary plasmodesmata may be similar to that which governs the continual branching and rearrangement of the cortical ER network (Lucas *et al.*, 1993).

This discussion has concentrated on the formation of secondary plasmodesmata in non-vascular tissue. However, a range of specialised plasmodesmata occur throughout the plant that may be primary or secondary in nature, but have distinct structures related to their function. One such case is the specialised plasmodesmata known as pore-plasmodesma units – PPUs (van Bel and Kempers, 1996) – between the sieve element and companion cells of the phloem which, while branched, have an asymmetric structure with one channel connecting the sieve element to the middle lamella, and multiple branches linking from there to the companion cell. Plasmodesmata in the phloem and other specialised plasmodesmata will be discussed in Chapters 7 and 8.

### *1.4.4 The future*

Advances in our understanding of plasmodesmal structure will be central to our understanding of their function, and there is still much to learn. At present, there is no emerging technology that seems likely to greatly improve our imaging resolution, but microscopy will doubtless improve over time. The greatest advances at present are emerging from new discoveries at the protein level. Once plasmodesmal proteins are cloned and characterised, their subsequent localisation will give new insights into plasmodesmal structure. Hopefully, in the future, we can start to understand the functions of the globular proteins seen in electron micrographs of plasmodesmata, embedded in both the plasma membrane and desmotubule, and find some of the many receptor, docking and regulatory proteins that are hypothesised to be part of the pore. Other welcome advances will be an understanding of whether transport occurs via the cytoplasmic sleeve, the desmotubule or the plasma membrane, or a combination of all three methods. The publication of both the *Arabidopsis* and rice genome sequences is allowing genomic approaches that will lead to the discovery of plasmodesmal proteins, and advanced proteomic techniques may increase the speed with which such proteins can be discovered. Once the first proteins are characterised, the doors will open to true 'functional genomics' where further analyses, such as overexpression and gene knockouts, can be used to study plasmodesmal structure, function and development.

## Acknowledgements

I would like to thank Ian Pitkethly and Shona Thompson for their help in preparing the figures, and Morven Pearson for help in compiling the reference list.

## References

Allen, R.D. and Bowen, C.C. (1966) 'Fine structure of *Psilotum nudumi* cells during division', *Caryologia* **19**, 299–342.

Allison, A.V. and Shalla, T.A. (1974) 'The ultrastructure of local lesions induced by potato virus X: a sequence of cytological events in the course of infection', *Phytopathology* **64**, 784–793.

Badelt, K., White, R.G., Overall, R.L. and Vesk, M. (1994) 'Ultrastructural specialisations of the cell wall sleeve around plasmodesmata', *American Journal of Botany* **81**(11), 1422–1427.

Baker, J.R. (1948) 'The cell theory: a restatement, history and critique', *Quarterly Journal of Microscopical Science* **89**, 103–125.

Baluška, F., Barlow, P.W. and Volkmann, D. (2000) Actin and myosin VIII in developing root cell apex cells, in *Actin: A Dynamic Framework for Multiple Plant Cell Functions* (eds C.J. Staiger, F. Baluška, D. Volkmann and P.W. Barlow), Kluwer, Dordrecht, pp. 457–477.

Baluška, F., Cvrckova, F., Kendrick-Jones, J. and Volkmann, D. (2001) 'Sink plasmodesmata as gateways for phloem unloading myosin VIII and calreticulin as molecular determinants of sink strength?', *Plant Physiology* **126**, 39–46.

Baluška, F., Samaj, J., Napier, R. and Volkmann, D. (1999) 'Maize calreticulin localizes preferentially to plasmodesmata in root apex', *The Plant Journal* **19**(4), 481–488.

Barclay, G.F., Peterson, C.A. and Tyree, M.T. (1982) 'Transport of fluorescein in trichomes of *Lycopersicon esculentum*', *Canadian Journal of Botany* **60**, 397–402.

Beebe, D.U. and Turgeon, R. (1991) 'Current perspectives on plasmodesmata: structure and function', *Physiologia Plantarum* **83**, 194–199.

Beffa, R.S., Hofer, R.-M., Thomas, M. and Meins, Jn.F. (1996) 'Decreased susceptibility to viral disease of β-1,3-glucanase-deficient plants generated by antisense transformation', *The Plant Cell* **8**, 1001–1011.

Belitser, N.V., Zaalishvili, G.V. and Sytnianskaja, N.P. (1982) '$Ca^{2+}$-binding sites and $Ca^{2+}$-ATPase activity in barley root tip cells', *Protoplasma* **111**, 63–78.

Blackman, L.M. and Overall, R.L. (1998) 'Immunolocalisation of the cytoskeleton to plasmodesmata of *Chara corallina*', *The Plant Journal* **14**(6), 733–741.

Blackman, L.M. and Overall, R.L. (2001) 'Structure and function of plasmodesmata', *Australian Journal of Plant Physiology* **28**, 709–727.

Blackman, L.M., Boevink, P., Santa Cruz, S., Palukaitis, P. and Oparka, K.J. (1998a) 'The movement protein of cucumber mosaic virus traffics into sieve elements in minor veins of *Nicotiana clevelandii*', *The Plant Cell* **10**, 525–537.

Blackman, L.M., Gunning, B.E.S. and Overall, R.L. (1998b) 'A 45 kDa protein isolated from the nodal walls of *Chara corallina* is localised to plasmodesmata', *The Plant Journal* **15**(3), 401–411.

Blackman, L.M., Harper, J.D.I. and Overall, R.L. (1999) 'Localization of a centrin-like protein to higher plant plasmodesmata', *European Journal of Cell Biology* **78**, 297–304.

Blatt, M.R., Leyman, B. and Geelen, D. (1999) 'Molecular events of vesicle trafficking and control by SNARE proteins in plants', *New Phytologist* **144**, 389–418.

Boevink, P., Oparka, K.J., Santa Cruz, S., Martin, B., Betteridge, A. and Hawes, C. (1998) 'Stacks on tracks: the plant Golgi apparatus traffics on an actin/ER network', *The Plant Journal* **15**(3), 441–447.

Botha, C.E.J., Cross, R.H.M., van Bel, A.J.E. and Peter, C.I. (2000) 'Phloem loading in the sucrose-export-defective (*SXD-1*) mutant maize is limited by callose deposition at plasmodesmata in bundle sheath-vascular parenchyma interface', *Protoplasma* **214**, 65–72.

Botha, C.E.J., Hartley, B.J. and Cross, R.H.M. (1993) 'The ultrastructure and computer-enhanced digital image analysis of plasmodesmata at the Kranz mesophyll-bundle sheath interface of *Themeda triandra* var. *imberbis* (Retz) A. Camus in conventionally-fixed leaf blades', *Annals of Botany* **72**, 255–261.

Boyko, V., Ferralli, J. and Heinlein, M. (2000) 'Cell-to-cell movement of TMV RNA is temperature-dependent and corresponds to the association of movement protein with microtubules', *The Plant Journal* **22**, 315–325.

Brown, R.C., Lemmon, B.E., Stone, B.A. and Olsen, O.-A. (1997) 'Cell wall (1-3)- and (1-3, 1-4)-β-glucans during early grain development in rice (*Oryza sativa* L.)', *Planta* **202**, 414–426.

Bucher, G.L., Tarina, C., Heinlein, M., Di Serio, F., Meins, F., Jr. and Iglesias, V.A. (2001) 'Local expression of enzymatically active class I β-1,3-glucanase enhances symptoms of TMV infection in tobacco', *The Plant Journal* **28**(3), 361–369.

Buvat, R. (1957) 'L'infrastructure des plasmodesmes et la continuité des cytoplasmes', *Comptes Rendue l'Academie des Sciences Paris* **245**, 198–201.

Cantrill, L.C., Overall, R.L. and Goodwin, P.B. (1999) 'Cell-to-cell communication via plant endomembranes', *Cell Biology International* **23**(10), 653–661.

Carr, D.J. (1976) Historical perspectives on plasmodesmata, in *Intercellular Communication in Plants: Studies on Plasmodesmata* (eds B.E.S. Gunning and A.W. Robards), Springer-Verlag, Berlin, pp. 291–295.

Chaffey, N. and Barlow, P. (2002) 'Myosin, microtubules and microfilaments: co-operation between cytoskeletal components during cambial cell division and secondary vascular differentiation in trees', *Planta* **214**, 526–536.

Chen, M.-H., Sheng, J., Hind, G., Handa, A.K. and Citovsky, V. (2000) 'Interaction between the tobacco mosaic virus movement protein and host cell pectin methylesterases is required for viral cell-to-cell movement', *The EMBO Journal* **19**(5), 913–920.

Chen, S., Das, P. and Hari, V. (1994) '*In situ* localization of ATPase activity in cells of plants infected by maize dwarf mosaic potyvirus', *Archives of Virology* **134**, 433–439.

Chisholm, S.T., Mahajan, S.K., Whitham, S.A., Yamamoto, M.L. and Carrington, J.C. (2000) 'Cloning of the *Arabidopsis RTM1* gene, which controls restriction of long-distance movement of tobacco etch virus', *Proceedings of the National Academy of Sciences USA* **97**, 489–494.

Chisholm, S.T., Parra, M.A., Anderberg, R.J. and Carrington, J.C. (2001) '*Arabidopsis RTM1* and *RTM2* genes function in phloem to restrict long-distance movement of tobacco etch virus', *Plant Physiology*, **127**, 1667–1675.

Choi, C.W. (1999) 'Modified plasmodesmata in Sorghum (*Sorghum bicolor* L. Moench) leaf tissues infected by maize dwarf mosaic virus', *Journal of Plant Biology* **42**, 63–70.

Cook, M.E., Graham, L.E., Botha, C.E.J. and Lavin, C.A. (1997) 'Comparative ultrastructure of plasmodesmata of *Chara* and selected bryophytes: toward an elucidation of the evolutionary origin of plant plasmodesmata', *American Journal of Botany* **84**, 1169–1178.

Crawford, K.M. and Zambryski, P.C. (2000) 'Subcellular localization determines the availability of non-targeted proteins to plasmodesmatal transport', *Current Biology* **10**, 1032–1040.

Cronshaw, J. (1980) 'ATPases in mature and differentiating phloem and xylem', *The Journal of Histochemistry and Cytochemistry* **28**, 375–377.

Dawson, J.H., Musselman, L.J., Wolswinkel, P. and Dörr, I. (1994) 'Biology and control of *Cuscuta*', *Review of Weed Science* **6**, 265–317.

Denecke, J. (2001) Plant endoplasmic reticulum, in *Nature Encyclopedia of Life Sciences*, Nature Publishing Group, http://www.els.net/, London.

Ding, B. (1998) 'Intercellular protein trafficking through plasmodesmata', *Plant Molecular Biology* **38**, 279–310.

Ding, B. and Lucas, W.J. (1996) Secondary plasmodesmata: biogenesis, special functions and evolution, in *Membranes: Specialised Functions in Plants* (eds M. Smallwood, J.P. Knox and D.J. Bowels), BIOS Scientific, Oxford, UK, pp. 489–506.

Ding, B., Haudenshield, J.S., Hull, R.J., Wolf, S., Beachy, R.N. and Lucas, W.J. (1992a) 'Secondary plasmodesmata are specific sites of localization of the tobacco mosaic virus movement protein in transgenic tobacco plants', *The Plant Cell* **4**, 915–928.

Ding, B., Kwon, M.-O. and Warnberg, L. (1996) 'Evidence that actin filaments are involved in controlling the permeability of plasmodesmata in tobacco mesophyll', *The Plant Journal* **10**(1), 157–164.

Ding, B., Turgeon, R. and Parthasarathy, M.V. (1991) Plasmodesmatal substructure in cryofixed developing tobacco leaf tissue, in *Recent Advances in Phloem Transport and Assimilate Partitioning* (eds J.L. Bonnemain, S. Delrot, J. Dainty and W.J. Lucas ), Ouest editions, Nantes, pp. 317–323.

Ding, B., Turgeon, R. and Parthasarathy, M.V. (1992b) 'Substructure of freeze-substituted plasmodesmata', *Protoplasma* **169**, 28–41.

Dollar, G., Struckhoff, E., Michaud, J. and Cohen, R.S. (2002) 'Rab11 polarization of the *Drosophila* oocyte: a novel link between membrane trafficking, microtubule organization and *oskar* mRNA localization and translation', *Development* **129**, 517–526.

Dorokhov, Y.L., Mäkinen, K., Frolova, O.Y. *et al.* (1999) 'A novel function for a ubiquitous plant enzyme pectin methylesterase: the host–cell receptor for the tobacco mosaic virus movement protein', *FEBS Letters* **461**, 223–228.

Dörr, I. (1987) The haustorium of *Cuscuta* – new structural results, in *Parasitic Flowering Plants. Proceedings of the 4th International Symposium on Parasitic Flowering Plants* (eds H.C. Weber and W. Forstreuter), Marburg, Germany, pp. 163–170.

Drake, G.A., Carr, D.J. and Anderson, W.P. (1978) 'Plasmolysis, plasmodesmata and the electrical coupling of oat coleoptyle cells', *Journal of Experimental Botany* **29**, 1205–1214.

Duckett, C.M., Oparka, K.J., Prior, D.A.M., Dolan, L. and Roberts, K. (1994) 'Dye-coupling in the root epidermis of *Arabidopsis* is progressively reduced during development', *Development* **120**, 3247–3255.

Ehlers, K. and Kollmann, R. (1996) 'Formation of branched plasmodesmata in regenerating *Solanum nigrum* protoplasts', *Planta* **199**, 126–138.

Ehlers, K. and Kollmann, R. (2001) 'Primary and secondary plasmodesmata: structure, origin and functioning', *Protoplasma* **216**, 1–30.

Ehlers, K., Schulz, M. and Kollmann, R. (1996) 'Subcellular localization of ubiquitin in plant protoplasts and the function of ubiquitin in selective degradation of outer-wall plasmodesmata in regenerating protoplasts', *Planta* **199**, 139–151.

Epel, B.L. (1994) 'Plasmodesmata: composition: structure and trafficking', *Plant Molecular Biology* **26**, 1343–1356.

Epel, B.L., Kuchuck, B., Kotlizky, G., Shurtz, S., Erlanger, M. and Yahalom, A. (1995) 'Isolation and characterization of plasmodesmata', *Methods in Cell Biology* **50**, 237–253.

Epel, B.L., Van Lent, J.W.M., Cohen, L., Kotlizky, G., Katz, A. and Yahalom, A. (1996) 'A 41 kDa protein isolated from maize mesocotyl cell walls immunolocalises to plasmodesmata', *Protoplasma* **191**, 70–78.

Erwee, M.G. and Goodwin, P.B. (1984) 'Characterisation of the *Egeria densa* leaf symplast: response to plasmolysis, deplasmolysis and to aromatic amino acids', *Protoplasma* **122**, 162–168.

Erwee, M.G., Goodwin, P.B. and van Bel, A.J.E. (1985) 'Cell–cell communication in the leaves of *Commelina cyanea* and other plants', *Plant Cell and Environment* **8**, 173–178.

Esau, K. (1963) 'Ultrastucture of differentiated cells in higher plants', *American Journal of Botany* **50**, 495–506.

Esau, K. (1977) *Anatomy of Seed Plants*, 2nd edn, John Wiley and Sons Inc., New York.

Esau, K. and Charvat, I.D. (1975) 'An ultrastructural study of acid phosphatase localization in cells of *Phaseolus vulgaris* phloem by the use of the azo dye method', *Tissue and Cell* **7**, 619–630.

Esau, K. and Thorsch, J. (1985) 'Sieve plate pores and plasmodesmata, the communication channels of the symplast – ultrastructural aspects and developmental relations', *American Journal of Botany* **72**, 1641–1653.

Fisher, D.B. (2000) Long-distance transport, in *Biochemistry and Molecular Biology of Plants* (eds B. Buchanan, W. Gruissem and R. Jones), American Society of Plant Physiologists, Rockville, pp. 730–784.

Franceschi, V.R., Ding, B. and Lucas, W.J. (1994) 'Mechanism of plasmodesmata formation in characean algae in relation to evolution of intercellular communication in higher plants', *Planta* **192**, 347–358.

Fry, S.C. (2001) Plant cell walls, in *Nature Encyclopedia of Life Sciences*, Nature Publishing Group, http://www.els.net/, London.

Gardiner, W. (1900) 'The genesis and development of the walls and connecting threads of the plant', *Proceedings of the Royal Society. Series B. Biological Sciences* **66**, 186–188.

Gardiner, W. (1907) 'On the mode of formation of the initial cell wall, the genesis and neogenesis of the connecting threads and the methods of connecting living tissues', *Proceedings of the Cambridge Philosophical Society. Biological Sciences* **14**, 209–210.

Gisel, A., Barella, S., Hempel, F.D. and Zambryski, P.C. (1999) 'Temporal and spatial regulation of symplastic trafficking during development in *Arabidopsis thaliana* apices', *Development* **126**, 1879–1889.

Goodwin, P.B. (1983) 'Molecular size limit for movement in the symplast of the *Elodea* leaf', *Planta* **157**, 124–130.

Grabski, S., de Feijter, A.W. and Schindler, M. (1993) 'Endoplasmic reticulum forms a dynamic continuum for lipid diffusion between contiguous soybean root cells', *The Plant Cell* **5**, 25–38.

Gu, X. and Verma, D.P.S. (1996) 'Phragmoplastin, a dynamin-like protein associated with cell plate formation in plants', *The EMBO Journal* **15**(4), 695–704.

Gubler, F., Ashford, A.E. and Jacobsen, J.V. (1987) 'The release of β-amylase through gibberellin-treated barley aleurone cell walls', *Planta* **172**, 155–161.

Guenoune-Gelbart, D. and Epel, B.L. (2004) $MP^{TMV}$ does not mediate an active gliding of ER through plasmodesmata, *Abstracts of the Fifth International Plasmodesmata Conference*, Pacific Grove, California, USA, p. 160.

Gunning, B.E.S. and Overall, R.L. (1983) 'Plasmodesmata and cell-to-cell transport in plants', *Bioscience* **33**, 260–265.

Gunning, B.E.S. and Robards, A.W. (1976) *Intercellular Communication in Plants: Studies on Plasmodesmata*, Springer-Verlag, Berlin.

Hashimoto, T. and Inze, D. (2003) 'Cell biology: how unique is the plant cell?', *Current Opinion in Plant Biology* **6**, 1–3.

Haupt, S., Oparka, K.J., Sauer, N. and Neumann, S. (2001) 'Macromolecular trafficking between *Nicotiana tabacum* and the holoparasite *Cuscuta reflexa*', *Journal of Experimental Botany* **52**, 173–177.

Heinlein, M., Epel, B.L., Padgett, H.S. and Beachy, R.N. (1995) 'Interaction of tobacco virus movement proteins with the plant cytoskeleton', *Science* **270**, 1983–1985.

Hepler, P.K. (1982) 'Endoplasmic reticulum in the formation of the cell plate and plasmódesmata', *Protoplasma* **111**, 121–133.

Huang, Z., Andrianov, V.M., Han, Y. and Howell, S.H. (2001) 'Identification of *Arabidopsis* proteins that interact with the cauliflower mosaic virus (CaMV) movement protein', *Plant Molecular Biology* **47**, 663–675.

Hughes, J.E. and Gunning, B.E.S. (1980) 'Glutaraldehyde-induced deposition of callose', *Canadian Journal of Botany* **58**, 250–258.

Hunte, C., Janßen, M., Schulz, M., Traub, O., Willecke, K. and Schnabl, H. (1993) 'Age-dependent modifications and further localization of the CX 26-like protein from *Vicia faba* L.', *Botanica Acta* **106**, 207–212.

Hunte, C., Schnabl, H., Traub, O., Willecke, K. and Schulz, M. (1992) 'Immunological evidence of connexin-like proteins in the plasma membrane of *Vicia faba* L.', *Botanica Acta* **105**, 104–110.

Iglesias, V.A. and Meins, F., Jr. (2000) 'Movement of plant viruses is delayed in a β-1,3-glucanase-deficient mutant showing a reduced plasmodesmatal size exclusion limit and enhanced callose deposition', *The Plant Journal* **21**(2), 157–166.

Itaya, A., Woo, Y.-M., Masuta, C., Bao, Y., Nelson, R.S. and Ding, B. (1998) 'Developmental regulation of intercellular protein trafficking through plasmodesmata in tobacco leaf epidermis', *Plant Physiology* **118**, 373–385.

Janssen, B.J., Lund, L. and Sinha, N. (1998) 'Overexpression of a homeobox gene, *LeT6*, reveals indeterminate features in the tomato compound leaf', *Plant Physiology* **117**, 771–786.

Jones, D.L., Gilroy, S., Larson, P.B., Howell, S.H. and Kochian, L.V. (1998) 'Effect of aluminium on cytoplasmic $Ca^{2+}$ homeostasis in root hairs of *Arabidopsis thaliana* (L.)', *Planta* **206**, 378–387.

Jones, M.G.K. (1976) The origin and development of plasmodesmata, in *Intercellular Communication in Plants: Studies on Plasmodesmata* (eds B.E.S. Gunning and A.W. Robards), Springer-Verlag, Berlin, pp. 81–105.

Juniper, B.E. (1977) 'Some speculations of the possible roles of the plasmodesmata in the control of differentiation', *Theoretical Biology* **66**, 583–592.

Juniper, B.E. and Barlow, P.W. (1969) 'The distribution of plasmodesmata in the root tip of maize', *Planta* **89**, 352–360.

Kauss, H. (1987) 'Some aspects of calcium-dependent regulation in plant metabolism', *Annual Review of Plant Physiology* **38**, 47–72.

Kirby, A.R., Gunning, A.P., Waldron, K.W., Morris, V.J. and Ng, A. (1996) 'Visualization of plant cell walls by atomic force microscopy', *Biophysical Journal* **70**, 1138–1143.

Knight, A.E. and Kendrick-Jones, J. (1993) 'A myosin-like protein from a higher plant', *Journal of Molecular Biology* **231**, 148–154.

Kollmann, R. and Glockmann, C. (1985) 'Studies on graft unions I. Plasmodesmata between cells of plants belonging to different unrelated taxa', *Protoplasma* **124**, 224–235.

Kollmann, R. and Glockmann, C. (1991) 'Studies on graft unions III. On the mechanism of secondary formation of plasmodesmata at the graft interface', *Protoplasma* **165**, 71–85.

Kollmann, R. and Glockmann, C. (1999) Multimorphology and nomenclature of plasmodesmata in higher plants, in *Plasmodesmata: Structure, Function, Role in Cell Communication* (eds A.J.E. van Bel and W.J.P. van Kesteren), Springer-Verlag, Berlin, pp. 149–172.

Kotlizky, G., Shurtz, S., Yahalom, A., Malik, Z., Traub, O. and Epel, B.L. (1992) 'An improved procedure for the isolation of plasmodesmata embedded in clean maize cell walls', *The Plant Journal* **2**(4), 623–630.

Kragler, F., Lucas, W.J. and Monzer, J. (1998) 'Plasmodesmata: dynamics, domains and patterning', *Annals of Botany* **81**, 1–10.

Lachaud, S. and Maurousset, L. (1996) 'Occurrence of plasmodesmata between differentiating vessels and other xylem cells in *Sorbus torminalis* (L.) Crantz and their fate during xylem maturation', *Protoplasma* **191**, 220–226.

Laporte, C., Vetter, G., Loudes, A.-M. *et al.* (2003) 'Involvement of the secretory pathway and the cytoskeleton in intracellular targeting and tubule assembly of *Grapevine fanleaf virus* movement protein in tobacco BY-2 cells', *The Plant Cell* **15**, 2058–2075.

Lee, J.-Y. and Lucas, W.J. (2001) 'Phosphorylation of viral movement proteins – regulation of cell-to-cell trafficking', *Trends in Microbiology* **9**(1), 5–8.

Leisner, S.M. and Turgeon, R. (1993) 'Movement of virus and photoassimilate in the phloem: a comparative analysis', *BioEssays* **15**(11), 741–748.

López-Sáez, J.F., Giménez-Martín, G. and Risueño, M.C. (1966) 'Fine structure of plasmodesmata', *Protoplasma* **61**, 81–84.

Lucas, W.J. and Gilbertson, R.L. (1994) 'Plasmodesmata in relation to viral movement within leaf tissues', *Annual Review of Phytopathology* **32**, 387–411.

Lucas, W.J. and Wolf, S. (1993) 'Plasmodesmata: the intercellular organelles of green plants', *Trends in Cell Biology* **3**, 308–315.

Lucas, W.J., Ding, B. and van der Schoot, C. (1993) 'Plasmodesmata and the supracellular nature of plants', *New Phytologist* **125**, 435–476.

Madore, M.A. and Lucas, W.J. (1986) Characterization of the source leaf symplast by means of lucifer yellow CH, in *Phloem Transport* (eds J. Cronshaw, W.J. Lucas and R.T. Giaquinta), Alan R. Liss, Inc., New York, pp. 129–133.

Mahajan, S.K., Chisholm, S.T., Whitham, S.A. and Carrington, J.C. (1998) 'Identification and characterisation of a locus (*RTM1*) that restricts long-distance movement of tobacco etch virus in *Arabidopsis thaliana*', *The Plant Journal* **14**(2), 177–186.

Mas, P. and Beachy, R.N. (2000) 'Role of microtubules in the intracellular distribution of tobacco mosaic virus movement protein', *Proceedings of the National Academy of Sciences USA* **97**, 12345–12349.

Mauseth, J.D. (1988) *Plant Anatomy*, Benjamin/Cummings, California.

Medina-Escobar, N., Haupt, S., Thow, G., Boevink, P., Chapman, S. and Oparka, K.J. (2003) 'High-throughput viral expression of cDNA-green fluorescent protein fusions reveals novel subcellular addresses and identifies unique proteins that interact with plasmodesmata', *The Plant Cell* **15**, 1507–1523.

Meiners, S. and Schindler, M. (1989) 'Characterization of a connexin homologue in cultured soybean cells and diverse plant organs', *Planta* **179**, 148–155.

Meiners, S., Xu, A. and Schindler, M. (1991) 'Gap junction protein homologue from *Arabidopsis thaliana*: evidence for connexins in plants', *Proceedings of the National Academy of Sciences USA* **88**, 4119–4122.

Mercer, F.V. (1956) 'Cytology and the electron microscope', *Proceedings of the Linnean Society of New South Wales* **81**, 4–19.

Mezitt Provencher, L., Miao, L., Sinha, N. and Lucas, W.J. (2001) '*Sucrose Export Defective1* encodes a novel protein implicated in chloroplast-to-nucleus signaling', *The Plant Cell* **13**, 1127–1141.

Moore, A.E. and Stone, B.A. (1972) 'Effect of infection with TMV and other viruses on the level of β-1,3-glucan hydrolase in leaves of *Nicotiana glutinosa*', *Virology*, **50**, 791–798.

Morozov, S.Y. and Solovyev, A.G. (2003) 'Triple gene block: modular design of a multifunctional machine for plant virus movement', *Journal of General Virology* **84**, 1354–1366.

Morvan, O., Quentin, M., Jauneau, A., Mareck, A. and Morvan, C. (1998) 'Immunogold localization of pectin methylesterases in the cortical tissues of flax hypocotyl', *Protoplasma* **202**, 175–184.

Münch, E. (1930) *Die Stoffbewegungen in der Pflanze*, G. Fisher, Jena, Germany.

Murillo, I., Cavallarin, L. and San Segundo, B. (1997) 'The maize pathogenesis-related PRms protein localizes to plasmodesmata in maize radicles', *The Plant Cell* **9**, 145–156.

Mushegian, A.R. and Koonin, E.V. (1993) 'The proposed plant connexin is a protein kinase-like protein (letter to the editor)', *The Plant Cell* **5**, 998–999.

Nägeli, C. and Cramer, L. (1855) *Pflanzenphysiologische Untersuchungen No.1*, Schultheiss, Zurich.

Nougarède, A., Landré, P., Rembur, J. and Hernandez, M.N. (1985) 'Are variations in the activities of 5′-nucleotidase and adenylate cyclase components in the release of inhibition in the pea cotyledonary bud?', *Canadian Journal of Botany* **63**, 309–323.

Olesen, P. (1979) 'The neck constriction in plasmodesmata', *Planta* **144**, 349–358.

Olesen, P. and Robards, A.W. (1990) The neck region of plasmodesmata: general architecture and some functional aspects, in *Parallels in Cell-to-Cell Junctions in Plants and Animals* (eds A.W. Robards, W.J. Lucas, J.D. Pitts, H.J. Jongsma and D.C. Spray), Springer-Verlag, Berlin, pp. 145–170.

Oparka, K.J. (2004) 'Getting the message across: how do plant cells exchange macromolecular complexes?', *Trends in Plant Science* **9**(1), 33–41.

Oparka, K.J. and Santa Cruz, S. (2000) 'The great escape: phloem transport and unloading of macromolecules', *Annual Review of Plant Physiology and Plant Molecular Biology* **51**, 323–347.

Oparka, K.J., Murphy, R., Derrick, P., Prior, D.A.M. and Smith, J.A.C. (1991) 'Modification of the pressure-probe technique permits controlled intracellular microinjection of fluorescent probes', *Journal of Cell Science* **98**, 539–544.

Oparka, K.J., Roberts, A.G., Boevink, P. *et al.* (1999) 'Simple, but not branched, plasmodesmata allow the nonspecific trafficking of proteins in developing tobacco leaves', *Cell* **97**, 743–754.

Orfila, C. and Knox, J.P. (2000) 'Spatial regulation of pectic polysaccharides in relation to pit fields in cell walls of tomato fruit pericarp', *Plant Physiology* **122**, 775–781.

Overall, R.L. (1999) Substructure of plasmodesmata, in *Plasmodesmata: Structure, Function, Role in Cell Communication* (eds A.J.E. van Bel and W.J.P. van Kesteren), Springer-Verlag, Berlin, pp. 129–130.

Overall, R.L. and Blackman, L.M. (1996) 'A model of the macromolecular structure of plasmodesmata', *Trends in Plant Science* **1**(9), 307–311.

Overall, R.L., Wolfe, J. and Gunning, B.E.S. (1982) 'Intercellular communication in *Azolla* roots: I Ultrastructure of plasmodesmata', *Protoplasma* **111**, 134–150.

Palevitz, B.A. and Hepler, P.K. (1985) 'Changes in dye coupling of stomatal cells of *Allium* and *Commelina* demonstrated by microinjection of Lucifer yellow', *Planta* **164**, 473–479.

Pennazio, S., D'Agostino, G., Appiano, A. and Redolfi, P. (1978) 'Ultrastructure and histochemistry of the resistant tissue surrounding lesions of tomato bushy stunt virus in *Gomphrena globosa* leaves', *Physiological Plant Pathology* **13**, 165–171.

Peterson, C.A. and Rauser, W.E. (1979) 'Callose deposition and photoassimilate export in *Phaseolus vulgaris* exposed to excess cobalt, nickel and zinc', *Plant Physiology* **63**, 1170–1174.

Pfluger, J. and Zambryski, P.C. (2001) 'Cell growth: the power of symplastic isolation', *Current Biology* **11**, 436–439.

Porter, K.R. and Machado, R.D. (1960) 'Studies on the endoplasmic reticulum IV, its form and distribution during mitosis in cells of onion root tip', *Journal of Biophysical and Biochemical Cytology* **7**, 167–180.

Quader, H., Hofmann, A. and Schnepf, E. (1987) 'Shape and movement of the endoplasmic reticulum in onion bulb epidermis cells: possible involvement of actin', *European Journal of Cell Biology* **44**, 17–26.

Radford, J.E. and White, R.G. (1998) 'Localization of a myosin-like protein to plasmodesmata', *The Plant Journal* **14**(6), 743–750.

Radford, J.E., Vesk, M. and Overall, R.L. (1998) 'Callose deposition at plasmodesmata', *Protoplasma* **201**, 30–37.

Reichelt, S. and Kendrick-Jones, J. (2000) Myosins, in *Actin: A dynamic Framework for Multiple Plant Cell Functions* (eds C.J. Staiger, F. Baluška, D. Volkmann and P. Barlow), Kluwer Academic Publishers, Dordrecht, The Netherlands, pp. 29–44.

Reichelt, S., Knight, A.E., Hodge, T.P. *et al.* (1999) 'Characterization of the unconventional myosin VIII in plant cells and its localization at the post-cytokinetic cell wall', *The Plant Journal* **19**, 555–567.

Rinne, P.L.H. and van der Schoot, C. (1998) 'Symplasmic fields in the tunica of the shoot apical meristem coordinate morphogenetic events', *Development* **125**, 1477–1485.

Rinne, P.L.H., Kaikuranta, P.M. and van der Schoot, C. (2001) 'The shoot apical meristem restores its symplastic organisation during chilling-induced release from dormancy', *The Plant Journal* **26**(3), 249–264.

Robards, A.W. (1968a) 'A new interpretation of plasmodesmatal ultrastructure', *Planta* **82**, 200–210.

Robards, A.W. (1968b) 'Desmotubule – a plasmodesmatal substructure', *Nature* **218**, 784.

Robards, A.W. (1971) 'The ultrastructure of plasmodesmata', *Protoplasma* **72**, 315–323.

Robards, A.W. (1976) Plasmodesmata in higher plants, in *Intercellular Communication in Plants: Studies on Plasmodesmata* (eds B.E.S. Gunning and A.W. Robards), Springer-Verlag, Berlin, pp. 15–57.

Robards, A.W. and Lucas, W.J. (1990) 'Plasmodesmata', *Annual Review of Plant Physiology* **41**, 369–419.

Roberts, A.G. and Oparka, K.J. (2003) 'Plasmodesmata and the control of symplasmic transport', *Plant, Cell and Environment* **26**, 103–124.

Roberts, A.G., Santa Cruz, S., Roberts, I.M., Prior, D.A.M., Turgeon, R. and Oparka, K.J. (1997) 'Phloem unloading in sink leaves of *Nicotiana benthamiana*: comparison of a fluorescent solute with a fluorescent virus', *The Plant Cell* **9**, 1381–1396.

Roberts, I.M., Boevink, P., Roberts, A.G., Sauer, N., Reichel, C. and Oparka, K.J. (2001) 'Dynamic changes in the frequency and architecture of plasmodesmata during the sink–source transition in tobacco leaves', *Protoplasma* **218**, 31–44.

Robinson-Beers, K. and Evert, R.F. (1991) 'Fine structure of plasmodesmata in mature leaves of sugarcane', *Planta* **184**, 307–318.

Roy, S., Watada, A.E. and Wergin, W.P. (1997) 'Characterization of the cell wall microdomain surrounding plasmodesmata in apple fruit', *Plant Physiology* **114**, 539–547.

Ruan, Y.-L., Llewellyn, D.J. and Furbank, R.T. (2001) 'The control of single-celled cotton fiber elongation by developmentally reversible gating of plasmodesmata and coordinated expression of sucrose and $K^+$ transporters and expansin', *The Plant Cell* **13**, 47–60.

Russin, W.A., Evert, R.F., van der Veer, P.J., Sharkey, T.D. and Briggs, S.P. (1996) 'Modification of a specific class of plasmodesmata and loss of sucrose export ability in the *sucrose export defective1* maize mutant', *The Plant Cell* **8**, 645–658.

Russow, E. (1883) 'Uber die Perforation der Zellwand und den Zusammenhang der Protoplasmakörper benechbarter Zellen', *Sitzungsberichte der Gesellschaft-naturforschender bei der Universität Dorpat* **6**, 562.

Schnepf, E. and Sych, A. (1983) 'Distribution of plasmodesmata in developing *Sphagnum* leaflets', *Protoplasma* **116**, 51–56.

Schulz, M., Traub, O., Knop, M., Willecke, K. and Schnabl, H. (1992) 'Immunofluorescent localization of a connexin 26-like protein at the surface of mesophyll protoplasts from *Vicia faba* L. and *Helianthus annuus* L.', *Botanica Acta* **105**, 111–115.

Sivaguru, M., Fujiwara, T., Samaj, J. *et al.* (2000) 'Aluminium-induced 1 – >3-β-D-glucan inhibits cell-to-cell trafficking of molecules through plasmodesmata: A new mechanism of aluminium toxicity in plants', *Plant Physiology* **124**, 991–1005.

Staehelin, L.A. and Hepler, P.K. (1996) 'Cytokinesis in higher plants', *Cell* **84**, 821–824.

Steinberg, G. and Kollmann, R. (1994) 'A quantitative analysis of the interspecific plasmodesmata in the non-division walls of the plant chimera *Laburnocytisus asamii* (Poit.) Schneid.', *Planta* **192**, 75–83.

Stone, B.A. and Clarke, A.E. (1992) *Chemistry and Biochemistry of (1-3)-β-Glucans*, La Trobe University Press, Melbourne, Australia.

Strasburger, E. (1901) 'Über plasmaverbindungen pflanzlicher Zellen', *Jahrbücher für Wissenschaftliche Botanik* **36**, 493–610.

Strugger, S. (1957) 'Der elektronenmikroskopische Nachweis von Plasmodesmen mit Hilfe der Uranylimprägnierung an Wurzelmeristemen', *Protoplasma* **48**, 365–367.

Susi, P. (2000) 'Dye-coupling in tobacco mesophyll cells surrounding growing tobacco mosaic tobamovirus-induced local lesions', *Journal of Phytopathology* **148**, 379–382.

Tangl, E. (1879) 'Ueber offene Communicationen zwischen den Zellen des Endosperms einiger Samen', *Jahrbüecher füer Wissenschaftliche Botanik* **12**, 170–190.

Tilney, L.G., Cooke, T.J., Connelly, P.S. and Tilney, M.S. (1991) 'The structure of plasmodesmata as revealed by plasmolysis, detergent extraction and protease digestion', *Journal of Cell Biology* **112**(4), 739–747.

Tucker, E.B. (1987) 'Cytoplasmic streaming does not drive intercellular passage in staminal hairs of *Setcreasea purpurea*', *Protoplasma* **137**, 140–144.

Tucker, J.E., Mauzerall, D. and Tucker, E.B. (1989) 'Symplastic transport of carboxyfluorescein in staminal hairs of *Setcreasea purpurea* is diffusive and includes loss to the vacuole', *Plant Physiology* **90**, 1143–1147.

Turner, A., Wells, B. and Roberts, K. (1994) 'Plasmodesmata of maize root tips: structure and composition', *Journal of Cell Science* **107**, 3351–3361.

van Bel, A.J.E. and Kempers, R. (1996) 'The pore/plasmodesm unit: key element in the interplay between sieve element and companion cell', *Progress in Botany* **58**, 278–291.

van Bel, A.J.E. and Knoblauch, M. (2000) 'Sieve element and companion cell: the story of the comatose patient and the hyperactive nurse', *Australian Journal of Plant Physiology* **27**, 477–487.

van der Schoot, C. and Rinne, P.L.H. (1999) The symplasmic organisation of the shoot apical meristem, in *Plasmodesmata: Structure, Function, Role in Cell Communication* (eds A.J.E. van Bel and W.J.P. van Kesteren), pp. 225–242.

van der Schoot, C., Dietrich, M.A., Storms, M., Verbeke, J.A. and Lucas, W.J. (1995) 'Establishment of a cell-to-cell communication pathway between separate carpels during gynoecium development', *Planta* **195**, 450–455.

van Gestel, K., Slegers, H., von Witsch, M., Samaj, J., Baluška, F. and Verbelen, J.-P. (2003) 'Immunological evidence for the presence of plant homologues of the actin-related protein Arp3 in tobacco and maize: subcellular localization to actin-enriched pit fields and emerging root hairs', *Protoplasma* **222**, 45–52.

Vaughn, K.C., Hoffman, J.C., Hahn, M.G. and Staehelin, L.A. (1996) 'The herbicide dichlobenil disrupts cell plate formation: immunogold characterization', *Protoplasma* **197**, 117–132.

Volk, G.M., Turgeon, R. and Beebe, D.U. (1996) 'Secondary plasmodesmata formation in the minor-vein phloem of *Cucumis melo* L. and *Cucurbita pepo* L.', *Planta* **199**, 425–432.

Volkmann, D., Mori, T., Tirlapur, U.K. *et al.* (2003) 'Unconventional myosins of the plant-specific class VIII: endocytosis, cytokinesis, plasmodesmata/pit-fields and the cell-to-cell coupling', *Cell Biology International* **27**, 289–291.

Waigmann, E. and Zambryski, P.C. (1995) 'Tobacco mosaic virus movement protein-mediated protein transport between trichome cells', *The Plant Cell* **7**, 2069–2079.

Waigmann, E., Chen, M.-H., Bachmaier, R., Ghoshroy, S. and Citovsky, V. (2000) 'Regulation of plasmodesmal transport by phosphorylation of tobacco mosaic virus cell-to-cell movement protein', *The EMBO Journal* **19**(18), 4875–4884.

Waigmann, E., Turner, A., Peart, J., Roberts, K. and Zambryski, P.C. (1997) 'Ultrastructural analysis of leaf trichome plasmodesmata reveals major differences from mesophyll plasmodesmata', *Planta* **203**, 75–84.

Wells, A.L., Lin, A.W., Chen, L.-Q. *et al.* (1999) 'Myosin VI is an actin-based motor that moves backwards', *Nature* **401**, 505–508.

Whaley, W.G., Mollenhauer, H.H. and Leech, J.H. (1960) 'The ultrastructure of the meristematic cell', *American Journal of Botany* **47**, 401–449.

White, R.G., Badelt, K., Overall, R.L. and Vesk, M. (1994) 'Actin associated with plasmodesmata', *Protoplasma* **180**, 169–184.

Wille, A.C. and Lucas, W.J. (1984) 'Ultrastructural and histochemical studies on guard cells', *Planta* **160**, 129–142.

Wolf, S., Deom, C.M., Beachy, R.N. and Lucas, W.J. (1989) 'Movement protein of tobacco mosaic virus modifies plasmodesmatal size exclusion limit', *Science* **246**, 377–379.

Wolf, S., Deom, C.M., Beachy, R.N. and Lucas, W.J. (1991) 'Plasmodesmatal function is probed using transgenic tobacco plants that express a virus movement protein', *The Plant Cell* **3**, 593–604.

Wu, J.H. and Dimitman, J.E. (1970) 'Leaf structure and callose formation as determinants of TMV movement in bean leaves as revealed by UV irradiation studies', *Virology* **40**, 820–827.

Yahalom, A., Lando, R., Katz, A. and Epel, B.L. (1998) 'A calcium-dependent protein kinase is associated with maize mesocotyl plasmodesmata', *Journal of Plant Physiology* **153**, 354–362.

Yahalom, A., Warmbrodt, R.D., Laird, D.W. *et al.* (1991) 'Maize mesocotyl plasmodesmata proteins cross-react with connexin gap junction protein antibodies', *The Plant Cell* **3**, 407–417.

Zambryski, P.C. and Crawford, K. (2000) 'Plasmodesmata: gatekeepers for cell-to-cell transport of developmental signals in plants', *Annual Review of Cell and Developmental Biology* **16**, 393–421.

Zee, S.-Y. (1969) 'Fine structure of the differentiating sieve elements of *Vicia faba*', *American Journal of Botany* **17**, 441–456.

# 2 Evolution of plasmodesmata

John A. Raven

## 2.1 Introduction

Consideration of the evolution of plasmodesmata (Raven, 1997; Cook and Graham, 1999) requires that the present-day distribution of these structures be known, and this chapter begins with a survey of the occurrence of plasmodesmata starting with the higher plants, on which most of the structural and functional studies have been performed. The distribution of plasmodesmata and structures such as pit plugs that may serve a similar function is then considered in relation to what is known about the evolution of photosynthetic organisms. In particular, the likely number of independent origins of plasmodesmata and related structures is estimated for the algae. The extent of intercellular differentiation in the various clades of organisms with plasmodesmata is then analysed to determine if the degree of complexity of structure is related to the presence of plasmodesmata, and hence to the selective pressures that may have been involved in their evolution. This necessarily involves a comparison of the minimum extent of differentiation found in organisms possessing plasmodesmata or their possible functional analogues, such as the pit plugs of red algae, with the maximum degree of differentiation found in three groups of large photosynthetic organisms lacking such intercellular connections. These three groups are: multicellular organisms lacking plasmodesmata, large differentiated acellular organisms such as the ulvophycean green alga *Caulerpa* and symbioses of cyanobacteria and algae with fungi and metazoans having their own intercellular transport connections.

## 2.2 The distribution of plasmodesmata among extant photosynthetic organisms

The taxonomic scheme used here differs from that employed by Raven (1997), which was derived mainly from Van den Hoek *et al.* (1995). For the Phylum (= Division) level and below, the scheme of Graham and Wilcox (2000) is used. This differs from Van den Hoek *et al.* (1995) mainly in implicitly assuming fewer independent origins of more complex morphological forms from unicellular flagellates among the green algae. The scheme of Graham and Wilcox (2000) has been supplemented by subsequent findings (Nozaki *et al.*, 2000; Draisma *et al.*, 2001; Rousseau *et al.*, 2001; Leonardi *et al.*, 2002; Lopez-Bautista and Chapman, 2002). Taxonomy above the

Phylum level follows Baldauf (2003), but with her 'Heterokonta' extended to include the Cryptophyta and Haptophyta, as well as the Heterokontophyta and the Chromista (Cavalier-Smith, 1986).

Plasmodesmata occur in all higher plants or embryophytes, i.e. bryophytes or tracheophytes (Table 2.1; Cook and Graham, 1999). The default, and plesiomorphic, developmental state is that all cells within a given phase of the life cycle are interconnected by plasmodesmata. Plasmodesmata do not occur between the sporophyte and the gametophyte phases of the life cycle; this again is the plesiomorphic condition. The apomorphic condition within a plant is the absence of plasmodesmata between certain cells due to loss during development, between mature guard cells and other cells, and between sieve tube–companion cell complexes and other cells in source organs of some flowering plants. Embryophytes evolved from charophycean green algae (Division/Phylum Chlorophyta); the immediate charophycean ancestors of the embryophytes (Charales and, more distantly, Coleochaetales: Karol *et al.*, 2001) possessed plasmodesmata, although most extant members of the Charophyceae lack plasmodesmata (Table 2.1). The Charophyceae is one of the five currently recognized Classes of the Chlorophyta (Graham and Wilcox, 2000). Of the other classes of green algae, plasmodesmata occur in some members (Orders Chaetophorales and Oedogoniales) of the Chlorophyceae, and in the Order Trentepohliales and the genus *Sporocladopsis* of the Ulvophyceae (Raven, 1997; Graham and Wilcox, 2000; Chapman *et al.*, 2001; Leonardi *et al.*, 2002; Lopez-Bautista and Chapman, 2003; Lopez-Bautista *et al.*, 2003; Table 2.1). These two Classes also have members with intercellular cytoplasmic connections which are structurally different from plasmodesmata, i.e. in colonial flagellate *Volvox* in the Chlorophyceae and in the unbranched filamentous *Ctenocladus* and *Smithsoniella* in the Ulvophyceae (Table 2.1). There are no plasmodesmata in the other two classes of green algae – the Prasinophyceae (with no multicellular members) and the Trebouxiophyceae (with unicellular and multicellular representatives) (Table 2.1).

The only other algae with plasmodesmata are the brown algae (Phaeophyceae) in the Division/Phylum Heterokontophyta (Raven, 1997; Graham and Wilcox, 2000; Table 2.2). All extant brown algae are multicellular and some attain considerable complexity (Bell and Mooers, 1997). Several other Classes of heterokonts have multicellular representatives [e.g. the Chrysophyceae, Pelagophyceae, Phaeothamniophyceae, Schizocladiophyceae and Tribophyceae (= Xanthophyceae)], but none have plasmodesmata (Raven, 1997; Graham and Wilcox, 2000; Kawai *et al.*, 2003; Table 2.2). Cavalier-Smith (1986) places the heterokonts in the Kingdom Chromista with two other algal Divisions/Phyla – the Cryptophyta and the Haptophyta. Although there are a few multicellular haptophytes, there are no reports of plasmodesmata in this higher taxon (Raven, 1997; Graham and Wilcox, 2000; Table 2.2). The other algae are either all small unicells (Chlorarachniophyta in the Cercozoa; Euglenophyta in the Discicristata) or have multicellular representatives lacking plasmodesmata (Dinophyta in the Alveolata) (Raven, 1997; Graham and Wilcox, 2000; Table 2.2).

**Table 2.1** Occurrence of structural features and intercellular connections in green algae and higher plants (Viridiplantae *sensu* Cavalier-Smith, 1981)

| Taxon | Extent of differentiation | Occurrence of plasmodesmata |
|---|---|---|
| **CHLOROPHYTA** | | |
| **Prasinophyceae** | Wall-less flagellate or walled coccoid unicells; phycoma (resting cysts) up to 1 mm in diameter | No |
| **Chlorophyceae** | | |
| Sphaeopleales | Small to large cells, colonies, unbranched filaments | No |
| Volvocales | Flagellate unicells or colonies | No[a] |
| Tetracystales | Unicells or colonies | No |
| Dunaliellales | Flagellate unicells or colonies | No |
| Chaetophorales | Branched or unbranched filaments | Yes |
| Oedogoniales | Branched or unbranched filaments | Yes |
| **Trebouxiophyceae** | Unicells, colonies, unbranched or branched filaments, foliose thalli | No |
| **Ulvophyceae** | | |
| Ulotrichales | Unbranched or branched filaments, foliose thalli | No |
| Ulvales | Unbranched or branched filaments, foliose thalli | No[b] |
| Siphonocladales | Unbranched or branched filaments, some with large cells | No |
| Caulerpales | Acellular, some very large and differentiated | No |
| Dasycladales | Acellular, some very large and differentiated | No |
| Trentepohliales | Branched filaments | Yes |
| *Sporocladopsis*[c] | Unbranched filaments | Yes |
| **Charophyceae** | | |
| Klebsormidiales | Unbranched filaments | No |
| Zygnematales | Unicells, unbranched filaments | No |
| Coleochaetales | Branched filaments | Yes |
| Charales | Branched filaments, large internodal cells, parenchymatous nodal cells | Yes |
| **Embryophyta** | Branched filaments with differentiated gametangia in some free-living gametophytes, usually (always in sporophytes) highly differentiated parenchymatous tissue | Yes |

[a]Intercellular cytoplasmic connections in *Volvox* are not true plasmodesmata.

[b]Intercellular cytoplasmic connections in *Ctenocladus* and *Smithsoniella* are not true plasmodesmata. The phylogenetic position of these two genera is not entirely clear. C.J. O'Kelly (http://www.bigelow.org/~cjo/ucdefb.html) regards *Ctenocladus* as basal in the Ulvales.

[c]*Sporocladopsis* has not been assigned to an Order.

*Source*: Based on Tables 1 and 2 of Raven (1997), but with green algal taxonomy based on Graham and Wilcox (2000) with additional data from Nozaki *et al.* (2000), Leonardi *et al.* (2002) and Lopez-Bautista and Chapman (2003).

**Table 2.2** Occurrence of structural features and intercellular connections in algae other than the Chlorophyta

| Taxon | Extent of differentiation | Occurrence of plasmodesmata |
|---|---|---|
| **PLANTAE**[a] | | |
| **GLAUCOPHYTA** | Unicellular, often flagellate | No |
| **RHODOPHYTA** | | |
| Bangiophyceae | Unicellular, unbranched and branched filaments | No[b] |
| Floridiophyceae | Unbranched and branched filaments, complex pseudoparenchymatous thalli | No[c] |
| **CERCOZOA** | | |
| **CHLORARACHNIOPHYTA** | Amoeboid or flagellate unicells | No |
| **ALVEOLATA** | | |
| **DINOPHYTA** | Unicellular, unbranched and branched filaments | No |
| **CHROMISTA**[d] | | |
| **CRYPTOPHYTA** | Unicellular, usually flagellate | No |
| **HAPTOPHYTA** | Unicellular, unbranched and branched filaments | No |
| **HETEROKONTOPHYTA** | | |
| Chrysophyceae | Unicellular, colonial, unbranched and branched filaments | No |
| Synurophyceae | Unicellular, colonial | No |
| Pedinellophyceae | Unicellular | No |
| Parmaphyceae | Unbranched and branched filaments | No |
| Xanthophyceae[e] | Unicellular, large cells, unbranched and branched filaments | No |
| Eustigmatophyceae | Unicellular | No |
| Bacillariophyceae | Unicellular, large cells, unbranched filaments | No |
| Bolidophyceae | Unicellular | No |
| Raphidophyceae | Unicellular | No |
| Dictyochophyceae | Unicellular | No |
| Pelagophyceae[f] | Unicellular, colonial, unbranched or branched filaments | No |
| Schizocladiophyceae | Branched filaments | No |
| Phaeophyceae[g] | Branched filaments, complex parenchymatous parenchymatous or pseudoparenchymatous thalli | Yes |
| **DISCICRISTATA** | | |
| **EUGLENOPHYTA** | Unicellular | No |

[a]Plantae *sensu* Cavalier-Smith (1981) includes the Chlorophyta and Embryophyta considered in Table 2.1.

[b]Some branched filamentous bangiophytes [e.g. *Conchocelis* (sporophytes) phase of *Bangia/Porphyra*] have pit plugs, others (e.g. *Stylonema*) do not (Brodie and Irvine, 2002).

[c]All floridiophytes have pit plugs.

[d]Chromista *sensu* Cavalier-Smith (1986), i.e. including cryptophytes and haptophytes with the heterokonts: cf. Baldauf (2003).

[e]= Tribophyceae.

[f]*sensu* Saunders *et al.* (1997), i.e. including Sarcinochrysidales (Sarcinochrysidophyceae) and Chrysomeridales.

[g]= Fucophyceae.

*Source*: Based on Table 3 of Raven (1997), with phylogeny modified to take into account Saunders *et al.* (1997), Graham and Wilcox (2000), Draisma *et al.* (2001), Rousseau *et al.* (2001), Baldauf (2003) and Kawai *et al.* (2003).

The emphasis on extant plants should not discourage further searches for plasmodesmata in fossil algae. There seem to be no reports of plasmodesmata-sized pores in the cell walls of fossil algae, although they have been reported in fossils of higher plants (Edwards, 2003). Pit plugs, which are an order of magnitude larger in diameter than plasmodesmata (Pueschel, 1980), have been reported for fossil red algae (Vannucci *et al.*, 2000).

## 2.3 The phylogeny of photosynthetic organisms and its relation to the occurrence of plasmodesmata

Molecular phylogenetic methods combined with cladistic techniques have put our understanding of the evolution of photosynthetic organisms (and, of course, other organisms) on a much firmer basis. However, many ambiguities remain, and this brief account of the phylogeny of algae and higher plants in relation to the occurrence of plasmodesmata should be regarded as provisional, especially in terms of the branching order at the Kingdom level and above in the Eukarya.

It is certain that oxygen-producing photosynthesis had a single evolutionary origin among the immediate ancestors of the extant cyanobacteria (Monera: Bacteria). The cyanobacteria are paraphyletic, i.e. are a group which arose from a common ancestor sharing all of the unique characteristics of the group, but which does not contain all of the members of that group. This comes about by the consideration of the occurrence, by vertical transmission after one or more sequential endosymbiotic events, of many cyanobacterial genes ($\sim$18% of the total gene complement in *Arabidopsis*) in the eukaryotic photosynthetic organisms (Martin *et al.*, 2002). The absolute number of cyanobacterium-derived genes found in *Arabidopsis* means that the plastid ancestor must have had a total number of genes similar to those found in such relatively complex cyanobacteria as *Nostoc* (Martin *et al.*, 2002). Such multicellular cyanobacteria have intercellular connections which are analogous to gap junctions rather than to plasmodesmata, so that it is unlikely that the origins of plasmodesmata in (eukaryotic) algae and plants are to be found in the plastid ancestor.

The majority of evidence favours a single primary endosymbiotic event by which a cyanobacterium became symbiotic in, and then genetically integrated with, the common ancestor of the Divisions/Phyla Glaucocystophyta, Rhodophyta and Chlorophyta plus Embryophyta. This group of Phyla is termed ‘Plants’ by Baldauf (2003), equivalent to the Plantae of Cavalier-Smith (1981). Among the plants, plasmodesmata are only found in the Chlorophyta and the Embryophyta (Tables 2.1 and 2.2). As we have already seen, it is highly likely that the plasmodesmata of the embryophytes were inherited from their charophycean ancestors. Since the Charophyceae, Chlorophyceae and Ulvophyceae were derived from different unicellular ancestors, it is very likely that the plasmodesmata of these three Classes had independent origins (Raven, 1997; Graham and Wilcox, 2000; Chapman *et al.*, 2001; Leonardi *et al.*, 2002; Lopez-Bautista and Chapman, 2003; Lopez-Bautista *et al.*, 2003). Within the Chlorophyceae, the branching order of the Orders Oedogoniales and Chaetophorales is unknown, as is the closeness of their relation; hence, it is not clear if the occurrence of plasmodesmata in both of these Orders can be explained by a single origin with no subsequent loss, or whether by two origins. Alternatively, there is the possibility of a single origin with loss of plasmodesmata from any Orders which are more closely related to the Oedogoniales or the Chaetophorales than these two Orders are to each other. Two origins of plasmodesmata within the Chlorophyceae would be consistent with the very different modes of cell division and cell expansion found in the Chaetophorales and in the Oedogoniales (Raven, 1997; Graham and Wilcox, 2000). For the Ulvophyceae, there is the possibility that plasmodesmata evolved independently in the ancestor of the Trentepohliales and in that of *Sporocladopsis*, since there is little morphological evidence other than the occurrence of plasmodesmata to link the two taxa, and there is no molecular phylogenetic evidence on *Sporocladopsis*.

The other photosynthetic eukaryotes attained their plastids by secondary endosymbioses involving either green algae (Chlorarachniophyta, Euglenophyta) or red algae (Cryptophyta, Dinophyta, Haptophyta, Heterokontophyta) as the plastid donors (Graham and Wilcox, 2000; Falkowski *et al.*, 2004). The Chlorarachniophyta are in the Cercozoa, while the Euglenophyta are members of the Discicristata (Baldauf, 2003), and there is agreement that the 'green' plastids in these two Phyla originated from two separate secondary endosymbioses. The Dinophyta are in the higher taxon Alveolata, while the Cryptophyta, Haptophyta and Heterokontophyta are in what Baldauf terms the 'Heterokonts', but which would be better referred to as Chromista (Cavalier-Smith, 1986). There is still controversy as to the number of secondary endosymbiotic events involved in the occurrence of 'red' plastids in the Chromista and the Alveolata. Cavalier-Smith (2003) favours a single secondary endosymbiotic event at the base of the Chromista plus Alveolata (the Chromalveolate Hypothesis), while others prefer separate origins for the plastids of (at least) the Alveolata and the Chromista.

Since the only algae from the six Phyla mentioned in the last paragraph which possess plasmodesmata are the Phaeophyceae in the Phylum Heterokontophyta (Tables 2.1 and 2.2), the only point at issue is whether these plasmodesmata evolved in the ancestral brown algae, or were inherited from the plastid ancestor or the host-cell ancestor. The rationale for transfer with the plastid ancestor is not, of course, that the information for producing plasmodesmata is thought to be encoded in the plastid genome. The reason is that many nuclear genes must also have been transferred to the host nucleus from the plastid ancestor because most genes needed for plastid function are nuclear-coded, and many other genes might also have been transferred in the secondary endosymbioses similar to what happened in the primary endosymbiosis (Martin *et al.*, 2002; Falkowski *et al.*, 2004). Both of the plastid ancestor and the host-cell possibilities are unlikely as vehicles for the genes required to produce plasmodesmata. The plastid ancestor of the heterokonts was a bangiophycean red alga, and although some members of the Bangiophyceae have pit plugs, none have plasmodesmata. The host cell for the secondary symbiosis leading to the heterokont (at least) plastids has not been clearly defined, but it was not a plasmodesmata-containing green alga.

These considerations suggest that there were certainly four origins of plasmodesmata, i.e. in the common ancestor of the Coleochaetales plus Charales in the Charophyceae, leading to the Embryophyta, in the common ancestor of the Chaetophorales plus Oedogoniales in the Chlorophyceae, in the ancestor of the Trentepohliales plus *Sporocladopsis* in the Ulvophyceae and in the ancestor of the Phaeophyceae. There is also the possibility of five separate origins, i.e. the first, third and fourth origins mentioned above plus separate origins in the Chaetophorales and the Oedogoniales. The molecular phylogenetic studies of Booton *et al.* (1998) who showed that the Chaetophorales and the Oedogoniales are not closely related favours the five separate origins of plasmodesmata in the Chlorophyta. An additional origin, i.e. six in all, would be required if plasmodesmata originated independently in the Trentepohliales and in *Sporocladopsis*.

Independent evidence as to the number of independent origins of plasmodesmata in green algae would be desirable, supplementing the molecular phylogenetic analyses based on nuclear-encoded genes which necessarily are not related to plasmodesmatal structure or function, e.g. 18S ribosomal RNA. One possible line of evidence could come from further fine-structural analysis indicating either shared characteristics among plasmodesmata which are of such complexity or peculiarity that they are unlikely to have evolved more than once, or a range of complex and/or peculiar structural features which differ among plasmodesmata in different taxa. Another, and perhaps more likely, source of evidence in the future is the identification of plasmodesmata-specific components at the molecular genetic rather than the purely structural data, although microscopy would be needed to determine, after suitable labelling for fluorescence microscopy or transmission electron microscopy, whether the putative components of plasmodesmata were indeed found in plasmodesmata. A start has been made with this in algae for the charalean *Chara corallina* (Blackman and Overall, 1998; Blackman *et al.*, 1998).

## 2.4 Functional aspects of the distribution and evolution of plasmodesmata

### *2.4.1 Background*

Plasmodesmata of higher plants have been known for several decades to function in symplasmic (Munch, 1930) transport of low relative molecular mass solutes, and more recent work showed a cut-off of approximately 1000 Da. The simple notion was that symplasmic communication, in the sense of transferring solutes between differentiated cells with different synthetic capacities and extents of access to apoplasmic solutes, could be contrasted with the lack of symplasmic communication for the informational macromolecules whose differential expression among cells underpinned differentiation. This view has been challenged even more recently with the finding that there is a regulated transfer of informational macromolecules through higher plant plasmodesmata, starting with the work of Kempers *et al.* (1993): [see Lee *et al.* (2000) and

Oparka *et al.* (2004)]. Furthermore, functional symplasmic domains occur which are subsets of the apparent size of the symplasm as determined by the occurrence of plasmodesmata. Transmission electron microscopy shows that the boundaries of such domains are usually characterized by plugs of osmophilic, electron-dense material in plasmodesmata in the cases of higher plants (Ehlers *et al.*, 1999) and the Charales (Kwiatkowska *et al.*, 2003). Thus, the view rightly persists that plasmodesmata are in some way related to differentiation in algae and plants with the regulated transfer of informational macromolecules permitting differentiation to be established, and the less restricted movement of small solutes allowing differentiated parts of the plant to interact at the level of nutrient solute sources and sinks (Raven, 1997; Cook and Graham, 1999; Graham *et al.*, 2000).

The structure of algal plasmodesmata differs from that of higher plants, not least by the general absence of a desmotubule or any obvious structural analogue in the algae (Cook *et al.*, 1997; Raven, 1997; Cook and Graham, 1999). However, Marchant (1976) reports desmotubules in *Uronema* and *Stigeoclonium* (Chlorophyceae: Chaetophorales), and Cook *et al.* (1997) comment on associations of endoplasmic reticulum with the plasmodesmata of *Chara*. The absence of a desmotubule from at least some plasmodesmata of charalean algae is in agreement with the only published estimate of the upper size limit for transfer of molecules through algal plasmodesmata which showed that molecules up to 45 kDa could move between cells of the charalean alga *Nitella* (Kituyama *et al.*, 1992). Despite this, the extent of differentiation, and hence the probable need for transfer of low-molecular mass solutes between cells of different position and/or metabolic potential, is a convenient (because readily observed: Bell and Mooers, 1997) datum to which the occurrence of plasmodesmata can be related.

### 2.4.2 *Cyanobacteria*

The cyanobacteria do not have plasmodesmata, but are briefly considered here because of their evolutionary relationship to organisms with plasmodesmata via the endosymbiotic origin of plastids, and the presence of many more cyanobacterial genes in higher plants than are involved in photosynthesis (Martin *et al.*, 2002). As has been indicated above, the complex cyanobacteria which were ancestral to the plastids had gap-junction-like intercellular connections between vegetative cells (Adams and Duggan, 1999). In addition, there would also have been cytoplasmic connections through cell walls between vegetative cells and a neighbouring heterocyst (a cell specialized in biological nitrogen fixation). These cytoplasmic connections are involved in the transport of organic carbon from photosynthetically competent photosynthetic cells to the heterocysts, and the movement of combined nitrogen resulting from nitrogen fixation into the vegetative cells (Adams and Duggan, 2002). These intercellular connections do not have the structure of plasmodesmata.

### 2.4.3 Chlorophyta

Five of the Orders of the Chlorophyta which have plasmodesmata (Charales and Coleochaetales; Chaetophorales and Oedogoniales; Trentepohliales) have members with branched filaments or more complex vegetative structures (Raven, 1997; Graham and Wilcox, 2000; Table 2.1). To this level of complexity of vegetative structure can be added differentiated antheridia and oogonia in the Coleochaetales and Charales, with the likelihood of symplasmic transfer of resources to the developing gametangia (Kwiatkowska *et al.*, 2003). Furthermore, there are colourless (non-photosynthetic) vegetative cells in the Chaetophorales (rhizoids in the soil-dwelling *Fritschiella*) and the Oedogoniales (hair cells in the epiphytic or epilithic *Bulbochaete*): Raven (1997), Graham and Wilcox (2000). In both of these cases, there is presumably symplasmic transport of photosynthate to the colourless cells, while in *Fritschiella* there could also be symplasmic transfer of soil-derived nutrients to the aerial photosynthetic filaments. Members of the Charales also have non-photosynthetic rhizoid cells growing in fine-grained, usually nutrient-rich, sediment; tracer experiments show that these rhizoids are responsible for the uptake of a large fraction of the $HPO_4^{2-}$ and $NH_4^+$ acquired by the whole alga, and that there is symplasmic transfer of the labelled nitrogen and phosphorus to the submerged photosynthetic shoot of the alga (Box *et al.*, 1984; Raven, 2003). Experiments with labelled inorganic carbon supplied to illuminated shoots of charalean algae show symplasmic distribution from sites at which photosynthesis exceeds the demand for photosynthate in growth to sites where the reverse is the case, such as growing apices; organic carbon is also supplied to the rhizoids (Raven and Smith, 1978; Raven, 1981, 2003; Andrews *et al.*, 1984). Finally, there is a taxonomically widespread occurrence of localised growth in these Orders, again raising the possibility of nutrient solute transfer from non-growing cells to dividing and expanding cells (Raven, 1997; Graham and Wilcox, 2000). The exception to these generalizations is *Sporocladopsis* which consists of unbranched filaments (Leonardi *et al.*, 2002; Table 2.1). These relatively complex structural and developmental features could be seen as representing a functional 'requirement' for plasmodesmata in terms of permitting differentiation, and in facilitating symplasmic transfer of the low relative molecular mass products of the 'division of labour' from source to sink cells. However, the two chlorophycean Orders also have unbranched filamentous members (i.e. *Uronema* in the Chaetophorales and *Oedogonium* in the Oedogoniales), so that it is possible that plasmodesmata in these two Orders evolved in unbranched filamentous organisms. Alternatively, plasmodesmata could have been retained in apomorphic (derived) unbranched forms if the plesiomorphic (ancestral) vegetative condition in which plasmodesmata evolved in these Orders was branched filaments. Further molecular phylogenetic work is needed to distinguish between these possibilities.

The non-plasmodesmatal intercellular connections in the unbranched filaments of the ulvophycean genera *Ctenocladus* and *Smithsoniella* are as functionally enigmatic as those of the unbranched members of the chlorophycean Orders Chaetophorales and

Oedogoniales considered above (Table 2.1). The intercellular connections in the polyphyletic colonial flagellate genus *Volvox* may be involved in signalling in relation to directional swimming of the colonies as well as to differentiation among cells, and especially the distinction between somatic and germ cells (Kirk, 1998; Nozaki *et al.*, 2000; Kaiser, 2001; Table 2.1).

### *2.4.4 Heterokontophyta*

The only heterokonts with plasmodesmata are in the Class Phaeophyceae (= Fucophyceae) where all species examined have been shown to have plasmodesmata lacking desmotubules (Table 2.1). The least complex morphological forms in the brown algae are branched filaments (Raven, 1997; Graham and Wilcox, 2000). The most complex parenchymatous and pseudoparenchymatous forms reach very large sizes (up to 50 m in length and 180 kg fresh weight; Raven, 1999). There has been a polyphyletic evolution of long-distance transport systems in these larger brown algae analogous to higher plant phloem, with a more or less pronounced increase in the diameter of plasmodesmata in the end walls of the conducting elements during ontogeny, reaching a diameter of 1 μm in *Macrocystis* (Raven, 2003). These phloem-like long-distance transport mechanisms have been subjected to very considerable physiological investigation (Raven, 2003). By contrast, as we shall see, unmodified plasmodesmata in more typical vegetative cells have received much less physiological attention.

Molecular phylogenetic analysis shows that the least derived of extant brown algae are the parenchymatous Dictyotales, while the Ectocarpales and Scytosiphonales with their predominantly branched filamentous growth form, and several Orders with complex parenchymatous or pseudoparenchymatous thalli, are more derived within the phylogenetic scheme (Draisma *et al.*, 2001; Rousseau *et al.*, 2001). The Dictyotales have relatively few cell types (Bell and Mooers, 1997) despite their large, dichotomous thalli with apical growth, although *Dictyopteris* has a midrib containing differentiated elongate cells whose structure is consistent with a role in long-distance transport (Katsaros and Galatis, 1988; Raven, 2003). However, nothing seems to be known of whether the rate of apical growth in the Dictyotales can be accounted for in terms of local photosynthesis in excess of local respiration. There seem to be no cells in the thallus of the Dictyotales which lack the capacity for photosynthesis, and most of the thallus area is only two-cell layers thick, although in *Dictyopteris* there is a slightly thickened midrib more than two cells thick. This indicates a limited requirement for radial symplasmic transport of photosynthate, and little more of nutrient solutes at low concentrations in the growth medium (e.g. $NH_4^+$, $NO_3^-$, $HPO_4^{2-}$) for which uptake at the outer surface of superficial cells is likely in terms of limitations on diffusion through diffusion boundary layers and apoplasm (Raven, 1984). Another possible requirement for symplasmic transport of resources in the Dictyotales is the growth of gametangia and sporangia in the isomorphic diplohaplontic life cycle. At all events, the thalli of the least derived brown algae as indicated by molecular phylogenetic studies, and even of more derived branched filaments with differentiated sporangia and gametangia, are more complex than the branched filaments of the sister-class Schizocladiophyceae (Kawai *et al.*, 2003).

### *2.4.5 Conclusions*

Consideration of the structure of the algal taxa in which plasmodesmata are found, and the probable involvement of symplasmic transport of resources in growth, development and reproduction, permits some rationalization of their occurrence in terms of likely function. However, organisms like *Sporocladopsis* with unbranched filaments, no differentiation of sporangia or gametangia, no evidence of localized growth and no obvious relatives with more complex morphology, also have plasmodesmata. This poses problems as to the function providing the increased fitness of organisms with plasmodesmata. This is perhaps the most extreme case of the occurrence of plasmodesmata in organisms with very little obvious differentiation, and which are not obviously derived from a plasmodesmata-containing organism of more complex morphology. The discussion above shows that most occurrences of plasmodesmata in structurally simple algae can be rationalised in terms of the probable evolution of plasmodesmata in known structurally more complex relatives. The same is true of simple structures in embryophytes, e.g. the branched filaments (rather than parenchymatous thalli) of the gametophytes of the fern *Schizaea pusilla* (Bierhorst, 1971; Table 2.1). Here, the branched filaments are presumed to be a derived condition from ancestral parenchymatous thalli, and plasmodesmata presumably have a role in provisioning the highly differentiated gametangia (antheridia and archegonia). Furthermore, the sporophyte phase is large in complex organisms with many cell types (Bell and Mooers, 1997).

## 2.5 Functioning of complex photosynthetic organisms which lack plasmodesmata

### *2.5.1 Introduction*

Some complex photosynthetic organisms lack plasmodesmata. Light can be thrown on the evolution of plasmodesmata by considering the functioning of complex multicellular photosynthetic organisms that lack plasmodesmata (many examples), or that have intercellular plasma-membrane-bounded structures (pit plugs of rhodophytes) which might function as selective transport conduits. Also considered arė large acellular algae, and large photosynthetic organisms with small photobionts associated with fungi or metazoans with their own cell-to-cell transport pathways.

### *2.5.2 Multicellular algae lacking plasmodesmata or analogues of plasmodesmata*

Here, we consider the level of complexity that can be achieved by multicellular algae that lack plasmodesmata or other plasma-membrane-bounded symplasmic connections. The limit of complexity, approximated by the number of cell types in the organism, seems to be branched filamentous structures, sometimes with localised growth, and foliose thalli one or two cells thick, again sometimes with localised growth. These are both polyphyletic traits.

Branched filaments lacking symplasmic connections occur in several green algae. In the Ulvophyceae, there is *Cladophora* in the Cladophorales (Siphonocladales) and *Acrosiphonia* and *Spongomorpha* in the Ulotrichales, as well as *Microthamnion* in the Trebouxiophyceae (Table 2.1). In the Heterokontophyta, there are *Phaeothamnion* (Phaeothamniophyceae), *Giraudiopsis* (Pelagophyceae) and *Schizocladium* (Schizocladiophyceae) (Table 2.2). The haptophyte *Pleurochrysis* has branched filaments, as do *Dinoclonium* and *Dinothrix* in the Dinophyta, and *Stylonema*, a bangiophyte red alga lacking pit plugs (Raven, 1997; Graham and Wilcox, 2000; Brodie and Irvine, 2003; Table 2.2). These algae generally lack differentiated gametangia or sporangia (*Stylonema* is an exception), and have no non-photosynthetic vegetative cells. Monostromatic thalli lacking symplasmic connections occur in the green algae in *Prasiola* (Prasiolales: Trebouxiophyceae) and *Monostroma* (Ulotrichales: Ulvophyceae), while distromatic or tubular monostromatic thalli lacking symplasmic connections are found in *Ulva* (including *Enteromorpha*) (Raven, 1997; Graham and Wilcox, 2000; Tables 2.1 and 2.2). Among red algae, the bangiophycean *Porphyra* has mono- or di-stromatic thalli lacking pit plugs as the gametophyta phase; pit plugs occur in the branched filamentous sporophyte (*Conchocelis*) phase. There are no colourless cells in these thalli, although *Porphyra* has superficial gametangia. Growth is generally diffuse and intercalary, but with some localisation; it is not known if cell-to-cell transport of resources is necessary to support this growth.

Branched filaments and foliose mono- or di-stromatic thalli seem to be the most complex that can be produced without intercellular connections, and the division of labour that such connections permit.

### *2.5.3 Rhodophyta and pit plugs: (trans)mission impossible?*

All extant floridiophyte red algae, and the sporophyte (*Conchocelis*) phase of *Bangia/Porphyra*, have 'pit plugs' between cell walls that were produced at a cell division (Graham and Wilcox, 2000; Table 2.2). In complex thalli (Bell and Mooers, 1997; Graham and Wilcox, 2000), secondary pit plugs are produced in walls not formed at the division forming the two cells linked by the pit plug. Here, one of the cells divides unequally, forming a pit plug in the new cross-wall; the smaller cell then fuses with the adjacent cell (Graham and Wilcox, 2000). A similar process of cell division is involved in the replacement of a damaged cell in a red algal thallus by division of the undamaged cell acropetal to the damaged cell. This provides a pit plug between the acropetal cell and the replacement cell, which then divides unequally to establish a pit plug linking it to the basipetal undamaged cell.

These pit plugs are all within a single phase of the life cycle (e.g. a female gametophyte) of a single organism. Secondary pit plugs are also formed between different phases of the life cycle. Thus, they are formed between the haploid female gametophyte and the diploid, matrotrophic, carposporophyte (Graham and Wilcox, 2000, p. 367). Secondary pit plugs also form between parasitic red algae and their more (adelphoparasites) or less (alloparasites) closely related red algal hosts, and in parasexual processes (Graham and Wilcox, 2000). The pit plug resembles plasmodesmata in that the two cells sharing the plug have plasma membrane continuity through the wall, surrounding the part of the pitplug that is in the wall. Pit plugs differ from plasmodesmata in size: they are an order of magnitude greater in diameter than are plasmodesmata. They also differ in their contents: pit plugs appear to be mainly proteinaceous, presumably of relatively few molecular species, while plasmodesmata contain cytosol, with or without a desmotubule. This summarises the structure of pit plugs in the early branching bangiophytes and in the more derived coralline floridiophyceans (Pueschel, 1990). However, many other floridiophyte red algae have a more complex pit plug structure, with a cap membrane separating the cytosol of each cell from the core of the plug (Pueschel, 1990). The cap membrane is continuous with the plasma membrane of each cell via a 'T' or 'Y' junction in which the plug-side leaflet of the cap membrane continues as the plug-side leaflet of the plasma membrane round the plug, and the cytosol-side leaflet of the plug is continuous with the cytosol-side leaflet of the plasma membrane of each cell. The wall-side leaflet of the plasma membrane around the plug is continuous with the wall-side leaflet of the plasma membrane of each cell. In addition, there are often polysaccharide 'caps' at each end of the plug. These caps can occur in the presence or absence of a cap membrane and, when a cap membrane is present, the caps can be entirely on the plug side of the membrane or on both sides.

The function of pit plugs has been much disputed, and the term 'pit plug' is used here rather than 'pit connection' as less of a pre-judgement as to the function which is associated with the structure (see Graham and Wilcox, 2000). Clearly, the plugs are a 'connection' in the physical sense since they physically link the cells, but there is no evidence as to whether they alter the mechanical properties of the walls or of the tissue. There are data on the possible role of pit plugs in cell-to-cell transfer of solutes without passage through the apoplasm. This circumlocution is used instead of 'symplasmic transport' in order to accommodate the need for solutes to traverse the plug, which is clearly not cytosol, in any such transfer, and also the need to traverse the cap membranes when these are present. The data on such intercellular transport phenomena are now briefly outlined (see Raven 1984, 1997, 2003).

There are electrophysiological and tracer carbon data consistent with the diffusive movement of low-molecular mass ions and neutral solutes through pit plugs (Raven, 1984, 1997, 2003). There is also circumstantial evidence for short-distance (a few hundreds of μm or even a few millimetres) transport of low-molecular mass solutes from photosynthetic cells to non-photosynthetic cells in large thalli, and from host photosynthetic cells to cells of red algal parasites. In these cases, there is a continuity of pit plugs linking cells in the transport pathway, and the transport does not seem to be apoplasmic (Raven, 1984, 1997, 2003). In all of these cases, diffusive movement of the solutes rather than mass flow of solution would suffice, granted certain assumptions which will be discussed below. Longer distance transport, i.e. many millimetres or more, of solutes from cell to cell through specialised cells elongated in the direction of transport is also known from red algae. Mass flow of solution is involved here, and there is partial or complete dissolution of pit plugs in the end walls of the conducting elements (Raven, 1984, 2003; Pueschel, 1990; Gonen *et al.*, 1996).

Is the structure of pit plugs and, where present, cap membranes and plug caps, consistent with diffusive transfer of solutes between the cytosols of cells linked by pit plugs (Raven, 1984, 1997)? The answer could be 'yes' in the case of pit plugs with no cap membranes or polysaccharide caps, provided the protein components of the pit plug are not so dense as to eliminate the occurrence of solvent water from the plug, or prevent a continuous solvent water phase through the plug that permits the diffusion of low-molecular mass solutes. To have the same conductance per unit area of cell wall, the plug should have an effective conductance (per area of cell wall) for low-molecular mass solutes which is no less than that of the cytosol sleeve (per area of cell wall) of higher plant plasmodesmata. Any shortfall in conductance per unit cross-sectional area of the plug relative to that of the plasmodesma sleeve, and the larger number of plasmodesmata per unit area of wall, is compensated for by a larger area of pit plug than of plasmodesma sleeve on a wall area basis. Where cap membranes occur, there is the additional requirement for some mechanism of transport of low-molecular mass solutes across the cap membranes. Such transfer could be nothing more than the channels, discriminating largely on molecular mass with a cut-off at 800–1000 Da, such as occurring in the outer membrane of mitochondria and the outer of the two invariant membranes of the plastid envelope (Raven, 1984). If such porins were targeted to the cap membrane, they would be unlikely to diffuse laterally into the plasma membrane surrounding the rest of the cell, or that separating plug from cell wall, because of the 'T' or 'Y' membrane junctions (Raven, 1984). The arguments outlined in the last paragraph clearly do not provide a proven mechanism for diffusion of solute from cytosol to cytosol via pit plugs, but they do suggest areas for future research on the composition and properties of pit plugs and of cap membranes. The conclusions arrived at above as to the correlation between the complexity (number of cell types) of algal thalli and the occurrence of plasmodesmata would be consistent with the functioning of pit *plugs* as pit *connections* in terms of cell-to-cell transport, since red algae contain some of the most complex of the algae according to the Bell and Mooers (1997) definition.

### *2.5.4 Morphologically complex acellular macroalgae*

At the other extreme from the limited complexity of the branched filaments or foliose thalli lacking plasmodesmata (or their functional equivalents) are the algae which have very significant differentiation of the thallus yet have no separation of the thallus into a multicellular structure, i.e. are acellular (Table 2.1). This complexity by its nature cannot be approximated to the number of cell types as used by Bell and Mooers (1997). Some of these large cells (see Raven, 1981, 1999) are sufficiently differentiated so as to be functionally equivalent in terms of the spatial variation in resource acquisition, storage and use as a multicellular organism, with perhaps 10 different cell types [as defined by Bell and Mooers (1997)]. However, they have no obvious mechanism which regulates the transport of informational macromolecules from a part of the cell where expression of the corresponding genes is appropriate to some other part of the thallus where those gene products would not be appropriate to the division of labour within the cell. Such regulation of the distribution of informational macromolecules clearly does occur. A recent example is the work of Serikawa and Mandoli (2001) on the asymmetric distribution of carbonic anhydrase activity, and of the carbonic anhydrase mRNA, in the thallus of the marine dasycladalean green alga *Acetabularia acetabulum*. At least, the occurrence of cytoplasmic streaming in the more differentiated members of the ulvophycean Orders Caulerpales and Dasycladales means that the non-informational products of the division of labour, such as photosynthate and inorganic nutrients, are readily transported around the organism. Probably, the streaming of cytoplasm functions as an indiscriminate carrier of small molecules, with source regions adding solutes to the stream and sink regions removing them.

The most complex acellular algae occur in the Chlorophyta. The freshwater and soil surface *Protosiphon* (Chlorophyceae) has a spherical photosynthesising part of the single cell with colourless rhizoids penetrating the soil or fine-grain sediment substrate. This organism is a rhizophyte (Raven, 1981, 1984) as are the charalean algae, and *Fritschiella,* that have already been mentioned. The differentiation here is into the photosynthetic part, and the rhizoids which presumably specialise in the uptake of nutrients from the soil or sediment where they are at higher concentrations than in the overlying water. However, it is in the rhizophytic marine members of the Caulerpales, such as many members of the type genus *Caulerpa*, that acellular differentiation is greatest. These algae have a branching, apically growing 'rhizome' system creeping over or through the fine-grained substrate. At intervals, this 'rhizome' bears positively gravitropic rhizoids (root equivalents) penetrating the sediment; these are known to specifically function in nutrient ($NH_4^+$, amino acids, $HPO_4^{2-}$) uptake (Raven, 1981, 1984, 2003; Chisholm *et al*., 1996). At other intervals, the 'rhizome' bears assimilators (leaf equivalents) involved in photosynthesis whose often complex morphology gives rise to specific names such as *Caulerpa cactoides, Caulerpa hypnoides* and *Caulerpa taxifolia*.

Granted the capacity, as yet incompletely understood, for differentiation within these acellular thalli, and apparently unrestricted distribution of the resource products of division of labour, why is this capacity not more widespread? There seem to be two structural constraints. One problem is that damage to a single cell means damage to the whole organism in these acellular algae; sealing off the plasmodesmata of the damaged cell, as can occur in multicellular algae and higher plants with plasmodesmata, is not an option. While these acellular algae have very effective wound-plugging capacities (Raven, 1981), it is likely that the investment in the 'puncture repair outfits' is relatively greater in the acellular algae than in corresponding multicellular algae or plants with plasmodesmata. The other, habitat-specific, problem is that of mechanical support of an acellular organism growing on land without the high-density, buoyancy-conferring aquatic medium (Raven, 1981).

As against these probable evolutionary disadvantages of the acellular habit among algae, a potential advantage is the reproductive effort that the organism can employ. The reproductive effort in this case is the fraction of the resources gathered in the lifetime of the vegetative organism which is used in reproduction. Some members of the Dasycladales, and most members of the then Caulerpales, are semelparous, i.e. have a single, terminal reproductive event (Graham and Wilcox, 2000). This is probably an ancient (found in the Late Palaeozoic) trait (Forsythe *et al.*, 2002). In the extreme case, almost all of the protoplasm (i.e. everything but the cell walls) becomes gametes in some members of the Caulerpales (Van den Hoek *et al.*, 1995; Clifton, 1997; Graham and Wilcox, 2000). This could be construed as a selective advantage of the acellular habit, increasing reproductive effort over what could be readily achieved with many small, symplasmically linked cells in an otherwise similar organism. How, for example, could gametes be released from cells of a multicellular organism which do not face onto the medium? Other algal clades have made less evolutionary use of the acellular habit than have the Chlorophyta. The only other taxon with significant developments of acellular organisms is the heterokont Class Tribophyceae (= Xanthophyceae) (Table 2.2). Here, *Botrydium* is a close morphological and ecological parallel to the chlorophycean *Protosiphon*. The other acellular genus of the Tribophyceae is the freshwater, estuarine and coastal *Vaucheria*, a close structural analogue of *Dichotomosiphon*, the only freshwater member of the ulvophycean Order Caulerpales. In both cases, the tribophycean genus is more widespread than the green algal analogue.

### *2.5.5 Symbioses of (mainly) unicellular algae with fungi and metazoan*

We have already considered two ways in which a unicellular algae can become a large, differentiated photosynthetic organism. One way is to become multicellular with (for organisms more complex than a branched filament or a foliose thallus with little differentiation of reproductive structures) plasmodesmata or (in red algae) pit plugs. The second way is to become a differentiated acellular organism.

The third major way in which a unicellular alga can become a large primary producer is to become symbiotic with a non-photosynthetic organism which has already achieved a considerable size and complexity (Raven, 1981, 1984). The non-photosynthetic host organisms include ascomycete and basidiomycete fungi in lichens, and sponges, cnidarians, trematodes, bivalves and ascidians in marine and freshwater symbioses with metazoans. The non-photosynthetic hosts already had symplasmic intercellular transport mechanisms when they became symbiotic (Raven, 1993). In the fungi, these are based on incomplete septa in the hyphae, whereas in the metazoa, they are the gap junctions. For longer distance transport of resources by mass flow, the metazoa use apoplasmic pathways such as flagella-driven flow through conduits in sponges, the coelenteron in cnidarians and the haemolymph in bivalves (Raven, 1981, 1984, 1993).

Regardless of whether the photobionts are extracellular (lichens) or intracellular (many metazoan symbioses), the transfer of photosynthate from photobiont cytosol to host cytosol, and of externally acquired nutrients from host cytosol to photobiont cytosol, involves transfer of the resources across the two membranes and the intervening space separating the two cytosols. This intermembrane space is phylogenetically and, for cases in which new photobionts are acquired in each new host generation, ontogenetically, an apoplasmic compartment. This need for transmembrane fluxes does not distinguish these symbioses from other, including multicellular or acellular, eukaryotic phototrophs, since the transfer of solutes into and out of the plastids involves an analogous pathway with two membranes separated by what was originally apoplasm, according to the hypothesis of the origin of plastids.

## 2.6 Conclusions

The comparative approach to the occurrence of plasmodesmata and analogous structures in photosynthetic organisms, combined with information from cladistic analysis of molecular biological and other data, has yielded a substantial amount of information on the evolution of plasmodesmata. Discounting, albeit without molecular genetic evidence, the possibility of extensive horizontal gene transfer as an explanation of the occurrence of plasmodesmata (see Raven, 1997), the most parsimonious explanation of the phylogenetic data requires four, or probably five, independent origins of structurally defined plasmodesmata in the green algae, and one in the heterokont line. The embryophyte plasmodesmata were derived from those of their charophycean algal ancestors. In five of these six cases, the occurrence of plasmodesmata in a clade is correlated with morphological complexity in that clade, which exceeds that of a branched filament with little differentiation of gametangia or sporangia and no non-photosynthetic cells. A similar correlation is found for the distribution of pit plugs in red algae. This correlation is in general agreement with the putative function of plasmodesmata in the symplasmic distribution of the resource products of the division of labour and in regulating the distribution of informational macromolecules which developmentally permits the division of labour. However, the absence of desmotubules from the plasmodesmata of most algae calls into question a role in restricting the distribution of mRNA and of soluble proteins. It is not yet clear if there is evolutionary significance in the functional similarities noted (Lee *et al.*, 2000) between the trafficking of macromolecules through plasmodesmata and through nuclear pores.

Despite the polyphyletic origin of plasmodesmata in algae, there are very significant ecological and evolutionary roles for algae without plasmodesmata *sensu stricto*. While the brown algae, with plasmodesmata, contain the largest algae known and dominate many marine benthic regions, there are very few marine green algae with plasmodesmata or close analogues, and the most differentiated marine green algae are acellular. The red algae, which are a predominant macroalgal group in the sea in terms of species number, have pit plugs rather than plasmodesmata.

While the algae evolved plasmodesmata several times, the same selective pressures were presumably involved in the evolution of large differentiated phototrophs, with the proven or likely capacity for symplasmic transport via acellular construction (plus cytoplasmic streaming), pit plugs or symbiosis with fungi or metazoan with symplasmic transport in host already established.

## Note

Niklas (2004) gives a good account of the occurrence and phylogeny of cell walls, a necessary but not sufficient condition for the occurrence of plasmodesmata.

Shepherd, Beilby and Bisson (2004) have analysed the relationship between the structure of coenocytes of the ulvophycean *Ventricaria* (= *Valonia*) *ventricosa* (Siphonocladales) and their unusual electrphysiology. The cytoplasm is alveolate, with uninucleate domains interconnected by fine cytoplasmic stands containing microtubules, and surrounded by sulpated polysaccharide which also occurs in the vacuole, in pores penetrating the cell wall, and on the surface of the cell wall. The finest of these strands are only 80–100 nm in diameter, providing analogies with the intercellular connections such as those found in basidiomycetes, if not directly with plasmodesmata.

## References

Adams, D.G. and Duggan, P.S. (1999) 'Heterocyst and akinete differentiation in cyanobacteria', *New Phytologist* **144**, 2–33.

Andrews, M., Box R., Fyson, A. and Raven, J.A. (1984) 'Source-sink characteristics of carbon transport in *Chara hispida*', *Plant, Cell and Environment* **7**, 683–687.

Baldauf, S. (2003) 'Deep roots of eukaryotes', *Science* **300**, 1703–1706.

Bell, G. and Mooers, A.O. (1997) 'Size and complexity among multicellular organisms', *Biological Journal of the Linnean Society* **60**, 345–363.

Bierhorst, D.W. (1971) *Morphology of Vascular Plants*, Macmillan, New York.

Blackman, L.M. and Overall, R.L. (1998) 'Immunolocalization of the cytoskeleton of *Chara corallina*', *Plant Journal* **14**, 733–741.

Blackman, L.M., Gunning, B.E.S. and Overall, R.L. (1998) 'A 45 kDa protein isolated from the nodal walls of *Chara corallina* is localized to plasmodesmata', *Plant Journal* **15**, 411–421.

Booton, G.C., Floyd, G.L. and Fuerst, P.A. (1998) 'Origins and affinities of the filamentous green algal orders Chaetophorales and Oedogoniales based on 18S rRNA sequences', *Journal of Phycology* **34**, 312–318.

Box, R., Andrews, M. and Raven, J.A. (1984) 'Intercellular transport and cytoplasmic streaming in *Chara hispida*', *Journal of Experimental Botany* **35**, 1016–1021.

Brodie, J.A. and Irvine, L.M. (2003) *Seaweeds of the British Isles. Volume 1 Rhodophyta, Part 3B Bangiophycideae*, Natural History Museum, London.

Cavalier-Smith, T. (1981) 'Eukaryotic kingdoms: seven or nine?', *Biosystems* **14**, 461–468.

Cavalier-Smith, T. (1986) 'The Kingdom Chromista: origin and systematics', in *Progress in Phycological Research* (eds F.E. Round and D.J. Chapman), Biopress, Bristol, UK, pp. 309–347.

Cavalier-Smith, T. (2003) 'Menome reduction and evolution of novel genetic membranes and protein-targetting machinery in eukaryote-eukaryote chimaeras (meta-algae)', *Philosophical Transactions of the Royal Society of London* **358**, 109–134.

Chapman, R.L., Borhsenious, O., Brown, R.C., Henk, M.C. and Waters, D.A. (2001) 'Phragmoplast-mediated cytokinensis in *Trentepohlia*: results of TEM and immunofluorescence cytochemistry', *International Journal of Systematic and Evolutionary Microbiology* **51**, 759–765.

Chisholm, J.R.M., Dauga, C., Ageron, E., Grimont, P.A.D. and Jaubert, J.M. (1996) ' "Roots" in mixotrophic algae', *Nature* **381**, 382.

Clifton, K.E. (1997) 'Mass spawning of green algae on coral reefs', *Science* **275**, 1116–1118.

Cook, M.E. and Graham, L.E. (1999) Evolution of plasmodesmata, in *Plasmodesmata: Nanochannels with megatasks* (eds A. van Bel and C. Kesteren), Springer Verlag, Berlin, pp. 101–117.

Cook, M.E., Graham, L.E., Botha, C.J. and Lavin, C.A. (1997) 'Comparative ultrastructure of *Chara* and selected bryophytes: toward an elucidation of the evolutionary origin of plant plasmodesmata', *American Journal of Botany* **84**, 1169–1178.

Draisma, S.G.A., Prud'Homme van Reine, W.F., Stam, W.T. and Olsen, J.L. (2001) 'A reassessment of phylogenetic relations within the Phaeophyceae based on RUBISCO large subunit and ribosomal DNA sequences', *Journal of Phycology* **37**, 586–603.

Edwards, D. (2003) 'Xylem in early tracheophytes', *Plant Cell and Environment* **26**, 57–72.

Ehlers, K., Binding, H. and Kollmann, R. (1999) 'The formation of symplasmic domains by plugging of plasmodesmata', *Protoplasma* **209**, 181–192.

Falkowski, P.G., Katz, M., Knoll, A.H. *et al.* (2004) 'The evolution of modern eukaryotic phytoplankton', *Science*, **305**, 354–360.

Forsythe, G.T.W., Wood, R. and Dickson, J.A.D. (2002) 'Mass spawning in ancient reef communities? Evidence from Late Palaeozoic phylloid algae', *Palaios* **17**, 615–621.

Gonen, Y., Kimmel, E., Tel-Or, E. and Friedlander, M. (1996) 'Intercellular assimilate transport in *Gracilaria cornea* (Gracilariales, Rhodophyta)', *Hydrobiologia* **326/327**, 421–428.

Graham, L.E., Cook, M.E. and Busse, J.S. (2000) 'The origin of plants: body plan changes contributing to major evolutionary radiation', *Proceedings of the National Academy of Sciences USA* **97**, 4535–4540.

Graham, L.J. and Wilcox, L.W. (2000) *Algae*, Prentice Hall, Upper Saddle River, NJ.

Kaiser, D. (2001) 'Building a multicellular organism', *Annual Review of Genetics* **35**, 103–123.

Karol, K.G., McCourt, R.M., Cimino, M.T. and Delwiche, C.F. (2001) 'The closest living relatives of plants', *Science* **294**, 2351–2353.

Katsaros, C. and Galatis, B. (1988) 'Thallus development in *Dictyopteris membranacea* (Phaeophyta, Dictyotales)', *British Phycological Journal* **23**, 71–88.

Kawai, H., Maeba, S., Sasaki, H., Okuda, K. and Henry, E.C. (2003) '*Schizocladia ischensis*: a new filamentous marine chromophyte belonging to a new class Schizocladiophyceae', *Protist* **154**, 211–228.

Kempers, R., Prior, D.A.M., van Bel, A.J.E. and Oparka, K.J. (1993) 'Plasmodesmata between sieve elements and companion cells of extrafascicular stem phloem of *Cucurbita maxima* permit passage of 3 kDa fluorescent probes', *Plant Journal* **4**, 567–575.

Kirk, D.L. (1998) *Volvox. Molecular–Genetic Origins of Multicellularity and Cellular Differentiation*, Cambridge University Press, Cambridge.

Kituyama, M., Hara, Y., Shimado, K., Yakomoto, H. and Hiramata, Y. (1992) 'Intercellular transport of macromolecules in *Nitella Plant*', *Cell and Environment* **33**, 413–417.

Kwiatkowska, M., Poplonska, K. and Wotczak, A. (2003) '*Chara tomentosa* antheridial plasmodesmata at various stages of spermatogenesis', *Biologia Plantarum* **46**, 233–238.

Lee, J.Y., Yoo, B.C. and Lucas, W.J. (2000) 'Parallels between nuclear pore and plasmodesmata trafficking of informational macromolecules', *Planta* **210**, 177–187.

Leonardi, P.I., Caceres, E.J. and Correa, J.A. (2002) 'Ultrastructure and taxonomy of *Sporocladopsis nove-zelandiae* (Ulvophyceae, Chlorophyta)', *Botanica Marina* **45**, 324–330.

López-Bautista, J.M. and Chapman, R.L. (2003) 'Phylogenetic affinities of the Trentepohliales inferred from small-subunit rDNA', *International Journal of Systematics and Evolutionary Microbiology* **53**, 2099–2106.

López-Bautista, J.M., Waters, D.A. and Chapman, R.L. (2003) 'Phragmoplastin, green algae and the evolution of cytokinesis', *International Journal of Systematics and Evolutionary Microbiology* **53**, 1715–1718.

Marchant, H.J. (1976) 'Plasmodesmata in algae and fungi', in *Communication in Plants: Studies in Plasmodesmata* (eds B.E.S. Gunning and A.W. Robards), Springer Verlag, Berlin, pp. 59–80.

Martin, W., Rojan, T., Richly, E. *et al.* (2002) 'Evolutionary analysis of *Arabidopsis*, cyanobacteria and chloroplast genomes reveals plastid phylogeny and thousands of cyanobacterial genes in the nucleus', *Proceedings of the National Academy of Sciences USA* **99**, 12246–12251.

Münch, E. (1930). *Die Stoffbewegung in der Pflanze*, Fischer, Jena.

Niklas, K.J. (2004) 'The cell walls that bind the tree of life', *Bioscience* **54**, 831–841.

Nozaki, H., Misazawa, K., Kajita, T., Kato, M., Nohara, S. and Watanabe, M.M. (2000) 'Origin and evolution of the colonial Volvocales (Chlorophyceae) as inferred from multiple chloroplast gene sequences', *Molecular Phylogenetics and Evolution* **17**, 256–268.

Oparka, K.J. (2004) 'Getting the message across: how do plant cells exchange macromolecular complexes?', *Trends in Plant Science* **9**, 33–44.

Pueschel, C.M. (1990) 'Cell structure', in *Biology of the Red Algae* (eds K.M. Cole and R.G. Sheath), Cambridge University Press, Cambridge, pp. 7–41.

Raven, J.A. (1981) 'Nutritional strategies of submerged benthic plants: the acquisition of C, N and P by rhizophytes and haptophytes', *New Phytologist* **88**, 1–30.

Raven, J.A. (1984) *Energetics and Transport in Aquatic Plants*, A.R. Liss, New York.

Raven, J.A. (1993) 'Energy and nutrient acquisition by autotrophic symbioses and their asymbiotic ancestors', *Symbiosis* **14**, 33–60.

Raven, J.A. (1997) 'Multiple origins of plasmodesmata', *European Journal of Phycology* **32**, 95–101.

Raven, J.A. (1999) 'The size of cells and organisms in relation to the evolution of embryophytes', *Plant Biology* **1**, 2–12.

Raven, J.A. (2003) 'Long-distance transport in non-vascular plants', *Plant, Cell and Environment* **26**, 73–85.

Raven, J.A. and Smith, F.A. (1978) 'Effects of temperature on ion content, ion fluxes and energy metabolism in *Chara corallina*', *Plant, Cell and Environment* **1**, 231–238.

Rousseau, F., Burrowes, R., Peters, A.F., Kuhlenkamp, R. and de Reviers, B. (2001) 'A comprehensive phylogeny of the Phaeophyceae based on mrDNA sequences resolves the earliest divergences', *Comptes Rendus de l'Academie des Sciences Serie III – Sciences de la Vie – Life Sciences* **324**, 305–319.

Saunders, G.W., Potter, D. and Andersen, R.A. (1997) 'Phylogenetic affinities of the Sarcinochrysidales and Chrysomeridales (Heterokonta) based on analysis of molecular and combined data', *Journal of Phycology* **33**, 310–318.

Serikawa, K.A. and Mandoli, D.F. (2001) 'Asymmetric subcellular mRNA distribution correlates with enzymatic activity for carbonic anhydrases from *Acetabularia acetabulum*', *Plant Physiology*, **125**, 900–911.

Shepherd, V.A., Beilby, M.J. and Bisson, M.A. (2004) 'When is a cell not a cell? A theory relating coenocytic structure to the unusual electrophysiology of *Ventricaria ventricosa (Valonia ventricosa)*', *Protoplasma* **223**, 79–91.

Van den Hoek, C., Mann, D.G. and Jahns, H.M. (1995) *Algae: An Introduction to Phycology*, Cambridge University Press, Cambridge.

Vannucci, G., Basso, D. and Fravega, P. (2000) 'New observations on the anatomy of the fossil alga *Subterraniphylum* Elliot', *Revista Italiano di Palaeontologia e Stratigraphia* **106**, 237–245.

# 3 Plasmodesmata: protein transport signals and receptors

Friedrich Kragler

## 3.1 Introduction

Exchange of signals between cells is suggested to be an evolutionary prerequisite to establish organ-bearing organisms. To build patterns and organs and to coordinate environmental inputs, it is vital to transfer signal molecules between cells. In plants, a cell-to-cell signaling pathway is provided by plasmodesmata (PD), an intercellular structure that facilitates cell-to-cell transport of regulatory macromolecules. A number of proteins, which are in general summarized as non-cell autonomous proteins (NCAPs; Table 3.1), utilize the PD transport pathway to gain access to neighboring cells. A function in cell-to-cell signaling by transport of proteins via PD is suggested for homeotic transcription factors (see Chapter 6). In addition, PD provide the means for RNA transport from cell to cell. With the aid of special non-cell autonomous RNA binding proteins, viral genomic RNA, messenger RNA and post-transcriptional gene silencing RNA molecules enter the PD transport pathway and act as non-cell autonomous signal molecules (Haywood *et al.*, 2002).

To comprehend the role of NCAPs requires the understanding of the pathway and the means by which cell-to-cell movement is achieved. The transport of macromolecules is thought to be specific and receptor-mediated. Amino acid sequence modifications within NCAPs result in their transport deficiency, and studies of fusion proteins consisting of NCAPs and reporter proteins indicate that NCAPs expose a signal that is sufficient and necessary to mediate cell-to-cell translocation. Obviously, specific structural motifs of NCAPs are recognized by special PD pathway receptor(s). Several plant-produced proteins were recently identified and suggested to act as part of the PD transport pathway. In this review, supporting evidence that underlies the function of PD transport pathway proteins and PD targeting signals are presented and discussed.

## 3.2 Components of the PD transport pathway

### *3.2.1 Principles of signal-mediated protein transport*

The signal hypothesis proposed by Günter Blobel more than two decades ago for protein transport into organelles states that nascent proteins or newly synthesized proteins contain intrinsic, organelle-specific targeting signals (Blobel, 1980). Such

**Table 3.1** Non-cell autonomous proteins and their characteristics

| Protein | Function | Tissue/Localization | References |
|---|---|---|---|
| *Transcription factors* | | | |
| *Zm*KNOTTED1 | Shoot apex meristem identity | Shoot apex/nucleus | Lucas *et al.*, 1995; Jackson, 2002; Kim *et al.*, 2002b; Kim *et al.*, 2003 |
| *At*LEAFY | Cell identity, flower development | Shoot apex/nucleus | Sessions *et al.*, 2000; Haywood *et al.*, 2002; Wu *et al.*, 2003 |
| *Am*DEFICIENS | Cell identity, flower development | Shoot apex/nucleus | Schwarz-Sommer *et al.*, 1992; Trobner *et al.*, 1992; Mezitt and Lucas, 1996; Perbal *et al.*, 1996; Efremova *et al.*, 2001 |
| *Am*GLOBOSA | Cell identity, flower development | Shoot apex/nucleus | Schwarz-Sommer *et al.*, 1992; Trobner *et al.*, 1992; Perbal *et al.*, 1996; Kragler *et al.*, 1998; Efremova *et al.*, 2001 |
| *At*SHORT ROOT[a] | Cell identity, root layer differentiation | Root stele – endodermis/ nucleus | Helariutta *et al.*, 2000; Nakajima *et al.*, 2001; Sena *et al.*, 2004 |
| *At*CAPRICE[a] | Cell identity, root hair differentiation | Root epidermis – hair cell/nucleus | Wada *et al.*, 1997; Schellmann *et al.*, 2002; Wada *et al.*, 2002 |
| *RNA transporting proteins* | | | |
| *Cm*RNB1 | Small ssRNA/ micro RNA transport | Phloem/N.D. | Yoo *et al.*, 2004 |
| *Cm*PP16 | RNA transport | Phloem/ cytosol – nucleus | Xoconostle-Cázares *et al.*,1999 |
| *Zm*KNOTTED1 | KNOTTED1 mRNA transport | Shoot apex/nucleus | Lucas *et al.*, 1995; Kragler *et al.*, 2000 |
| *Phloem proteins* | | | |
| *Cm*HSC70-1, 2 | HSP70 chaperone | Phloem/cytosol | Aoki *et al.*, 2002 |
| *Cm*PP2 | Dimeric lectin | Phloem/filaments | Balachandran *et al.*, 1997 |
| *Cm*PP36 | Cytochrome $b_5$ reductase | Phloem/cytosol | Xoconostle-Cázares *et al.*, 2000 |
| *Os*RBB13-1 | Thioredoxin h | Phloem/N.D. | Ishiwatari *et al.*, 1998 |
| *Rc*Cystatin | Protease-inhibitor | Phloem/N.D. | Balachandran *et al.*, 1997 |
| *Rc*Glutaredoxin | Redox regulation | Phloem/N.D. | Balachandran *et al.*, 1997 |
| Pumpkin kinases[a] | Phosphorylation | Phloem/N.D. | Yoo *et al.*, 2002 |
| Phloem sap proteins | N.D. | N.D. | Balachandran *et al.*, 1997 |

N.D., not determined; ER, endoplasmic reticulum.

[a]Not verified in microinjection assays to move and/or to increase PD size exclusion limit.

targeting signal(s) are recognized by selective signal receptors that initiate specific translocation of proteins across boundaries. Formulated for protein transport into the ER, these two features seem to be a fundamental principle for all protein translocation systems. In all translocation pathways, an initial receptor has to interact physically with a signal/targeting motif. The recognition by a pathway-specific receptor provides the means to initiate protein import into organelles such as the nucleus, mitochondria, plastids, ER and peroxisomes/glyoxysomes (Smith and Schnell, 2001; Schnell and Hebert, 2003; Soll and Schleiff, 2004; Wiedemann *et al.*, 2004).

A signal-mediated translocation system involves more than two components, and thus has more than two steps. Here, based on the known intracellular transport mechanisms, a generally accepted four-step model is presented for signal-mediated protein translocation across a boundary (Figure 3.1). First, the initial transport receptor (IniR) recognizes and interacts specifically with a cargo protein exposing a transport signal. Next the IniR-cargo complex binds either directly to a docking complex (DoC) located at the entry site of the channel/pore or to a transfer complex (TraC). An activated IniR-cargo-TraC complex moves to the DoC. The DoC recognizes the presence of a cargo and initiates a physical modification of the channel/pore and of the cargo, which facilitates the entry of the cargo. Finally, the cargo is translocated from the *cis* to the *trans* site through the channel/pore complex.

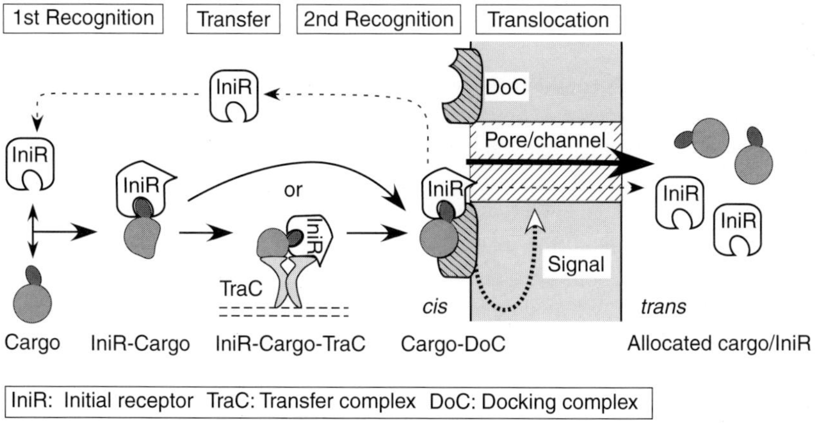


**Figure 3.1** An integrative model of signal-mediated protein translocation pathways across cellular barriers. A cargo exposes a specific translocation signal that physically interacts with an initial receptor (IniR). First, the cargo signal–IniR compatibility decides whether a molecule enters a targeted translocation pathway. Following initial recognition and structural rearrangements, the IniR-cargo complex is recognized either by a transfer complex (TraC) or directly by a docking complex (DoC) associated with the microchannel/pore. The presence of a cargo-IniR complex activates DoC and initiates structural modifications of both the cargo and the microchannel/pore complexes. Finally, these structural modifications allow the cargo to enter the microchannel/pore and to translocate in an energy consuming manner from the *trans*- across the barrier to the *cis*-site. Note that in intracellular translocation pathways, an HSP70-related chaperone activity can target all four steps: the initial 1st recognition, transfer, 2nd recognition/docking and translocation.

In general, the first protein physically interacting with a targeting signal exposed by a potential cargo protein resembles the transport receptor. Thus, the IniR is by definition the actual transport pathway receptor, which specifies the cargo and initiates the translocation of the cargo. The other downstream transport pathway components such as TraC, DoC and the structural channel/pore proteins are essential, but not necessarily selective toward a signal motif. Notably, according to intracellular transport pathways such as the peroxisomal import pathway, recognition and entry of a cargo into a translocation pathway does not require the permanent presence of an initial transport receptor at the transport channel/pore (Smith and Schnell, 2001). In most intracellular transport pathways, the formation of a cargo-IniR complex is necessary to activate a DoC present at the channel/pore to initiate translocation across a membranous border (Gorlich, 1997; Gorlich and Kutay, 1999; Smith and Schnell, 2001; Schnell and Hebert, 2003).

In a transport system with multiple hierarchical steps and components, we are confronted with the challenge to define the hierarchical order of involved components and to assign distinct function(s). For example, every single component may constitute an essential part of the system and, consequently, a mutation in one particular transport factor will not disclose its function. In genome-wide localization screens, the IniR or TraC, whose presence at the DoC and channel/pore is either transient or not required, may not be distinguished from a potential cargo. In interaction screens based on cargo proteins, a TraC or DoC might not be isolated. These downstream transport pathway components may associate exclusively with IniR-cargo complexes and not to a free cargo molecule. Similarly, in binding studies involving candidate DoC and TraC factors, interaction may depend on the combined presence of IniR and a cargo. For intracellular transport systems, such experimental challenges were solved mainly by utilizing biochemical approaches and *in vitro* import systems. However, an *in vitro* system is, in general, not applicable for intercellular transport pathways such as the PD transport pathway.

Our current insights of the intercellular PD protein transport system conform to the targeting signal hypothesis. Having outlined a general model based on intracellular transport pathways, we assume that the PD transport pathway is not fundamentally different from intracellular pathways presented in Figure 3.1. Here, the general features of intracellular transport pathways serve as a blueprint for the intercellular PD protein transport pathway and should facilitate the functional assignment of PD pathway components.

### *3.2.2 Putative PD pathway components*

Plasmodesmatal microchannels are enclosed by a continuous plasma membrane cylinder and embedded in the cellulosic cell wall (Esau and Charvat, 1975; Ehlers and Kollmann, 2001). This unique morphology of the plant-specific transport microchannel might demand a mechanism not mirrored in intracellular transport pathways. In evolutionary terms, plasmodesmal-related structures appeared relatively late, after a sophisticated intracellular protein transport system was already in place. Our current view regarding mechanistic aspects and components of the PD transport system implies that higher plants did not create a novel mechanism for cell-to-cell protein transport following other principles. Several lines of evidence suggest that existing intracellular transport mechanisms may have been adopted to provide the means of transport from cell to cell. For example, cytoskeletal factors such as actin-, myosin- and kinesin-related proteins are found within or in close proximity to the PD microchannel (Blackman and Overall, 1998; Blackman *et al.*, 1999; Reichel *et al.*, 1999; Samaj *et al.*, 2000). The cytoskeletal elements may provide the intracellular transport pathway towards PD, the driving force for dilation of the PD microchannel, and/or transfer along the PD microchannel. Also, similar to intracellular translocation mechanisms, structural modifications of cargo proteins such as altering the phosphorylation patterns (known to regulate import into nuclei or chloroplasts; Hood and Silver, 1999; Soll and Schleiff, 2004), seem to modulate protein transport via PD. The viral movement protein produced by tobacco mosaic virus (TMV-MP) is essential for viral cell-to-cell spread and is targeted by a kinase (Citovsky *et al.*, 1993), and a mutant TMV-MP mimicking phosphorylation is significantly impaired in its trafficking capacity (Waigmann *et al.*, 2000). Also, similar to organelle-bound protein import systems such as the peroxisomal, chloroplastic, mitochondrial and ER translocation machineries (Schnell and Hebert, 2003), a HSP70-related chaperone activity is found in association with the PD complex (Kragler *et al.*, 1998; Aoki *et al.*, 2002; Haywood *et al.*, 2002).

One would expect a high number of distinct receptors and translocation signals considering the structural complexity and dynamic nature of PD. A PD contains electron-dense protein complexes located between the plasma membrane and an appressed ER tube that forms a helical three-dimensional symplasmic channel (Ehlers and Kollmann, 2001). The number of structurally unrelated and actively transported NCAPs (Table 3.1) implies the presence of distinct PD receptors mediating transport. So far, however, due to the experimental challenges of the system, we identified a relatively small number of factors acting as regulatory components of the PD transport system. Here, the role of the four most promising PD-pathway components will be discussed and an attempt will be made to assign a function in a PD-pathway model (Table 3.2; Figure 3.2): the *N. tabacum* NON-CELL AUTONOMOUS PATHWAY PROTEIN 1 (*Nt*NCAPP1; Lee *et al.*, 2003) potentially an initial PD receptor (PD-IniR); the cytoskeleton with its associated factors, potentially regulating the access of cargos to the PD-IniR or PD transfer complex (PD-TraC); HSP70-related chaperones (Aoki *et al.*, 2002); and an yet-unidentified protein named INCREASED SIZE EXCLUSION LIMIT 1 (ISE1; Kim *et al.*, 2002a). The last of these is suggested to be either a structural component of the PD docking complex (PD-DoC) or of the PD microchannel (see Chapter 5). Although potential components of the PD transport pathway may be found among the viral movement protein interacting proteins (Soellick *et al.*, 2000; Chisholm *et al.*, 2001; Huang *et al.*, 2001; Kellmann, 2001; Matsushita *et al.*, 2001; Matsushita *et al.*, 2002; Fridborg *et al.*, 2003; Okinaka *et al.*, 2003; Dunoyer *et al.*, 2004; Mariano *et al.*, 2004), or among proteins localized to the PD complex (Dorokhov *et al.*, 1999; Chen *et al.*, 2000; Cutler

**Table 3.2** NCAP-interacting proteins/receptors regulating access to the PD transport pathway

| Protein | Binding partners | Suggested transport function | Subcellular localization | References |
|---|---|---|---|---|
| CmNCAPP1 | TMV-MP; *Cm*PP16, phloem proteins | Positive regulator | ER/phragmoplast | Lee *et al.*, 2003 |
| NtMPB2C | TMV-MP; *Zm*KNOTTED1 | Negative regulator | Microtubules | Kragler *et al.*, 2003; Kragler, unpublished |
| CmHSC70-1, 2 | Nascent proteins | Unfolding of NCAPs | Cytosolic | Aoki *et al.*, 2002 |
| AtISE | N.D. (not cloned) | Regulator of PD size exclusion limit | N.D. | Kim *et al.*, 2002a |

N.D., not determined; ER, endoplasmic reticulum.

*et al.*, 2000; Medina-Escobar *et al.*, 2003; Oparka, 2004), these proteins were not functionally characterized with respect to intercellular transport of plant-produced NCAPs and, thus, with the exception of phloem HCP70-1 (Aoki *et al.*, 2002; Haywood *et al.*, 2002), are not discussed here.

### *3.2.3 NCAPP1 – the initial PD receptor*

The NCAPP1 characteristics are consistent with a PD-IniR (Jackson and Kim, 2003; Lee *et al.*, 2003). The 40-kDa protein NCAPP1 specifically interacts with a number of PD cargo proteins present in the phloem, such as the RNA-binding phloem protein CmPP16, and also with the viral movement protein TMV-MP. As expected for a PD transport pathway component, plants silenced for NCAPP1, or overproducing a dominant-negative truncated variant, develop a phenotype consistent with obstructed cell-to-cell communication. Mutant plants have disorganized epidermal cells and flower tissues. Consistent with the notion that different PD receptors regulate the transport of distinct NCAPs, a dominant-negative form of NCAPP1 abolishes the transport of TMV-MP and CmPP16 but not of KN1 and cucumber mosaic virus movement protein. NCAPP1 is immuno-detected at the surface of ER vesicles in proximity of PD but not at PD orifices or within PD microchannels. NCAPP1 does not seem to act as a PD-DoC because of two reasons. First, the subcellular distribution of NCAPP1 is not consistent with a PD-DoC thought to be present at the orifice of the PD microchannel to provide the means of transport via the PD microchannel. Second, a PD-Doc is suggested to have the capacity to interact with cargo-IniR complexes but not with cargo proteins alone. In summary, the observed mutant phenotype and intracellular localization of NCAPP1 is consistent with a functional role as a membrane-associated initial PD receptor (Figure 3.2).

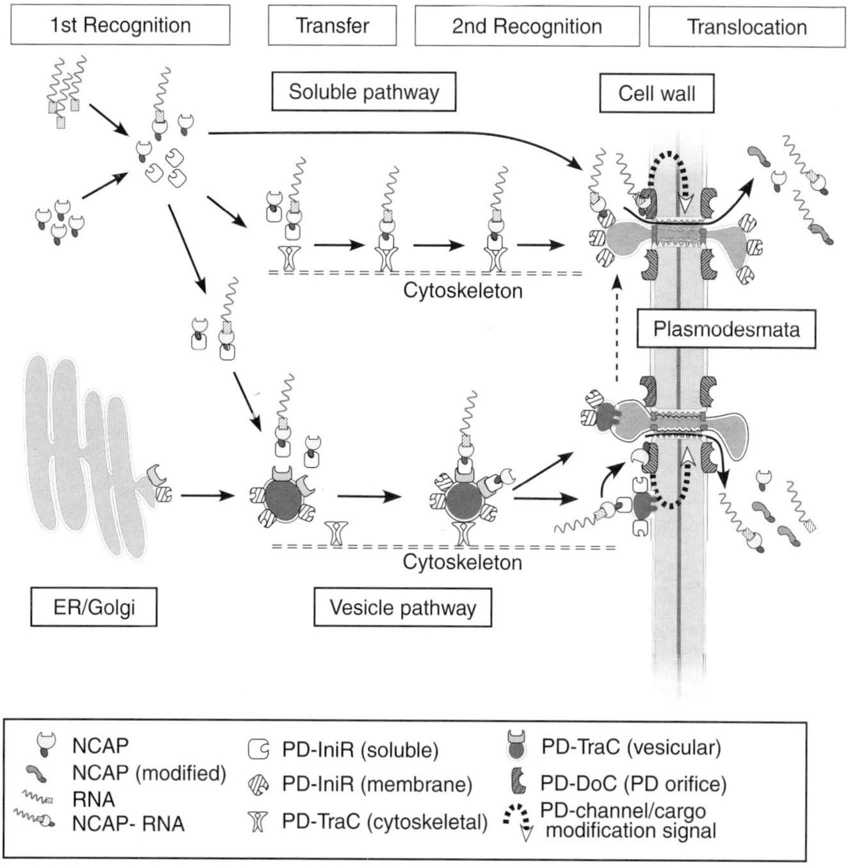


**Figure 3.2** Intracellular transport system likely involved in NCAP delivery from cell to cell via PD. NCAP or NCAP – nucleic acid cargo complexes that expose a PD targeting signal assemble either as soluble proteins in the cytosol (upper panel) or at ER-derived vesicles (lower panel). A PD initial receptor (PD-IniR) recognizes specifically a PD-cargo complex and is delivered to the cell periphery via interaction with a transfer complex (PD-TraC) associated to the cytoskeleton motor system. At the proximity of the PD orifice, a docking complex (PD-DoC) regulates the cargo delivery into and through the PD microchannel. Receptor(s) located on the ER or plasma membrane, in the immediate proximity to PD, may constitute the PD-DoC. The PD-DoC recognizes the presence of a cargo and mediates the PD size exclusion limit increase and potentially modifies the cargo. As in intracellular translocation pathways, structural modifications of the cargo by an HSP70-related chaperone may be essential at all steps. Note that a vesicular pathway for transport of a membrane-associated IniR to the ER structures associated to PD is indicated in the lower panel.

### *3.2.4 Potential role of the cytoskeleton in the PD transport pathway*

In most PD pathway models, the cytoskeleton plays a crucial role (Heinlein *et al.*, 1995; McLean *et al.*, 1995; Reichel *et al.*, 1999; Boyko *et al.*, 2000; Chaffey and Barlow, 2001; Haywood *et al.*, 2002; Oparka, 2004). As presented in Figure 3.2, the cytoskeleton could serve as the intracellular track for soluble complexes or

Rab-bearing vesicles serving as PD-TraC. The vesicle-bound transport complexes, transferred along the cytoskeleton via special classes of kinesin- and/or myosin-related proteins, allocate PD-targeted cargos to PD-associated membranes (Oparka, 2004). This view finds support in the observation that substances that depolymerize filamentous actin cables interfere with the intercellular transport of TMV-MP (Boyko *et al.*, 2002; Gillespie *et al.*, 2002) and alter the PD size exclusion limit (PD SEL; Ding *et al.*, 1996). However, with the exception of *Nt*MPB2C (Kragler *et al.*, 2003; Kragler, unpublished), no observations indicate a function of the cytoskeleton or cytoskeleton-associated factors in regulation of intercellular transport of nonviral endogenous NCAPs (as listed in Table 3.1). Also, the function of microtubules and/or filamentous actin in intercellular trafficking of viral TMV-MP remains equivocal. Functional microtubules and microtubule interaction seem not to be necessary for TMV-MP cell-to-cell transport (Gillespie *et al.*, 2002; Haywood *et al.*, 2002; Heinlein, 2002; Ding *et al.*, 2003; Oparka, 2004).

Currently, no gene products have been identified that unequivocally fulfil all requirements for a PD transfer complex protein (PD-TraC) that traffics NCAPs to the PD microchannels. However, a potential candidate protein that modulates PD-TraC function may be found in the microtubulin-associated protein, *Nt*MPB2C. The *Nt*MPB2C interacts physically with TMV-MP (Kragler *et al.*, 2003) and KNOTTED1 (Kragler, unpublished) and constitutes a plant-unique class of proteins with significant similarity to Rab-like and cytoskeleton-associated proteins. A coiled-coil structural motif mediates *Nt*MPB2C interaction with TMV-MP and association to microtubules. In overexpression studies, *Nt*MPB2C was observed to divert the subcellular distribution of TMV-MP (Kragler *et al.*, 2003) and KN1 (Kragler, unpublished results) toward microtubules-associated structures of unknown nature. A negative regulatory role is suggested on the basis of overexpression experiments. Here, *Nt*MPB2C seems to negatively regulate the access of NCAPs to either the PD-IniR or PD-TraC transport components. However, overloading a cell with PD pathway components together with NCAPs blocks NCAP cell-to-cell transport in a dominant negative fashion (Kragler *et al.*, 1998). Thus, we cannot rule out the possibility that *Nt*MPB2C functions as a positive regulator by acting as a PD-IniR or PD-TraC. In the PD-transport model presented in Figure 3.2, *Nt*MPB2C or related proteins could either limit the number of NCAPs accessing the PD transport pathway or facilitate transport via PD.

### *3.2.5 HSP70-related proteins and the PD transport pathway*

Chaperones of the 70-kDa heat shock proteins (HSP70) family assist in the transport of proteins into peroxisomes, ER, mitochondria and chloroplasts, as well as influence clathrin-coated vesicle dynamics, viral replication and transcriptional activation. HSP70s are ATP-consuming factors that, together with DNAJ-type HSP40 co-chaperones, modulate protein folding (Hartl and Hayer-Hartl, 2002) and mediate intracellular translocation of proteins across membranes (Schnell and

Hebert, 2003). HSP70 chaperones bind hydrophobic pockets of nascent proteins to delay/support folding and protein import into organelles.

An interesting link between the cytoskeleton formation and HSP70 chaperones is found in the HSP70-assisted modification of microtubles (Green and Liem, 1989; Liang and MacRae, 1997; Warner *et al.*, 2004). A 70-kDa HSP in the green alga *Chlamydomonas* supports the formation of flagellar microtubules core bundles (Sanchez and Cohen, 1994). HSP70 can attach to polymerized tubulin at a small (12 amino acid) *C*-terminal region, which overlaps with the site recognized by MAPs (Bloch and Johnson, 1995). Moreover, HSP70 and MAP1B contain structurally related tubulin-binding motifs, and the beet yellow virus produces a 65-kDa HSP70-related chaperone essential for cell-to-cell spread via PD that interacts with purified microtubules (Agranovsky *et al.*, 1991; Karasev *et al.*, 1992; Alzhanova *et al.*, 2001). With respect to the actin-based cytoskeleton, evidence exists that HSP70 stabilizes actin filaments in mammalian cells (Macejak and Luftig, 1991; Haus *et al.*, 1993; Weeds and Maciver, 1993).

In summary, members of the HSP70 heat shock protein family seem to facilitate cytoskeleton function. Because tubulin, actins and HSP70s are well-conserved proteins across kingdoms, a similar role could be suggested for HSP70 chaperones towards the plant cytoskeleton, and by this means also for the PD transport pathway. Here, chaperones acting on the cytoskeleton may be necessary to release cytoskeleton-associated cargos for transport and to increase the SEL of PD microchannels.

In all steps of the PD transport pathway, HSP70-related chaperones could act directly on PD cargo proteins to facilitate the exposure of a targeting signal and/or the transfer of NCAPs via the PD channels. Support for a role of chaperones in unfolding cargo proteins comes from studies with structurally stabilized KN1 molecules. Chemically cross-linked KN1 proteins are recognized by the PD translocation machinery but lack the capacity to move from cell to cell (Kragler *et al.*, 1998). Furthermore, HSP70-related proteins are detected in PD-enriched cell fractions (Aoki *et al.*, 2002). Here, according to intracellular transport systems, they would facilitate transport via the PD microchannel. In analogy to protein import into mitochondria or chloroplasts, chaperones could bind to nascent NCAPs and facilitate recognition by a PD-IniR, and/or as a component of the PD-DoC mediate transport by threading a large cargo into the PD microchannels.

### *3.2.6 Potential role of ISE1 – a mutant plant with impaired PD function*

During embryogenesis, a developmental transition occurs in the physical properties of the PD microchannel (Gisel *et al.*, 1999; Zambryski, 2004). *Arabidopsis* embryos in the torpedo stage limit the transport capacity of larger symplasmic tracers by decreasing the PD SEL. This observation facilitated a screen for embryo-defective mutants named *increased size exclusion limit* (*ise*), which are disorganized in PD-mediated diffusion of symplasmic tracers (Gisel *et al.*, 1999; see Chapter 5). The *ise1* mutant develops ectopic root hairs in all epidermal cells instead of distinct cell files. As CAPRICE, a regulator of root hair identity (Wada *et al*., 1997), is transported from cell to cell (Wada *et al*., 2002), this would indicate that ISE1 has a regulatory role in transport of NCAPs involved in cell differentiation. In mutant plants, the PD structures appear normal; thus, we can assume that the gene product is not a major structural PD component. Although the identity of the mutated *ISE1* gene is not yet known, ISE1 seem to function as a regulatory PD component tuning the dynamics of the PD microchannels. In the PD pathway model, ISE1 could function as part of PD-DoC or as a structural component of PD microchannels.

### *3.2.7 A PD pathway model*

Based on observed structural modifications of PD cargo proteins and the presence of specific PD receptor proteins that recognize structurally distinct NCAPs, a PD transport pathway model can be built (Figure 3.2) based on the intracellular translocation pathways (Figure 3.1). The potential PD pathway components have a similar role in both models. The transport from cell to cell via PD is achieved in four steps. The first step is the initial recognition by a PD-IniR pathway constituent. If this interaction occurs distant from the PD orifices a second step is necessary, which transfers the cargo to the proximity of the PD orifice via a PD-TraC associated with the cytoskeleton. In this scenario, a third step, which involves recognition of a cargo-InR complex present at the PD orifices, is mandatory. The PD-DoC, which senses the presence of a cargo, initiates gating of the PD microchannel and unfolding of the transported macromolecules. Translocation of NCAPs requires gating of the PD microchannel and structural protein modifications. Thus, the last and fourth step of the pathway is the translocation of the cargo across the cellular boundary (Figure 3.2). In the model shown, the PD-TraC is either constituted as a cytoskeleton-associated soluble complex, or as a hydrophobic vesicle-bound complex. Transport of membrane-associated PD pathway components and cargos is suggested to be achieved via an ER-based vesicular pathway. Such a membrane-bound pathway would be especially necessary to support the *de novo* formation of intercellular connected PD structures formed after cytokinesis.

The above HSP70-related chaperones and the cytoskeleton seem to act together with the three suggested PD pathway components NCAPP1, NtMPB2C and ISE1 to constitute the intercellular NCAP transport pathway. A membrane-associated PD-IniR is realized by the ER-associated NCAPP1. The NtMPB2C seems to function as a regulatory cofactor involved in the initial NCAP recognition by PD-IniR, or transfer of the cargo by PD-TraC. ISE1 could mediate the transport and initiates modifications of PD microchannels as part of the PD-DoC. Although the exact functional role of these PD pathway components has to be elucidated, if improperly expressed or mutated they interfere with NCAP transport from cell to cell. Also, sufficient evidence exists to prove that cytoskeletal elements mediate the intracellular transfer of cargos to PD. HSP70-like chaperones, however, can be correlated with all stages of PD cargo transport: recognition, delivery and modification of PD microchannel components and of transported macromolecules.

## 3.3 Identifying PD transport signal(s)

### *3.3.1 Definition of a targeting signal*

Two attributes of NCAPs define their identity as PD transported proteins. First, *NCAPs induce a PD SEL increase* and, second, *NCAPs are actively transported via PD* (Haywood *et al.*, 2002). Although disputed, no strong evidence exists for transport via PD of *endogenous proteins* in a nonspecifically and passive fashion (Zambryski and Crawford, 2000; Haywood *et al.*, 2002). Interestingly, like the nuclear pore complex, a dilated PD microchannel permits the unspecific diffusion of small heterologous-expressed proteins such as the jellyfish GFP (Oparka *et al.*, 1999; Crawford and Zambryski, 2000, 2001). This is in contrast to results obtained in microinjection assays. Here, a number of small proteins were tested for their transport capacity. The *Ricinus communis* 8-kDa ubiquitin (Balachandran *et al.*, 1997), soybean 22.5-kDa trypsin inhibitor (Lucas *et al.*, 1995) or heterologous proteins such as mouse 27.8-kDa GST (Kragler *et al.*, 1998) and jellyfish 27-kDa GFP (Kragler, unpublished) show no movement via PD or effect on the plasmodesmatal SEL. On the other hand, GST or GFP fused to the NCAP KN1 results in transport of the recombinant protein to neighboring cells (Kragler *et al.*, 1998, 2000; Kim *et al.*, 2002, 2003). Thus, KN1 is identified as a specific cargo and seems to carry a PD targeting motif. As in the intracellular protein import system, a specific PD translocation system seems to identify specific targeting signals.

In intracellular protein transport pathways, a targeting signal is essential for a protein to enter a translocation system. Conventionally, a targeting signal is defined as *specific, sufficient and necessary* for a protein to enter a given translocation pathway (Schell and Hebert, 2003). PD targeting signal(s), in analogy to the intracellular targeting signals, should be essential for an NCAP to enter the PD translocation pathway, and sufficient to mediate transport of cell-autonomous proteins. Here, the potential PD-targeting signals of four NCAPs are discussed (Table 3.3). These examples demonstrate the structural heterogeneity and intricacies presented by PD-targeted proteins: the homeodomain transcription factor KNOTTED1 (Lucas *et al.*, 1995; Kim *et al.*, 2002b, 2003), rice phloem Thioredoxin h (Balachandran *et al.*, 1997), pumpkin cytochrome $b_5$ reductase CmPP36 (Xoconostle-Cázares *et al.*, 2000) and pumpkin phloem chaperone HSC70-1 (Aoki *et al.*, 2002).

### *3.3.2 KNOTTED1 – existence of a PD-targeting and SEL increase signal*

The non-cell autonomous nature of KN1 was confirmed *in vivo* and *in vitro* (see also Chapter 6). In microinjection experiments, as measured with fluorescent tracer molecules, the presence of KN1 increases the PD SEL (Lucas *et al.*, 1995). In transgenic plants, KN1–GFP fusion proteins expressed in shoot apical L1 cells moved to the inner layers of the shoot apex (Kim *et al.*, 2002b, 2003). Transport studies performed with mutant KN1 (M6) highlights one short amino acid sequence located between the KNOX domain and the homeodomain, which is necessary for transport

**Table 3.3** NCAPs with experimentally evaluated PD signal motif(s)

| Protein | Function | Motif(s) |
|---|---|---|
| *Zm*KNOTTED1 | Transcription factor | *N*-terminus for SEL increase; the region between the ELK domain and the homeodomain resembles a monopartite NLS (cluster of lysine residues) |
| *Cm*HSC70-1, 2 | Chaperone | Plant SVR (a short nonconserved region close to *C*-terminus) sufficient to mediate transport of human HSP70; potentially proline-directed phosphorylation of neighboring threonine abolishes transport |
| *Cm*PP36 | Cytochrome $b_5$ reductase | Structurally similar to CmPP16 and RCNMV-MP; *N*-terminal hydrophobic 59 amino acids proteolytically cleaved off for transport |
| *Os*RBB13-1 | Thioredoxin h | Two charged *N*-terminal glutamic acid residues and a charged *C*-terminal amino acid cluster |

via PD (Lucas *et al.*, 1995; Kim *et al.*, 2002b). A charged lysine cluster region resembling a monopartite nuclear localization signal (NLS) was substituted by three alanines in the KN1 M6 mutant. The simplest explanation could have been that such a monopartite NLS-like signal is sufficient and necessary for KN1 movement. In such a model, the PD transport system adopted parts of the nuclear transport system to allocate proteins with an NLS. However, fusing a GFP protein to a comparable monopartite NLS abolished GFP cell-to-cell transport, which indicates that an NLS is not sufficient to act as a PD-targeting signal (Crawford and Zambryski, 2000). The combined results support the notion of a more extensive three-dimensional structural motif resembling the KN1 PD-targeting signal. The KNOX dimerization domain and/or the homeodomain neighboring the KN1 NLS motif may provide the structure essential for KN1 cell-to-cell movement.

As defined above, NCAPs have two distinct signaling features, increase of SEL and targeting. These two features could result from a single signal motif or two physically separated motifs. Evidence for the latter is found in transport studies employing *in vitro* isolated antagonistic peptides that interfere with KN1 transport. An antagonistic peptide, which competes with binding of KN1 to proteins in a PD-enriched cell fraction, blocked PD SEL increase and KN1 mRNA transport. Nevertheless, the KN1 antagonist still allowed limited cell-to-cell transfer of KN1 protein (Kragler *et al.*, 2000). Obviously, the opening of the PD microchannel seems essential for transport of a large KN1-protein–KN1-mRNA complex but not of KN1 protein alone. Regardless of whether these results assign a function to the

PD SEL increase for RNA transport, they also suggest the presence of two distinct KN1 PD signals: a targeting signal and a SEL increase signal.

### *3.3.3 Phloem Thioredoxin h – charged amino acids as PD motifs*

Supporting evidence that NCAPs in general may carry two signal motifs is found in a mutant rice phloem Thioredoxin h, OsRBB13-1 (Ishiwatari *et al.*, 1998). Deletion of the five *N*-terminal amino acids (MT1), or introduction of alanine substitutions in a charged *C*-terminal cluster (MT8), disabled the cell-to-cell transport activity of the enzyme. However, in a limited number of microinjection experiments, the structurally altered OsRBB13-1 MT8 mutant still induced an increase of the PD SEL. The known three-dimensional structure of Thioredoxin h enzymes closely related to RPP13-1 facilitated *in silico* structural comparison studies of the wild-type protein with mutant RPP13-1. Both mutations, MT1 and MT8, alter the charge distribution on the protein surface. The combined results led to the conclusion that the accessibility of charged residues for a PD transport receptor mediates the entrance of RBP13-1 into the PD transport pathway.

### *3.3.4 Proteolytic processing facilitates transport of CmPP36*

The pumpkin phloem 36-kDa protein, CmPP36, belongs to the cytochrome $b_5$ reductase family and was identified by its structural similarity to *Ricinus communis necrotic mosaic virus* movement protein (RCNMV) and CmPP16 (Xoconostle-Cázares *et al.*, 2000). The *N*-terminal membrane-targeting domain is cleaved to produce a truncated DN-CmPP36 found in the phloem sap. In microinjection experiments, DN-CmPP36, but not CmPP36, is able to mediate its cell-to-cell transport. DN-CmPP36 interacts with RNA, but does not (in contrast to RCNMV movement protein and CmPP16) mediate transport of RNA. The authors proposed that intercellular movement of CmPP36 produced in the vascular tissue requires proteolytic processing to produce a soluble, movement-competent protein to enter the sieve tube long-distance transport pathway via PD. An interesting aspect is that a secondary modification of the protein provides the means for translocation. Here, similar to the downstream events observed in protein import into ER, mitochondria and chloroplast translocated proteins are proteolytically cleaved during or after translocation to make the translocated proteins available for downstream events. By analogy, the hydrophobic *N*-terminal sequence of CmPP36 could serve as a special sorting motif, which utilizes the suggested vesicular PD translocation pathway to move toward the PD microchannel as indicated in the model presented in Figure 3.2. Once such CmPP36-tagged vesicle arrives at a PD-DoC, a PD-associated protease releases a processed CmPP36 which is then available for transport.

### *3.3.5 A short PD-targeting motif in phloem HSC70*

The high degree of structural and functional conservation of the HSP70 chaperones facilitated the identification of a PD-targeting motif used by a subclass of pumpkin phloem HSC70-1 (Aoki *et al.*, 2002). Sequence comparison and transport studies of truncated and mutated HSC70-1 protein suggested that a substrate-binding domain in combination with a short variable region (SVR) is essential for HSC70 trafficking. Replacing a threonine ($T_{637}$) present in the SVR of the cell-autonomous pumpkin HSC70-3 with proline was sufficient to constitute a structure that triggered HSC70-3 cell-to-cell transport. The SVR motif, however, was insufficient to mediate transport of GFP, but facilitated the transport of an otherwise cell-autonomous human HSP70. These results are consistent with the notion that a structurally complex targeting motif facilitates entrance of the pumpkin HSC70-1 into the PD translocation pathway. It seems that the selective pressure on organisms to maintain functional HSC70 chaperones permitted only alterations in the short variable region of cytosolic HSP70 chaperones to establish a PD-targeting motif embedded in a larger structural context.

### *3.3.6 The elusive nature of PD-targeting signals*

A simple amino acid stretch that resembles a PD signal motif, and is both necessary and sufficient to mediate the transport of a protein, has not yet been identified. All current results point to a complex structural motif including major parts of PD-transported proteins. The challenges in identifying a PD signal motif are manifold. The overall three-dimensional structure of NCAPs might function as a PD signal motif. It is also feasible that an NCAP may carry a number of distinct PD signal motifs that are recognized in parallel or in a hierarchical order by different initial PD receptors. It is also conceivable that the PD signal motif is of such low complexity that it is easily missed in mutant and bioinformatic studies.

Although simply composed targeting signals appear to mediate translocation into organelles, each explanation is exemplified by intracellular protein transport systems, e.g. peroxisomal firefly luciferase or yeast catalase A harbor two peroxisomal targeting motifs – a conventional *C*-terminal PTS1 and an as yet unidentified motif in the *N*-terminal third of the protein (Gould *et al.*, 1987; Kragler *et al.*, 1993). The unconventional peroxisomal targeting motif of catalase A, although extensively analyzed by point mutations and bioinformatical approaches, has not been resolved (A. Hartig, personal communication). Another example of the complexity of targeting signals can be found in attempts to resolve an NLS of zinc-binuclear cluster transcription factors. The nuclear localization of the yeast transcription factors AlcR, PrnA and Gal4 is mediated by noncanonical redundant NLS-like motifs recognized by distinct NLS receptor complexes (Nikolaev *et al.*, 2003). Moreover, a single NLS motif can be recognized by multiple receptors. A number of NLS motifs with lengths of up to 38 amino acids, or shorter unusual NLS motifs that bear no obvious resemblance to the classical SV40-like NLS, establish a complex and dynamic receptor–cargo interaction system. Studies covering nuclear transport mechanisms of yeast and vertebrates suggest that 16 categories of import signals and receptors, 10 categories of export signals and receptors, and 5 shuttling signals and receptors are used to establish regulated nuclear protein allocation (Nakielny and Dreyfuss, 1999; Allen *et al.*, 2000).

We seem to face a complexity of PD-targeting motifs similar to that observed with some peroxisomal and nuclear imported proteins. In the future, the PD-targeting motif structure may be resolved once we have identified a sufficient number of the three-dimensional structures of NCAPs and tested their function in transport assays. On the other hand, random GFP–cDNA fusion screens revealed 13 fusion proteins specifically accumulating like viral movement proteins at PD (Blackman *et al.*, 1998; Heinlein *et al.*, 1998; Cutler *et al.*, 2000; Kim *et al.*, 2003; Medina-Escobar *et al.*, 2003). Such systematic PD protein localization studies, combined with bioinformatic approaches, may reveal common PD-targeting motifs or, in the best-case scenario, identify PD pathway components that are co-transported with cargo proteins.

## 3.4 Conclusions and future prospects

The PD transport pathway parallels in many ways the nuclear import/export of macromolecules. For example, hundreds of different proteins with distinct structural features and functions are translocated (Table 3.1), and secondary modifications such as phosphorylation regulate the transfer of macromolecules (Waigmann *et al.*, 2000; Haywood *et al.*, 2002). An intriguing question is why the PD transport system evolved to such complexity? Controlling the exchange of homeotic signals in the form of transcription factors like KNOTTED1, LEAFY, CAPRICE or SHORT ROOT could provide a strong selective pressure on plants to evolve a sophisticated PD transport system. A default intercellular transport pathway for all cytosolic soluble proteins would impose a challenge for tissue differentiation and pattern formation. Several lines of evidences exist in favor of regulated cell-to-cell transport of NCAPs and non-cell autonomous RNA molecules (Haywood *et al.*, 2002; Ding *et al.*, 2003; Kim *et al.*, 2003; Qi *et al.*, 2004).

Despite several attempts to identify protein structures that act as PD-targeting signals, a common and simple sequence motif necessary and sufficient to guide heterologous proteins through PD could not be determined. This is not a unique situation, considering the complexity of other intracellular translocation signal motifs, and definitely needs more attention. Potentially, genome-wide GFP tagging screens might bring classes of consensus sequences that resemble PD-targeting motifs in the near future. Based on knowledge gained by high-throughput screenings, *in silico* facilitated identification of PD-targeting motifs should be feasible and allow us to assign non-cell-autonomous functions to novel as well as to already known gene products. On the mechanistic side, two questions remain to be answered: what is the driving force within the PD structures that translocates cargos across the PD microchannels? How is cargo transport between the *cis*- and *trans*-sites coordinated to avoid traffic jams within the PD microchannels?

## Acknowledgements

I offer my apologies to the colleagues whose work could not be properly discussed due to space limitations. I thank my colleagues Andreas Hartig, Christoph Schueller at the Institute of Biochemistry and Molecular Cell Biology and Helga Weisz from the Institute of Interdisciplinary Research and Education for contributions and discussions. The work was supported by a grant from the Austrian Science Foundation (FWF); Project#: P16928-Kragler.

## References

Agranovsky, A.A., Boyko, V.P., Karasev, A.V., Koonin, E.V. and Dolja, V.V. (1991) 'Putative 65 kDa protein of beet yellows closterovirus is a homologue of HSP70 heat shock proteins', *Journal of Molecular Biology* **217**, 603–610.

Allen, T.D., Cronshaw, J.M., Bagley, S., Kiseleva, E. and Goldberg, M.W. (2000) 'The nuclear pore complex: mediator of translocation between nucleus and cytoplasm', *Journal of Cell Science* **113**, 1651–1659.

Alzhanova, D.V., Napuli, A.J., Creamer, R. and Dolja, V.V. (2001) 'Cell-to-cell movement and assembly of a plant closterovirus: roles for the capsid proteins and Hsp70 homolog', *EMBO Journal* **20**, 6997–7007.

Aoki, K., Kragler, F., Xoconostle-Cázares, B. and Lucas, W.J. (2002) 'A subclass of plant heat shock cognate 70 chaperones carries a motif that facilitates trafficking through plasmodesmata', *Proceedings of the National Academy of Sciences USA* **99**, 16342–16347.

Balachandran, S., Xiang, Y., Schobert, C., Thompson, G.A. and Lucas, W.J. (1997) 'Phloem sap proteins from *Cucurbita maxima* and *Ricinus communis* have the capacity to traffic cell to cell through plasmodesmata', *Proceedings of the National Academy of Sciences USA* **94**, 14150–14155.

Blackman, L.M. and Overall, R.L. (1998) 'Immunolocalisation of the cytoskeleton to plasmodesmata of *Chara corallina*', *Plant Journal* **14**, 733–741.

Blackman, L.M., Boevink, P., Santa Cruz, S., Palukaitis, P. and Oparka, K.J. (1998) 'The movement protein of cucumber mosaic virus traffics into sieve elements in minor veins of *Nicotiana clevelandii*', *The Plant Cell* **10**, 525–538.

Blackman, L.M., Harper, J.D. and Overall, R.L. (1999) 'Localization of a centrin-like protein to higher plant plasmodesmata', *European Journal of Cell Biology* **78**, 297–304.

Blobel, G. (1980) 'Intracellular protein topogenesis', *Proceedings of the National Academy of Sciences USA* **77**, 1496–1500.

Bloch, M.A. and Johnson, K.A. (1995) 'Identification of a molecular chaperone in the eukaryotic flagellum and its localization to the site of microtubule assembly', *Journal of Cell Science* **108**, 3541–3545.

Boyko, V., Ashby, J.A., Suslova, E. *et al.* (2002) 'Intramolecular complementing mutations in tobacco mosaic virus movement protein confirm a role for microtubule association in viral RNA transport', *Journal of Virology* **76**, 3974–3980.

Boyko, V., Ferralli, J. and Heinlein, M. (2000) 'Cell-to-cell movement of TMV RNA is temperature-dependent and corresponds to the association of movement protein with microtubules', *Plant Journal* **22**, 315–325.

Chaffey, N. and Barlow, P. (2001) 'The cytoskeleton facilitates a three-dimensional symplasmic continuum in the long-lived ray and axial parenchyma cells of angiosperm trees', *Planta* **213**, 811–823.

Chen, M.H., Sheng, J., Hind, G., Handa, A.K. and Citovsky, V. (2000) 'Interaction between the tobacco mosaic virus movement protein and host cell pectin methylesterases is required for viral cell-to-cell movement', *EMBO Journal* **19**, 913–920.

Chisholm, S.T., Parra, M.A., Anderberg, R.J. and Carrington, J.C. (2001) '*Arabidopsis* RTM1 and RTM2 genes function in phloem to restrict long-distance movement of tobacco etch virus', *Plant Physiology* **127**, 1667–1675.

Citovsky, V., McLean, B.G., Zupan, J.R. and Zambryski, P. (1993) 'Phosphorylation of tobacco mosaic virus cell-to-cell movement protein by a developmentally regulated plant cell wall-associated protein kinase', *Genes and Development* **7**, 904–910.

Crawford, K.M. and Zambryski, P.C. (2000) 'Subcellular localization determines the availability of non-targeted proteins to plasmodesmatal transport', *Current Biology* **10**, 1032–1040.

Crawford, K.M. and Zambryski, P.C. (2001) 'Non-targeted and targeted protein movement through plasmodesmata in leaves in different developmental and physiological states', *Plant Physiology* **125**, 1802–1812.

Cutler, S.R., Ehrhardt, D.W., Griffitts, J.S. and Somerville, C.R. (2000) 'Random GFP::cDNA fusions enable visualization of subcellular structures in cells of *Arabidopsis* at a high frequency', *Proceedings of the National Academy of Sciences USA* **97**, 3718–3723.

Ding, B., Itaya, A. and Qi, Y. (2003) 'Symplasmic protein and RNA traffic: regulatory points and regulatory factors', *Current Opinion in Plant Biology* **6**, 596–602.

Ding, B., Kwon, M.O. and Warnberg, L. (1996) 'Evidence that actin filaments are involved in controlling the permeability of plasmodesmata in tobacco mesophyll', *Plant Journal* **10**, 157–164.

Dorokhov, Y.L., Makinen, K., Frolova, O.Y. *et al.* (1999) 'A novel function for a ubiquitous plant enzyme pectin methylesterase: the host–cell receptor for the tobacco mosaic virus movement protein', *FEBS Letters* **461**, 223–228.

Dunoyer, P., Thomas, C., Harrison, S., Revers, F. and Maule, A. (2004) 'A cysteine-rich plant protein potentiates potyvirus movement through an interaction with the virus genome-linked protein VPg', *Journal of Virology* **78**, 2301–2309.

Efremova, N., Perbal, M.C., Yephremov, A., Hofmann, W.A., Saedler, H. and Schwarz-Sommer, Z. (2001) 'Epidermal control of floral organ identity by class B homeotic genes in *Antirrhinum* and *Arabidopsis*', *Development* **128**, 2661–2671.

Ehlers, K. and Kollmann, R. (2001) 'Primary and secondary plasmodesmata: structure, origin, and functioning', *Protoplasma* **216**, 1–30.

Esau, K. and Charvat, I.D. (1975) 'An ultrastructural study of acid phosphatase localization in cells of *Phaseolus vulgaris* phloem by the use of the azo dye method', *Tissue and Cell* **7**, 619–630.

Fridborg, I., Grainger, J., Page, A., Coleman, M., Findlay, K. and Angell, S. (2003) 'TIP, a novel host factor linking callose degradation with the cell-to-cell movement of Potato virus X', *Molecular Plant–Microbe Interactions* **16**, 132–140.

Gillespie, T., Boevink, P., Haupt, S. *et al.* (2002) 'Functional analysis of a DNA-shuffled movement protein reveals that microtubules are dispensable for the cell-to-cell movement of tobacco mosaic virus', *The Plant Cell* **14**, 1207–1222.

Gisel, A., Barella, S., Hempel, F.D. and Zambryski, P.C. (1999) 'Temporal and spatial regulation of symplastic trafficking during development in *Arabidopsis thaliana* apices', *Development* **126**, 1879–1889.

Gorlich, D. (1997) 'Nuclear protein import', *Current Opinion in Cell Biology* **9**, 412–419.

Gorlich, D. and Kutay, U. (1999) 'Transport between the cell nucleus and the cytoplasm', *Annual Review of Cell and Developmental Biology* **15**, 607–660.

Gould, S.G., Keller, G.A. and Subramani, S. (1987) 'Identification of a peroxisomal targeting signal at the carboxy terminus of firefly luciferase', *Journal of Cell Biology* **105**, 2923–2931.

Green, L.A. and Liem, R.K. (1989) 'Beta-internexin is a microtubule-associated protein identical to the 70-kDa heat-shock cognate protein and the clathrin uncoating ATPase', *Journal of Biological Biochemistry* **264**, 15210–15215.

Hartl, F.U. and Hayer-Hartl, M. (2002) 'Molecular chaperones in the cytosol: from nascent chain to folded protein', *Science* **295**, 1852–1858.

Haus, U., Trommler, P., Fisher, P.R. *et al.* (1993) 'The heat shock cognate protein from *Dictyostelium* affects actin polymerization through interaction with the actin-binding protein cap32/34', *EMBO Journal* **12**, 3763–3771.

Haywood, V., Kragler, F. and Lucas, W.J. (2002) 'Plasmodesmata: pathways for protein and ribonucleoprotein signaling', *The Plant Cell* **14**, S303-S325.

Heinlein, M. (2002) 'Plasmodesmata: dynamic regulation and role in macromolecular cell-to-cell signaling', *Current Opinion in Plant Biology* **5**, 543–552.

Heinlein, M., Epel, B.L., Padgett, H.S. and Beachy, R.N. (1995) 'Interaction of tobamovirus movement proteins with the plant cytoskeleton', *Science* **270**, 1983–1985.

Heinlein, M., Padgett, H.S., Gens, J.S. *et al.* (1998) 'Changing patterns of localization of the tobacco mosaic virus movement protein and replicase to the endoplasmic reticulum and microtubules during infection', *The Plant Cell* **10**, 1107–1120.

Helariutta, Y., Fukaki, H., Wysocka-Diller, J. *et al.* (2000) 'The SHORT-ROOT gene controls radial patterning of the *Arabidopsis* root through radial signaling', *Cell* **101**, 555–567.

Hood, J.K. and Silver, P.A. (1999) 'In or out? Regulating nuclear transport', *Current Opinion in Cell Biology* **11**, 241–247.

Huang, Z., Andrianov, V.M., Han, Y. and Howell, S.H. (2001) 'Identification of *Arabidopsis* proteins that interact with the cauliflower mosaic virus (CaMV) movement protein', *Plant Molecular Biology* **47**, 663–675.

Ishiwatari, Y., Fujiwara, T., McFarland, K.C. *et al.* (1998) 'Rice phloem thioredoxin h has the capacity to mediate its own cell-to-cell transport through plasmodesmata', *Planta* **205**, 12–22.

Jackson, D. (2002) 'Double labeling of KNOTTED1 mRNA and protein reveals multiple potential sites of protein trafficking in the shoot apex', *Plant Physiology* **129**, 1423–1429.

Jackson, D. and Kim, J.Y. (2003) 'Dispatch. Intercellular signaling: an elusive player steps forth', *Current Biology* **13**, R349–R350.

Karasev, A.V., Kashina, A.S., Gelfand, V.I. and Dolja, V.V. (1992) 'HSP70-related 65 kDa protein of beet yellows closterovirus is a microtubule-binding protein', *FEBS Letters* **304**, 12–14.

Kellmann, J.W. (2001) 'Identification of plant virus movement–host protein interactions', *Zeitschrift für Naturforschung* **56**, 669–679.

Kim, I., Hempel, F.D., Sha, K., Pfluger, J. and Zambryski, P.C. (2002a) 'Identification of a developmental transition in plasmodesmatal function during embryogenesis in *Arabidopsis thaliana*', *Development* **129**, 1261–1272.

Kim, J.Y., Yuan, Z. and Jackson, D. (2003) 'Developmental regulation and significance of KNOX protein trafficking in *Arabidopsis*', *Development* **130**, 4351–4362.

Kim, J.Y., Yuan, Z., Cilia, M., Khalfan-Jagani, Z. and Jackson, D. (2002b) 'Intercellular trafficking of a KNOTTED1 green fluorescent protein fusion in the leaf and shoot meristem of *Arabidopsis*', *Proceedings of the National Academy of Sciences USA* **99**, 4103–4108.

Kragler, F., Curin, M., Trutnyeva, K., Gansch, A. and Waigmann, E. (2003) 'MPB2C, a microtubule-associated plant protein binds to and interferes with cell-to-cell transport of tobacco mosaic virus movement protein', *Plant Physiology* **132**, 1870–1883.

Kragler, F., Langeder, A., Raupachova, J., Binder, M. and Hartig, A. (1993) 'Two independent peroxisomal targeting signals in catalase A of *Saccharomyces cerevisiae*', *Journal of Cell Biology* **120**, 665–673.

Kragler, F., Monzer, J., Shash, K., Xoconostle-Cázares, B. and Lucas, W.J. (1998) 'Cell-to-cell transport of proteins: Requirement for unfolding and characterization of binding to a putative plasmodesmal receptor', *Plant Journal* **15**, 367–381.

Kragler, F., Monzer, J., Xoconostle-Cázares, B. and Lucas, W.J. (2000) 'Peptide antagonists of the plasmodesmal macromolecular trafficking pathway', *EMBO Journal* **19**, 2856–2868.

Lee, J.Y., Yoo, B.C., Rojas, M.R., Gomez-Ospina, N., Staehelin, L.A. and Lucas, W.J. (2003) 'Selective trafficking of non-cell-autonomous proteins mediated by NtNCAPP1', *Science* **299**, 392–396.

Liang, P. and MacRae, T.H. (1997) 'Molecular chaperones and the cytoskeleton', *Journal of Cell Science* **110**, 1431–1440.

Lucas, W.J., Bouche-Pillon, S., Jackson, D.P. *et al.* (1995) 'Selective trafficking of KNOTTED1 homeodomain protein and its mRNA through plasmodesmata', *Science* **270**, 1980–1983.

Macejak, D.G. and Luftig, R.B. (1991) 'Stabilization of actin filaments at early times after adenovirus infection and in heat-shocked cells', *Virus Research* **19**, 31–45.

Mariano, A.C., Andrade, M.O., Santos, A.A. *et al.* (2004) 'Identification of a novel receptor-like protein kinase that interacts with a geminivirus nuclear shuttle protein', *Virology* **318**, 24–31.

Matsushita, Y., Deguchi, M., Youda, M., Nishiguchi, M. and Nyunoya, H. (2001) 'The tomato mosaic tobamovirus movement protein interacts with a putative transcriptional coactivator KELP', *Molecular Cell* **12**, 57–66.

Matsushita, Y., Miyakawa, O., Deguchi, M., Nishiguchi, M. and Nyunoya, H. (2002) 'Cloning of a tobacco cDNA coding for a putative transcriptional coactivator MBF1 that interacts with the tomato mosaic virus movement protein', *Journal of Experimental Botany* **53**, 1531–1532.

McLean, B.G., Zupan, J. and Zambryski, P.C. (1995) 'Tobacco mosaic virus movement protein associates with the cytoskeleton in tobacco cells', *The Plant Cell* **7**, 2101–2114.

Medina-Escobar, N., Haupt, S., Thow, G., Boevink, P., Chapman, S. and Oparka, K. (2003) 'High-throughput viral expression of cDNA-green fluorescent protein fusions reveals novel subcellular addresses and identifies unique proteins that interact with plasmodesmata', *The Plant Cell* **15**, 1507–1523.

Mezitt, L.A. and Lucas, W.J. (1996) 'Plasmodesmal cell-to-cell transport of proteins and nucleic acids', *Plant Molecular Biology* **32**, 251–273.

Nakajima, K., Sena, G., Nawy, T. and Benfey, P.N. (2001) 'Intercellular movement of the putative transcription factor SHR in root patterning', *Nature* **413**, 307–311.

Nakielny, S. and Dreyfuss, G. (1999) 'Transport of proteins and RNAs in and out of the nucleus', *Cell* **99**, 677–690.

Nikolaev, I., Cochet, M.F. and Felenbok, B. (2003) 'Nuclear import of zinc binuclear cluster proteins proceeds through multiple, overlapping transport pathways', *Eukaryotic Cell* **2**, 209–221.

Okinaka, Y., Mise, K., Okuno, T. and Furusawa, I. (2003) 'Characterization of a novel barley protein, HCP1, that interacts with the brome mosaic virus coat protein', *Molecular Plant–Microbe Interactions* **16**, 352–359.

Oparka, K.J. (2004) 'Getting the message across: how do plant cells exchange macromolecular complexes?', *Trends in Plant Science* **9**, 33–41.

Oparka, K.J., Roberts, A.G., Boevink, P. *et al.* (1999) 'Simple, but not branched, plasmodesmata allow the nonspecific trafficking of proteins in developing tobacco leaves', *Cell* **97**, 743–754.

Perbal, M.C., Haughn, G., Saedler, H. and Schwarz-Sommer, Z. (1996) 'Non-cell-autonomous function of the *Antirrhinum* floral homeotic proteins DEFICIENS and GLOBOSA is exerted by their polar cell-to-cell trafficking', *Development* **122**, 3433–3441.

Qi, Y., Pelissier, T., Itaya, A., Hunt, E., Wassenegger, M. and Ding, B. (2004) 'Direct role of a viroid RNA motif in mediating directional RNA trafficking across a specific cellular boundary', *The Plant Cell* **16**, 1741–1752.

Reichel, C., Mas, P. and Beachy, R.N. (1999) 'The role of the ER and cytoskeleton in plant viral trafficking', *Trends in Plant Science* **4**, 458–462.

Samaj, J., Peters, M., Volkmann, D. and Baluŝka, F. (2000) 'Effects of myosin ATPase inhibitor 2,3-butanedione 2-monoxime on distributions of myosins, F-actin, microtubules, and cortical endoplasmic reticulum in maize root apices', *Plant Cell Physiology* **41**, 571–582.

Sanchez, I. and Cohen, W.D. (1994) 'Assembly and bundling of marginal band microtubule protein: role of tau', *Cell Motility and the Cytoskeleton* **29**, 57–71.

Schellmann, S., Schnittger, A., Kirik, V. *et al.* (2002) 'TRIPTYCHON and CAPRICE mediate lateral inhibition during trichome and root hair patterning in *Arabidopsis*', *EMBO Journal* **21**, 5036–5046.

Schnell, D.J. and Hebert, D.N. (2003) 'Protein translocons: multifunctional mediators of protein translocation across membranes', *Cell* **112**, 491–505.

Schwarz-Sommer, Z., Hue, I., Huijser, P. *et al.* (1992) 'Characterization of the *Antirrhinum* floral homeotic MADS-box gene *deficiens*: evidence for DNA binding and autoregulation of its persistent expression throughout flower development', *EMBO Journal* **11**, 251–263.

Sena, G., Jung, J.W. and Benfey, P.N. (2004) 'A broad competence to respond to SHORT ROOT revealed by tissue-specific ectopic expression', *Development* **131**, 2817–2826.

Sessions, A., Yanofsky, M.F. and Weigel, D. (2000) 'Cell–cell signaling and movement by the floral transcription factors LEAFY and APETALA1', *Science* **289**, 779–782.

Smith, M.D. and Schnell, D.J. (2001) 'Peroxisomal protein import – the paradigm shifts', *Cell* **105**, 293–296.

Soellick, T., Uhrig, J.F., Bucher, G.L., Kellmann, J.W. and Schreier, P.H. (2000) 'The movement protein NSm of tomato spotted wilt tospovirus (TSWV): RNA binding, interaction with the TSWV N protein, and identification of interacting plant proteins', *Proceedings of the National Academy of Sciences USA* **97**, 2373–2378.

Soll, J. and Schleiff, E. (2004) 'Protein import into chloroplasts', *Nature Reviews Molecular Cell Biology* **5**, 198–208.

Trobner, W., Ramirez, L., Motte, P. *et al.* (1992) 'GLOBOSA: a homeotic gene which interacts with DEFICIENS in the control of *Antirrhinum* floral organogenesis', *EMBO Journal* **11**, 4693–4704.

Wada, T., Kurata, T., Tominaga, R. *et al.* (2002) 'Role of a positive regulator of root hair development, CAPRICE, in *Arabidopsis* root epidermal cell differentiation', *Development* **129**, 5409–5419.

Wada, T., Tachibana, T., Shimura, Y. and Okada, K. (1997) 'Epidermal cell differentiation in *Arabidopsis* determined by a Myb homolog, CPC', *Science* **277**, 1113–1116.

Waigmann, E., Chen, M.H., Bachmaier, R., Ghoshroy, S. and Citovsky, V. (2000) 'Regulation of plasmodesmal transport by phosphorylation of tobacco mosaic virus cell-to-cell movement protein', *EMBO Journal* **19**, 4875–4884.

Warner, A.H., Brunet, R.T., MacRae, T.H. and Clegg, J.S. (2004) 'Artemin is an RNA-binding protein with high thermal stability and potential RNA chaperone activity', *Archives of Biochemistry and Biophysics* **424**, 189–200.

Weeds, A. and Maciver, S. (1993) 'F-actin capping proteins', *Current Opinion in Cell Biology* **5**, 63–69.

Wiedemann, N., Frazier, A.E. and Pfanner, N. (2004) 'The protein import machinery of mitochondria', *Journal of Biological Biochemistry* **279**, 14473–14476.

Wu, X., Dinneny, J.R., Crawford, K.M. *et al.* (2003) 'Modes of intercellular transcription factor movement in the *Arabidopsis* apex', *Development* **130**, 3735–3745.

Xoconostle-Cázares, B., Ruiz-Medrano, R. and Lucas, W.J. (2000) 'Proteolytic processing of CmPP36, a protein from the cytochrome b(5) reductase family, is required for entry into the phloem translocation pathway', *Plant Journal* **24**, 735–747.

Xoconostle-Cázares, B., Xiang, Y., Ruiz-Medrano, R. *et al.* (1999) 'Plant paralog to viral movement protein that potentiates transport of mRNA into the phloem', *Science* **283**, 94–98.

Yoo, B.-C., Kragler, F., Varkonyi-Gasic, E. *et al.* (2004) 'A systemic small RNA signaling system in plants', *The Plant Cell* **16**, 1979–2000.

Yoo, B.-C., Lee, J.Y. and Lucas, W.J. (2002) 'Analysis of the complexity of protein kinases within the phloem sieve tube system. Characterization of *Cucurbita maxima* calmodulin-like domain protein kinase 1', *Journal of Biological Biochemistry* **277**, 15325–15332.

Zambryski, P.C. (2004) 'Cell-to-cell transport of proteins and fluorescent tracers via plasmodesmata during plant development', *Journal of Cell Biology* **164**, 165–168.

Zambryski, P.C. and Crawford, K. (2000) 'Plasmodesmata: gatekeepers for cell-to-cell transport of developmental signals in plants', *Annual Review of Cell and Developmental Biology* **16**, 393–421.

# 4 Comparative structures of specialised monocotyledonous leaf blade plasmodesmata

C.E.J. Botha, R.H.M. Cross and L. Liu

## 4.1 Introduction

Much has been written over the past 30 years about the structure of plasmodesmata and their functional relationships in higher plants. In view of their small size, the ultrastructure of plasmodesmata is highly significant and will continue to dominate many structure–function studies. We (arguably) do not yet fully comprehend how these dynamic quasi-organelles (Epel, 1994) function in symplasmic transport, or just how they alter or change their dynamics structurally, depending upon their state of transport (Schulz, 1999, and references cited). Structurally, the most studied aspect of plasmodesmata is the neck region within which callose and actin appear to be implicated in regulation or gating (see Overall and Blackman, 1996; Overall, 1999; Schulz, 1999; Botha and Cross, 2001, and literature cited). However, apart from the meticulous work presented by Schulz (1995), which showed that 350 mM mannitol stimulated dilation of the cytoplasmic annulus in *Pisum sativum* roots, little is known about how plasmodesmata alter their structure, depending on their functional state. Actin and myosin have been localized to plasmodesmata in *Chara* (Blackman and Overall, 1998) and it would seem that actin and/or myosin may be involved in the regulation of the cytoplasmic annulus.

The first comprehensive reviews of plasmodesmatal structure by Robards (1968b) and Robards and Lucas (1990) focused considerable attention on unravelling the structure of these intercellular communication pathways. Subsequently, Evert *et al.* (1996b) demonstrated that there are several ways in which plasmodesmatal structure may become modified, depending on the transport requirements of the cells, as well as by physical barriers that the cell wall and its associated substructures present (see also Robinson-Beers and Evert, 1991; Botha, 1992a,b; Russin *et al.*, 1996; Botha *et al.*, 2000).

There is some agreement about the structure of these plasmodesmata, especially so in leaves along the loading pathway from mesophyll to sieve tube in the angiosperms. Yet, it is their subtle variations (usually associated with particular cell interfaces, or particular functions) that makes them as intriguing now as they were to early researchers, and allows them to be involved in many more functions than originally thought possible (López-Sáez *et al.*, 1966; see Chapter 1). In some instances, plasmodesmata are so specialized as to allow only one-way traffic (Haritatos *et al.*, 2000;

Turgeon *et al.*, 2001), whist in others no such specialization or transport directionality apparently exists. There are many reports in the literature suggesting that plasmodesmata may become modified along the loading or unloading pathway, either normally through aging, as growth and thickening of the cell walls force alteration of their structure to accommodate increased wall thickening, or through the specific action of viruses (Robards, 1968a,b; Robinson-Beers and Evert, 1991; Epel, 1994; Oparka *et al.*, 1999; Botha *et al.*, 2000; Lee *et al.*, 2000). Blackman and Overall (2001) noted that little is known about the molecular architecture of plasmodesmata. Many researchers have provided excellent electron micrographs which show remarkably consistent features, but these images do not – perhaps cannot – reveal enough information about the dynamic nature of plasmodesmata. Blackman and Overall (2001) argue that the generally accepted size exclusion limit of plasmodesmata of 1 kDa, determined through methods involving the microinjection of fluorescent probes, is perhaps not an accurate measure of the size of molecules that can move from cell to cell. Overall (1999) argues convincingly for the presence of a molecular sieve-like structure situated at the neck region which is envisaged to prevent diffusion of membrane components between cells via plasmodesmata. Apparently, proteins commonly found in the plasma membrane are excluded from the regions around plasmodesmata (see Overall, 1999, and references cited). Blackman and Overall (2001) go on to state that some plasmodesmata may allow the non-selective movement of proteins up to 50 kDa, whereas others allow the selective transport of proteins such as transcription factors. Plasmodesmata are now known to traffic plasmodesmatally targeted and non-targeted proteins involved in signal transduction and other growth-critical functions (Crawford and Zambryski, 2000; Ehlers and Kollmann, 2001). Further factors, such as whether or not the plasmodesmata are simple or branched, may influence, or even govern, non-specific protein movement (Oparka *et al.*, 1999). Crawford and Zambryski (2001) suggest that leaf age dramatically affects the ability of cells to exchange proteins non-selectively. Epel (1994), therefore, correctly refers to plasmodesmata as dynamic quasi-organelles whose conductivity can be regulated by environmental and developmental signals. Their role in intercellular communication, in addition to the function associated with the movement of relatively simple assimilated molecules, therefore ensures that plasmodesmata remain somewhat mysterious, and that the state of this enigma surrounding their functional state guarantees that interest in solving structure in relation to function issues remains high.

Thus, the early research on the fine structure of plasmodesmata by Robards (1968a) and López-Sáez *et al.* (1966) remains as important today, despite the number of ultrastructural studies that have followed (Botha and Cross, 2001).

## 4.2 Maturity-related changes in plasmodesmatal structure

Evert *et al.* (1996a) made a comprehensive study of the anatomical and ultrastructural changes that accompany the sink-to-source transition in developing maize (*Zea mays* L. cv. W273) leaves. Sink, transition and source regions were positively identified by whole-leaf autoradiography. In the leaves examined, a structural gradient was found from non-importing to importing regions of the leaf blade. Although all components, except metaxylem elements, of the large bundles reached maturity before their counterparts in intermediate and small bundles, the final events in structural maturation were uniform for all bundle types across the blade. Among the very last structures to mature were the plasmodesmata at the interfaces between mesophyll cells and between mesophyll and bundle sheath (BS) cells. Evert *et al.* (1996a) reported that the maturation of the plasmodesmata coincided with maturation of the thick-walled sieve tubes, which are the last components of the vascular bundles to mature. Significantly, the vasculature reached structural maturity in advance of the cessation of import, and maturation of bundles involved in phloem loading was found not to be closely correlated with initiation of export from the blade. An important observation from the Evert *et al.* (1999a) study was that the deposition of suberin lamellae in the walls of the BS cells coincided with the deposition of secondary walls in the metaxylem vessels. It was suggested that a primary role of the suberin lamellae might be to prevent leakage of sucrose from the bundles in *Z. mays* (see also Evert *et al.*, 1985).

## 4.3 The plasmodesmatal cell wall interfaces in monocot leaves

Few laboratories have focused attention on the ultrastructure of the plasmodesmata of monocotyledons (see López-Sáez *et al.*, 1966; Evert *et al.*, 1977; Robinson-Beers and Evert, 1991; Botha, 1992a; Evert *et al.*, 1996a,b, and literature cited). In this chapter, we re-focus attention on plasmodesmatal structures with emphasis on those that exist in the monocotyledonous foliage leaf and, where appropriate, draw parallels with plasmodesmata in dicotyledonous plants.

Within monocotyledonous leaves, all interfaces from the mesophyll are interconnected by plasmodesmata internally, across different interfaces, at least to the vascular parenchyma (VP) cells. In many cases, a degree of plasmodesmatal isolation occurs between the VP cells and the phloem (Botha 1992a; Botha and van Bel, 1992b and literature cited). The anatomy of the monocotyledonous foliage leaf is variable, with the structure being dependent largely on the photosynthetic pathway that operates in the species. Within the lamina of the leaf in $C_3$ plants such as the grasses, the vascular bundles are surrounded by a BS, in which chloroplasts (if present) will be structurally similar to those occurring in the mesophyll cells. The bundle may be associated with an inner sheath, called the mestome sheath (MS), and beneath this are found the xylem and phloem. In the Cyperaceae, the transport pathway to the phloem becomes more complex, with a three-sheathed appearance. It comprises an outer parenchymatous BS and immediately beneath this, an encompassing MS, which is usually thick-walled and lignified at maturity. On the inside of this is a ring (sometimes incomplete) of parenchymatous cells that usually contain well-developed chloroplasts. The structural relationships between these sheaths is illustrated in the low magnification electron micrograph of a small

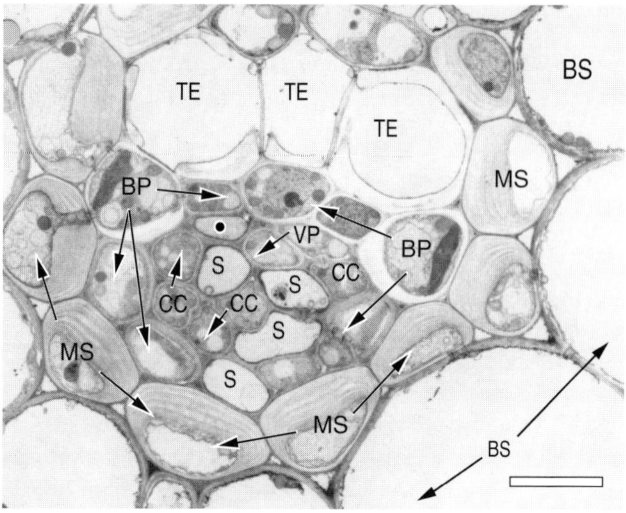


**Figure 4.1** *Mariscus congestus*. Transmission electron micrograph of part of an intermediate leaf blade vascular bundle showing arrangement of border parenchyma (BP) which encloses the underlying phloem tissue. The phloem is composed of parenchyma cells and thin-walled sieve tubes (S) which are associated with companion cells (CC). This bundle contains a solitary thick-walled sieve tube (solid dot), which is associated with and connected to vascular parenchyma cells. The mestome sheath (MS) has conspicuously thickened inner tangential walls. The MS has a suberized compound middle lamella in the inner radial and tangential walls of all sheath cells, separating it from the underlying BP cells. Symplasmic connectivity between the MS and the underlying BP is via deep-seated primary pit fields in this $C_4$ species. Bar = 5 μm.

vascular bundle from the leaf blade of *Mariscus congestus* (Figure 4.1). The anatomy of $C_4$ monocotyledonous leaves is generally complexed by the photosynthetic biochemistry, with a radiating Kranz mesophyll that surrounds a BS whose cells contain large agranal chloroplasts. An MS may occur internal to the BS and, if present, may surround the underlying vascular tissue entirely or partly (see Hattersley and Watson, 1975, for further details). The phloem, with its unique thin- and thick-walled sieve elements, is variously connected by both simple plasmodesmata for mesophyll interfaces and by branched pore-plasmodesmata between companion cells and thin-walled sieve-element–companion-cell complex. Companion cells are often connected to surrounding parenchymatous cells by branched plasmodesmata. Beebe and Russin (1999) suggest that the inner tangential walls of all BS cells (regardless of whether they are BS or MS cells) are in contact with VP cells. However, this is not the case in all monocots, specifically the Cyperaceae.

Of significant interest are the plasmodesmata that connect the mesophyll cells of leaves with their underlying vascular bundles. In many instances, a suberin lamella is present in at least one of the following interfaces – BS to MS and MS to VP. In the grasses, some are $C_3$ and others $C_4$, with all having a BS separating the mesophyll from underlying vascular tissues beneath the BS layer. In $C_4$ grasses, this sheath is the BS, containing as it does photosynthetic carbon reduction (PCR) chloroplasts. In contrast, Cyperaceae show greater structural variability, and three of four $C_4$ anatomical types of the Cyperaceae (Fimbristyloid, Chlorocyperoid, Eliocharoid) have an unusual version of Kranz anatomy, with one or two cell layers intervening between the primary carbon assimilation (PCA), or Kranz sheath, which is equivalent to the Kranz mesophyll in $C_4$ grasses, and an inner PCR layer. This inner layer is equivalent to the BS in $C_4$ grass leaf-blade bundles, but, because of its spatial separation from the PCA cells in the $C_3$ Cyperaceae, it is termed border parenchyma (BP). In many Cyperaceae, a layer of thick-walled cells separates the PCA from PCR layers. This layer has been called either an 'MS' or an 'endodermoid sheath'. As in the grasses, the MS in Cyperaceae abuts the BS, and in $C_4$ species, the MS separates the PCA from PCR layers. In the Cyperaceae, there is an increase in the number of cell interfaces across which transport of metabolites to the phloem must occur, with up to three sheath layers to cross in $C_3$ and $C_4$ species (see Soros and Dengler, 1998, and references cited) – from the mesophyll to BS→MS→BP→VP. It is important to note that the BP layer may contain PCR-supporting chloroplasts in the $C_4$ species, or PCA-chloroplasts in the $C_3$ species.

## 4.4 Plasmodesmata crossing the suberin lamella – constrictions and asymmetry

Many of the grasses and members of the Cyperaceae examined in our laboratories to date are associated with a suberin lamella, which is located either in the BS or the MS layers. In transectional view, leaf-blade vascular bundles may contain a suberin lamella in the outer or inner (or sometimes both) tangential walls. This suberin lamella may occur only on one side of the middle lamella between cells. For example, it occurs on the MS cell side in sugar cane, and on the BS cell side in maize. In maize, the only portions of the BS cell wall that is consistently suberized are those that are crossed by plasmodesmatal aggregates interconnecting BS with VP cells. Alternatively, the suberin lamella may occur on both sides of the middle lamella between associated cell interfaces, forming a 'tramline' or 'sandwich-like' appearance, with a polylamellate layer of suberin and waxy material delimiting the middle lamella and thus forming a compound middle lamella in the process. This double suberin lamella is common in many of the grasses that we have investigated to date, occurring in the BS–BS or MS–MS interfaces. In regions traversed by plasmodesmata, the compound middle lamella may be quite substantial – up to 80–100 nm in width. There are several reports in the literature which have indicated that the plasmodesmata traversing the suberin lamella are constricted where they pass through the suberin lamella (see Evert *et al.*, 1977; Robinson-Beers and Evert, 1991, and references cited; Botha, 1992a; Evert *et al.*, 1996a; Beebe and Russin, 1999). Plasmodesmatal structure across suberin lamellae thus share common attributes across the monocotyledonous species, some of which are highlighted here.

Where plasmodesmata cross the suberin lamella, containment and constriction to approximately 50 nm usually occurs. On either side of the restricting suberin layer, plasmodesmata are dilated and wider – usually 100–120 nm in width (including the plasma membrane). In almost all cases, these dilated regions show evidence of a desmotubule in the unconstricted suberin-free wall areas in mature leaf tissue. In their study, Evert *et al.* (1996a) described the complex processes involved in the maturing maize leaf blade in great detail. Their study focused attention on the formation of plasmodesmata in cell wall regions, where the prosuberin lamella and, subsequently the suberin lamella, formed in outer tangential and radial walls and inner tangential walls of the BS cells. Within this cell interface, plasmodesmata increased in length, keeping pace with the increase in the width of the new wall material deposited outside of the suberin lamella. In maize, the inner tangential walls become suberized only where plasmodesmatal aggregates occur (Robinson-Beers and Evert, 1991). Aspects of plasmodesmatal distribution and structure and the modifications associated with the presence of suberin lamellae in common walls are illustrated in Figures 4.2–4.16. In many cases, the plasmodesmatal aggregates that cross-interface, such as BS–MS or MS–MS, traverse a single or double suberin lamella lining the middle lamella in these regions. In many instances, plasmodesmata are structurally asymmetrical on either side of the suberin lamella. A case in point is the distribution of sphincters especially at the BS cell side in some maize examples, such as the sucrose transport deficient mutant (Beebe and Russin, 1999). Evert *et al.* (1996a) reported that development of the mesophyll-BS plasmodesmata coincides with the appearance of the sphincters on the mesophyll cell side of BS cells in maize. In contrast, sugarcane (Robinson-Beers *et al.*, 1991) develops sphincters on both orifices of plasmodesmata between bundles sheath and mesophyll plasmodesmata, and on both sides of plasmodesmata at other interfaces that are suberized. Our research supports the concept

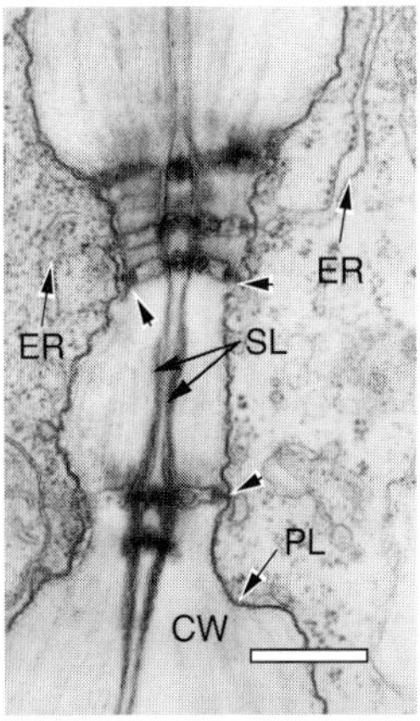


**Figure 4.2** Transection through common wall between two mestome sheath (MS) cells in *Saccharum officinarum* Var 293. Plasmodesmata occur in large pit fields in the primary wall between these cells. The compound middle lamella is delimited by polylamellate suberin lamellae (SL) on either side. Note that neck and median regions of these plasmodesmata are constricted. Plasmodesmata are linked by endoplasmic reticulum, which is ribosomal within the cytoplasm of the MS cells. Sphincters (arrowheads) are associated with both sides of individual plasmodesmata. Bar = 300 nm. Abbreviations: ER, endoplasmic reticulum; CW, cell wall; PL, plasmalemma.

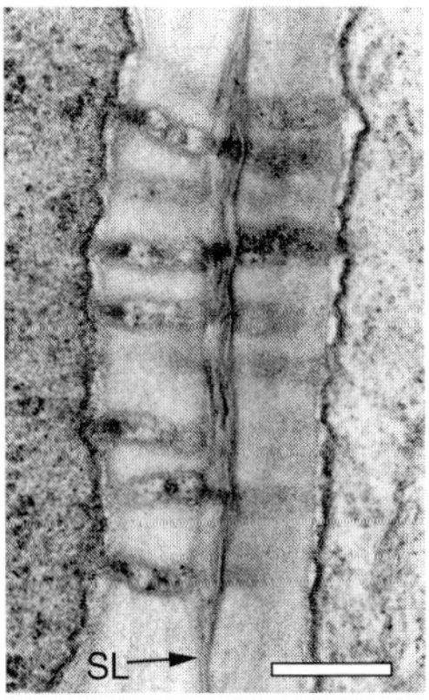


**Figure 4.3** *Panicum maximum*. Transection through the wall between a bundle sheath (BS) (left) and a Kranz mesophyll cell (right) in this $C_4$ species. The median regions of plasmodesmata are constricted where they pass through the compound middle lamella-polylamellate suberin–lamella complex. Plasmodesmatal orifices appear constricted on the BS cell-side. Note decreased electron density in the KMS cell wall and in the plasmodesmata exiting at this interface. Bar = 200 nm. Abbreviations: SL, suberin lamella.

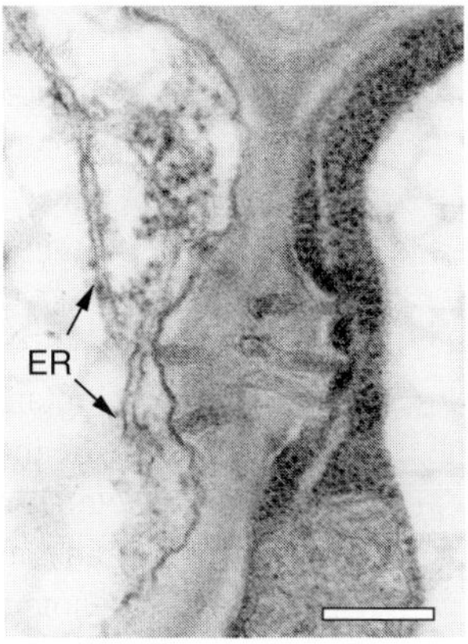


**Figure 4.4** *Triticum aestivum*. Transection through the common wall between a bundle sheath cell (left) and a vascular parenchyma cell (right). Plasmodesmata occur in a raised wall section between these two cell types. Endoplasmic reticulum (ER) is apparently in close contact with plasmodesmata in both cells. Bar = 300 nm.

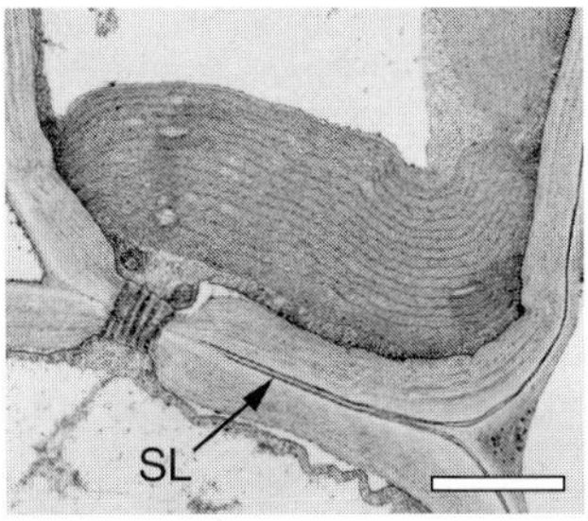


**Figure 4.5** *Saccharum officinarum* Var 376, showing a small plasmodesmatal pit field between two bundle sheath (BS) cells at low magnification. The compound middle lamella takes on a characteristic 'tramline' appearance, because of the presence of suberin lamellae (SL) in both BS cell walls. Suberin lamellae occur in the outer tangential and outer radial walls of BS cells, but stop short of the plasmodesmatal pit field illustrated here. Bar = 550 nm.

of variability of sphincter distribution in the monocotyledonous species that we have examined.

Figure 4.5 is a low magnification micrograph, showing the conspicuous 'tramline' appearance, where suberin lamellae outline the middle lamella between BS cells in sugarcane. A plasmodesmal pit field is shown near the inner radial wall,

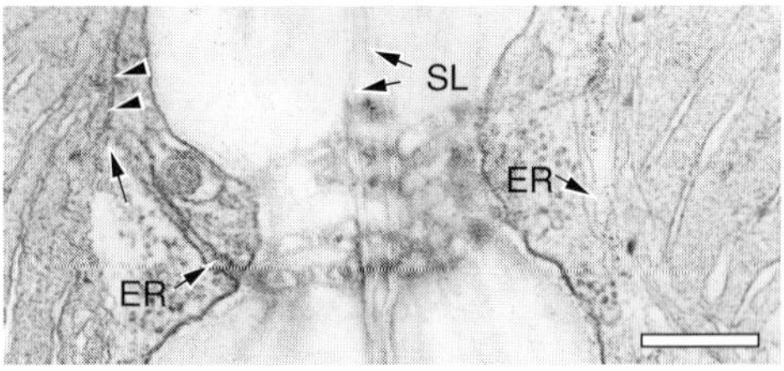


**Figure 4.6** *Saccharum officinarum* Var 376. Detail showing a small primary pit field between bundle sheath (BS) cells in a region of the wall which contains suberin lamellae (SL). Four plasmodesmata are visible in this view. The compound middle lamella constricts plasmodesmata which traverse this region. Note what appear to be direct plasmodesmata to chloroplast links via smooth endoplasmic reticulum in the BS cell to the left and to the right. Although quite electron-lucent, neck regions are constricted at either orifice. Bar = 200 nm. Abbreviations: ER, endoplasmic reticulum.

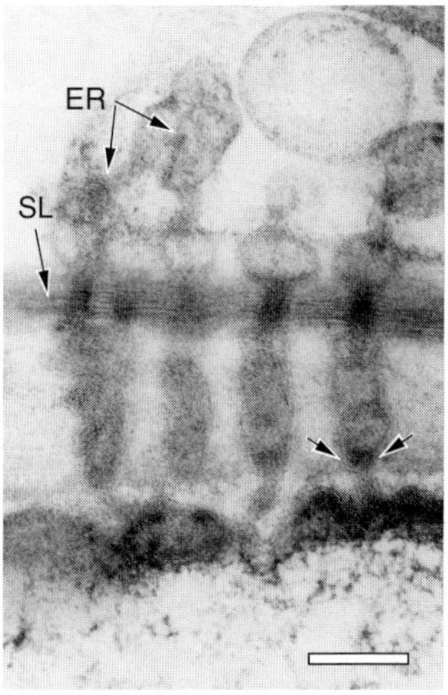


**Figure 4.7** *Cyperus pulcher*. Detail, showing plasmodesmata between mestome sheath (MS) (above) and vascular parenchyma (VP) cell (below). Plasmodesmata are constricted where they cross the polylamellate suberin lamella, but not either side of this structure. The asymmetrically thickened wall results in an unusual plasmodesmatal morphology, with a balloon-like outline commonly associated with the plasmodesmatal part in the MS cell wall side. Neck regions on the VP cell side are electron-dense and appear to be constricted and one contains a sphincter. Treatment in tannic acid enhanced the contrast of sphincters (paired arrowheads) on the VP side, but no such structure is visible on the MS-side of these plasmodesmata. Endoplasmic reticulum connects directly to these plasmodesmata on the MS cell side. Bar = 150 nm. Abbreviations: ER, endoplasmic reticulum; SL, suberin lamella.

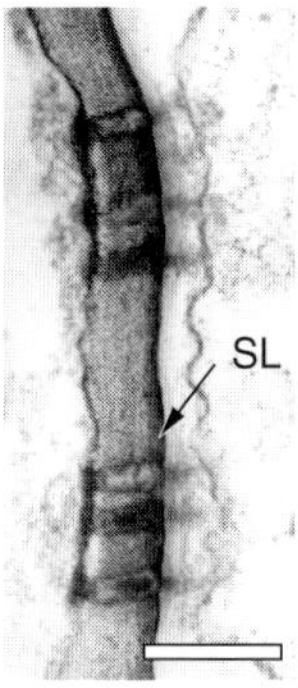


**Figure 4.8** *Cyperus fastigiatus*. Plasmodesmata between vascular parenchyma (VP) cells (left) and a mestome sheath cell (right). Note asymmetric outline and electron-dense appearance of plasmodesmata on VP side. All plasmodesmata in this and other examples treated in tannic acid reveal a very electron-dense extraplasmodesmal layer and sphincter-like structures associated with the neck regions on the VP-cell side and a generally more electron-lucent appearance on the MS-cell side. Desmotubules and cytoplasmic sleeves are discernible in this micrograph. Bar = 150 nm. Abbreviation: SL, suberin lamella.

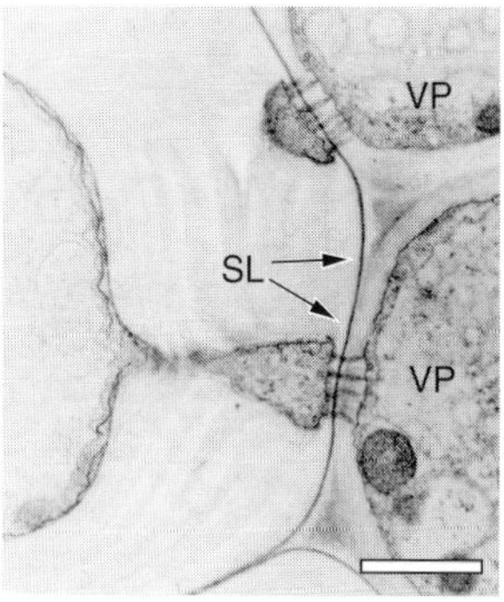


**Figure 4.9** *Cyperus pulcher*. Shows deep pits in the walls of very thick inner tangential mestome sheath (MS) cell wall, interfacing with an underlying vascular parenchyma (VP) cell (right) in part of an intermediate vein size class. A single suberin lamella (SL) is associated with the MS cell. Bar = 2.5 μm.

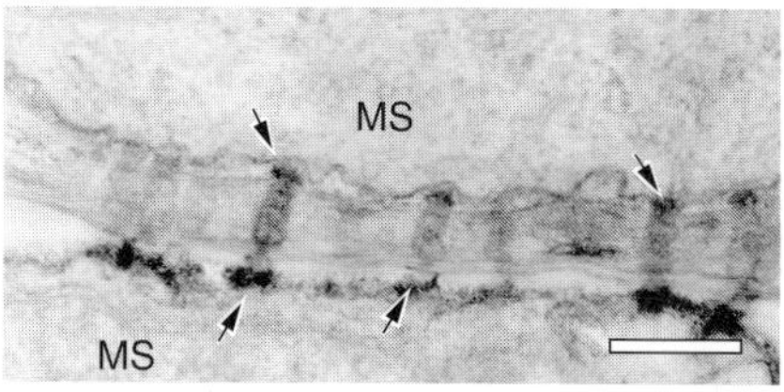


**Figure 4.10** *Mariscus congestus*. Electronmicrograph shows a plasmodesmatal field in the common radial wall between two mestome sheath cells. A polylamellate suberin lamella, showing the characteristic 'tramline' appearance where two lamellae occur either side of the middle lamella, is evident and plasmodesmata are constricted where they traverse the compound middle lamella. Neck regions in this tannic-acid-treated specimen are associated with electron-dense material. Bar = 150 nm. Abbreviation: MS, mestome sheath.

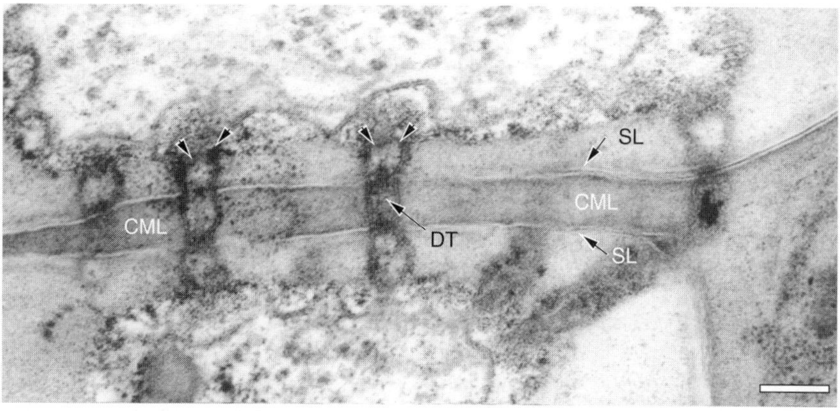


**Figure 4.11** *Cyperus sexangularis*. High-resolution electronmicrograph, showing plasmodesmatal field in the radial wall between two mestome sheath cells. Note compound middle lamella, delimited by suberin-lamellae. Plasmodesmata are only slightly constricted where they pass through the suberin-lamella-lined (SL) compound middle lamella (CML). Neck regions are dilated, with variable amount of electron-dense material associated with the plasmodesma–endoplasmic reticulum junctions, specifically associated with the outer plasmalemma leaflet. Plasmodesmata contain desmotubules (DTs). Bar = 100 nm.

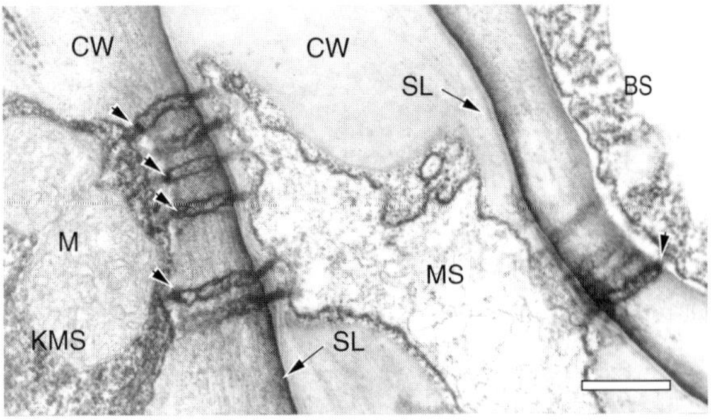


**Figure 4.12** *Mariscus congestus*. Glancing section showing plasmodesmatal fields between Kranz mesophyll cell (left), an endodermoid cell (middle) and a bundle sheath cell (BS) (right). Suberin lamellae constrict plasmodesmata at both endodermoid interfaces and neck regions appear to be associated with sphincters (arrowheads) at Kranz mesophyll and BS cell-sides. Bar = 1 μm. Abbreviations: CW, cell wall; SL, suberin lamella; MS, mestome sheath; KMS, krang merophyll sheath; M, mitochondrion; BS, bundle sheath.

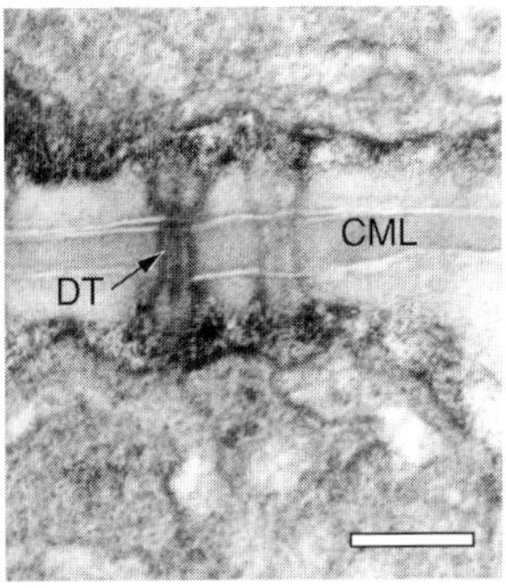


**Figure 4.13** *Cyperus pulcher*. This high-resolution electronmicrograph shows a pair of plasmodesmata that traverse a region of wall between mestome sheath cells containing compound middle lamella. Note desmotubule, surrounded by cytoplasmic annulus in the plasmodesma in plasmodesmata to the left. Neck regions are dilated, but associated with electron-dense, particulate material. Bar = 80 nm. Abbreviations: DT, desmotubule; CML, compound middle lamella.

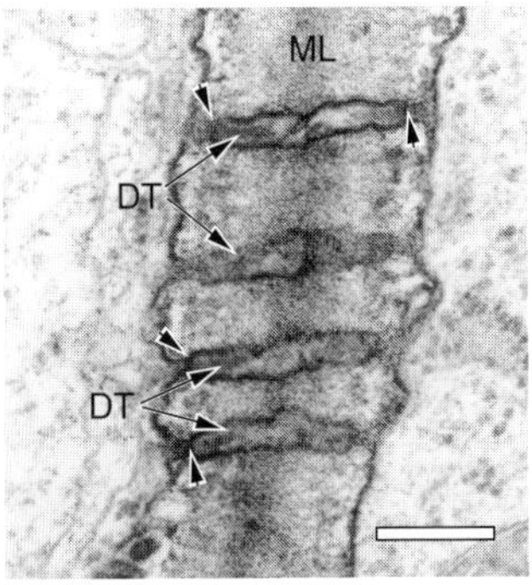


**Figure 4.14** *Cyperus sexangularis*. Section showing plasmodesmata traversing the common wall between two adjacent border parenchyma cells. Plasmodesmata are not constricted where they cross the middle lamella (ML). Note endoplasmic reticulum in close association with neck regions. Neck regions are variously dilated and plasmodesmata contain desmotubules. Bar = 100 nm. Abbreviations: DT, desmotubule; ML, middle lamella.

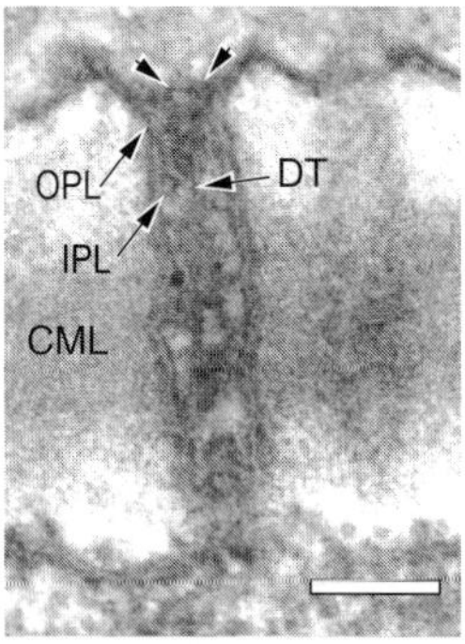


**Figure 4.15** *Cyperus pulcher*. Pair of plasmodesmata between a mestome sheath cell (above) and a vascular parenchyma cell (below). Neck regions are dilated. Note desmotubule and cytoplasmic annulus. Bar = 80 nm. Abbreviations: DT, desmotubule; OPL, outer plasmalemma leaflet; IPL, inner plasmalemma leaflet; CML, compound middle lamella.

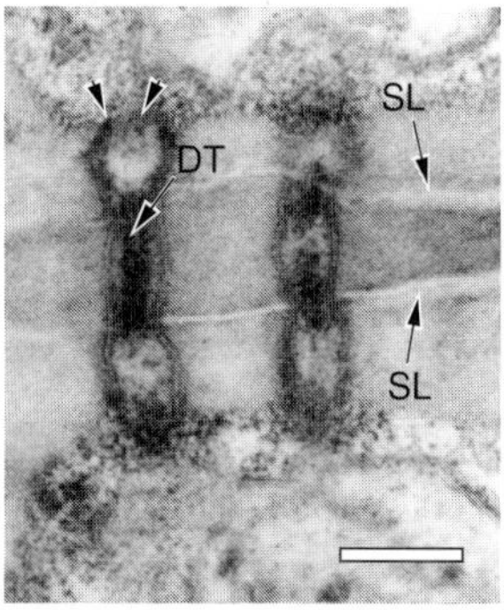


**Figure 4.16** *Cyperus sexangularis*. Plasmodesmata between adjacent mestome sheath cells, traversing compound middle lamella. Neck regions are variously stained and note desmotubule in the neck of plasmodesma to the lower left. Bar = 100 nm. Abbreviations: DT, desmotubule; SL, suberin lamella.

in an area where reduced suberin deposition occurs. Figures 4.2 and 4.6 show plasmodesmata traversing the common wall between concomitant MS cells (Figure 4.2) and BS cells (Figure 4.6) in sugarcane. Both images were taken from mature intermediate leaf-blade vascular bundles. Note that the plasmodesmata are constricted where they pass through the compound middle lamella. The neck regions are constricted and seem to be associated with sphincters at either orifice. The ER links plasmodesmata across this interface and in many other instances, as in the detail showing plasmodesmata between BS cell and a VP cell (Figure 4.4).

Our observations demonstrate that asymmetric plasmodesmatal structure is often associated with dissimilar cell associations and is co-associated with a marked difference in plasmodesmal and membrane transparency across the unlike interfaces, such as between the KMS–BS interface in *Panicum maximum* (Figure 4.3) and the BS–VP cell interface in *Cyperus pulcher*, *Cyperus fastigiatus* and *Mariscus congestus* (Figures 4.7, 4.8 and 4.10). A single suberin lamella is shown between MS and VP cells in the inner tangential and radial walls of the MS–VP interface in an intermediate vein in *C. pulcher* (Figure 4.9).

Figure 4.3 shows part of a large primary pit field between a Kranz mesophyll cell (right) and a BS cell (left) in *P. maximum*. As mentioned, the plasmodesmata across the KMS–BS interface often stain differentially. In this case, the plasmodesmata were generally more electron-dense on the BS cell side than on the KMS cell side. Plasmodesmata appear to be tightly constricted and encased where they pass through the suberin lamella and dilate on the KMS side. They contain clearly defined desmotubules and there appear to be sphincters at their orifices. In wheat leaves, plasmodesmata between BS and VP cells (Figure 4.4) appear to be equally constricted at either orifice. Interestingly, in paraformaldehyde–glutaraldehyde-fixed tissues, no sphincter or sphincter-like structures are evident. In many cases, plasmodesmata between BS cells in sugarcane are associated with a clearly defined ER system, which is either smooth or associated with ribosomal material.

Figures 4.2 and 4.6 show plasmodesmata at MS–MS and BS–BS interfaces, respectively, in leaf-blade tissues in sugarcane. Note the constricted appearance of the plasmodesmata, as well as the increase in electron density associated with the median constrictions. The ER in Figure 4.6 is closely associated with the chloroplast envelope (arrowheads) in the cell to the right.

Figures 4.6–4.9 illustrate aspects of plasmodesmatal structure in *C. pulcher* (Figures 4.7 and 4.9) and *C. fastigiatus* (Figures 4.8 and 4.10), respectively, between MS cells and associated VP cells. Figure 4.9 is a low magnification micrograph, which illustrates the asymmetric structure of plasmodesmata traversing the MS (left) and VP (right) interfaces in *C. pulcher*. Note that the MS cell wall is greatly thickened and possibly lignified. Plasmodesmata occur in deep pits and the compound middle lamella is associated with a single suberin lamella at this interface, which occurs on the MS-cell-wall side. The asymmetry of MS–VP plasmodesmata in *C. pulcher* is clearly illustrated at higher magnification in Figure 4.7. Treatment with tannic acid enhanced the sphincters (paired arrowheads) on the VP side, but no such structure was visible on the MS-side of these plasmodesmata. Plasmodesmata traversing the MS–VP interface in *C. fastigiatus* (Figure 4.8) are also constricted on the MS side (right), but not on the VP side (left). All plasmodesmata in this and other examples treated in tannic acid reveal a very electron-dense extra-plasmodesmal layer and sphincter-like structures associated with the VP-cell side, targeted with a generally more electron-lucent appearance on the MS-cell side.

Plasmodesmata are constricted where they pass through the suberin lamella but apparently not where they pass through the intervening middle lamella region in *M. congestus*, as illustrated in Figure 4.10.

Figures 4.11–4.16 illustrate aspects of the fine structural detail of plasmodesmata within selected leaf-blade vascular tissue from members of the Cyperaceae. All images shown here are from material in which tannic acid (0.01% w/v) was added as a mordant to the first glutaraldehyde fixation step. Details of this fixation procedure are described elsewhere (Botha *et al.*, 1993). The electron micrographs illustrate plasmodesmatal structure where these cross a double or single suberin lamella at the MS–MS or MS–VP interfaces. The fine structure of plasmodesmata involved in connections along the KMS–MS–BS (Figure 4.12) and the radial wall connections in the BP (Figure 4.14) interface are shown for comparison. High-resolution microscopy reveals a clearly defined outer and inner plasma membrane leaflet as the delimiter between the plasmodesmata and the cell walls. In many instances, 1–2 nm diameter, spherical structures are visible in association with both outer and inner plasma membrane leaflets. The desmotubules are separated from the inner plasma membrane leaflet by an annulus. The plasma membrane generally appears to be separated from the wall layers by about 20–40 nm. Some plasmolysis possibly occurred during fixation, but equally possible is the notion that this region of the wall–plasmalemma interface may be electron-lucent or even fluid. MS–MS plasmodesmata in the Cyperaceae are generally unoccluded with their exit aperture to the cytoplasm of concomitant cells free of dense 'plug-like' material (see Figures 4.11, 4.13, 4.15 and 4.16), but these plasmodesmata contain numerous granular substructures with varying electron-density. MS–VP and BP–BP plasmodesmata, too, are clear of obstruction, but sphincters are visible in some of these (unlabelled arrowheads point to sphincters in Figure 4.14). The extra-plasmodesmal neck regions show varying degrees of electron-lucence and these areas are potentially sites of extra-plasmodesmal collars, which may or may not be where callose is deposited. Figure 4.15 illustrates the point clearly where an opaque, dumbell-shaped region is located immediately outside the outer plasmalemma leaflet, extending towards the compound middle lamella. In MS–MS plasmodesmata, a collar is not as apparent, but in these cases, granular electron-dense particles are located in the extra-plasmodesmal space (unlabelled arrowheads point to these structures in Figures 4.11, 4.13 and 4.16).

## 4.5 Regulation at the neck – structural considerations

Myosin, actin and centrin have been reported to be involved in the regulation of the neck regions of plasmodesmata (White *et al.*, 1994; Radford and White, 1998; Blackman *et al.*, 1999). Radford and White (2001) have reported that treatment of plasmodesmata with 2,3-butanedione monoxime (BDM) resulted in strong constriction of the neck region of plasmodesmata in *Allium cepa, Z. mays* and *Hordeum vulgare*. Blackman *et al.* (1999) suggest several models for the involvement of centrin in plasmodesmatal functionality, each case proposing an active role for centrin in attachment and disposition of the ER in the neck region, thereby regulating transport through the desmotubule. Elevated calcium is thought to contract the centrin, thereby causing neck closure. In the Blackman *et al.* (1999) model, opening of the channels requires the relaxation of the centrin, presumably through phosphorylation reactions. Radford and White (1998) suggest that myosin may have a structural role in the preservation of the form of plasmodesmatal desmotubules, with the electron-dense strands reportedly linking the desmotubule to the plasma membrane (refer to the structures in Ding *et al.*, 1992; Botha *et al.*, 1993; Overall, 1999). These strands are potentially composed of myosin, which may in turn link actin to the plasma membrane, or to another, as yet unidentified integral membrane protein. Overall (1999) produced a series of electron micrographs that allow better interpretation of the structural components of plasmodesmata and these images highlight the complexity alluded to by Ding *et al.* (1992) and Botha *et al.* (1993). Published work by the Evert and Botha groups show neck modifications such as collars and sphincters in plasmodesmata of monocotyledonous plants, especially those associated with the plasmodesmata traversing the BS to VP interfaces. Collars and sphincters assume significance, in that the route taken by assimilates from the mesophyll through the BS to VP cell interfaces, is complicated by varying wall structures and the presence or absence of compound middle lamellae (see review by Beebe and Russin, 1999, and literature cited). As has been stated elsewhere, the suberin lamella is thought to regulate traffic, by forcing a symplasmic route across cell interfaces that contain this structure. Up- and down-regulation via sphincters, or collars, could thus be responsible for controlling flow rates through the suberin-encased wall areas.

Overall's model (1999) envisages that sphincter-like structures occur on the outside of plasmodesmata, whilst Evert *et al.* (1977) and others also interpret as sphincters the electron-dense structure routinely observed in plasmodesmatal neck regions in BS–cell associations. The issue at this stage in not about which of the two views is correct. What is important is that we recognise the internalized sphincters that have been noted in many grass leaves as significant structures that must play a role in the transport process.

## 4.6 Changes in wall structure and plasmodesmatal form – secondary modification

Unlike plasmodesmata that interconnect parenchymatous cells elsewhere in grass monocotyledonous leaves, those that traverse the route from mesophyll to VP and, specifically, those that cross the KMS–BS–MS route, differ as they develop in cell walls that undergo a considerable degree of change in structure (including partial lignification and development of suberin lamellae) during the maturation stages involved in the transition from the juvenile to the mature cell. Detailed examination of the structural changes that occur in plasmodesmata during the transition from sink to source led Evert *et al.* (1996a) to conclude that plasmodesmata are amongst the last structures to mature, along with the thick-walled sieve tubes, within developing leaves in *Z. mays*. In developing maize leaves, elements of the large transport bundles (with the exception of the metaxylem elements) mature first, before the intermediate and small (offloading) bundles. Their investigation revealed that the vasculature reaches structural maturity in advance of the cessation of import, and that the maturation of bundles involved with the phloem loading process in the leaf blades is not closely correlated with initiation of export from the blade. Evert *et al.* (1996a) further reported that deposition of the suberin lamellae within the walls of the BS cells coincides with the deposition of secondary walls in the metaxylem vessels.

Based upon the very detailed study by Evert *et al.* (1996a), one could speculate that plasmodesmata, traversing a compound middle lamella (composed of a middle lamella–suberin–lamella complex), will undergo similar changes during the cell and organ maturation process. Initially, plasmodesmata will be relatively narrow structures, filled with an electron-dense matrix, as illustrated for maize in the Evert *et al.* (1996a) paper. These juvenile plasmodesmata, ensheathed by either a middle lamella or a compound middle lamella, are restricted and the plasmodesmatal diameter is limited at this point in the still-differentiating cell wall. Increased addition to the matrix of the cell wall on either side of the middle lamella requires that the plasmodesmata extend in length to maintain their respective cellular links. The micrographs in the maize study by Evert *et al.* (1996a) show unequivocally that the desmotubule only becomes apparent closer to the time when the vascular tissue matures. It remains tempting to speculate that the desmotubule itself condenses out of the ER associations near the neck regions of plasmodesmata, and that only then do these specialised plasmodesmata become functional. Suberization seems linked to the final maturation process, the suberin lamellae limiting loss and thereby regulating the passage of assimilate or water (see Botha *et al.*, 1982; Evert *et al.*, 1985; Botha and Evert, 1986; Evert *et al.*, 1996a). The deposition of additional wall matrix outside of the compound middle lamella-encased plasmodesmata implies that the wall-building process, or stage of development of the plasmodesmata, is secondary to that which resulted in the initial formation of the plasmodesma during cytokinesis. However, it is important to qualify that they are not therefore *secondary plasmodesmata* but could be considered to be *secondarily modified* as a result of the continued wall-matrix deposition and resultant wall thickening that occurs during the overall leaf and vascular maturation process (Evert *et al.*, 1996a; Ehlers and Kollmann, 1999). Quite why the plasmodesmata increase in total diameter from the 50–80 nm routinely seen in the middle lamella or compound middle lamella region to 100 nm or more remains an intriguing question. Perhaps under normal circumstances, when the neck regions are not gated by collars or sphincters (see Figures 4.11, 4.13 and 4.16), the dilated portions of the plasmodesmata act as 'funnels', and the cytoplasmic sleeve between the plasma membrane and desmotubule is of sufficient diameter to allow normal cell-to-cell trafficking of assimilates?

## References


Beebe, D.U. and Russin, W.A. (1999) Plasmodesmata in the phloem-loading pathway, in *Plasmodesmata. Structure, Function and Role in Cell Communication* (eds A.J.E. van Bel and W.J.P. van Kesteren), Springer, Berlin, pp. 261–293.

Blackman, L.M. and Overall, R.L. (1998) 'Immunolocalisation of the cytoskeleton to plasmodesmata of *Chara corallina*', *The Plant Journal* **14**(6), 733–741.

Blackman, L.M. and Overall, R.L. (2001) 'Structure and function of plasmodesmata', *Australian Journal of Plant Physiology* **28**, 709–727.

Blackman, L.M., Harper, J.D.I. and Overall, R.L. (1999) 'Localization of a centrin-like protein to higher plant plasmodesmata', *European Journal of Cell Biology* **78**, 297–304.

Botha, C.E.J. (1992a) 'Plasmodesmatal distribution, structure and frequency in relation to assimilation in $C_3$ and $C_4$ grasses in southern Africa', *Planta* **187**(3), 348–358.

Botha, C.E.J. and Cross, R.H.M. (2001) 'Regulation within the supracellular highway – plasmodesma are the key', *South African Journal of Botany* **67**, 1–9.

Botha, C.E.J. and Evert, R.F. (1986) 'Free-space marker studies on the leaves of *Saccharum officinarum* and *Bromus unioloides*', *South African Journal of Botany* **52**, 335–342.

Botha, C.E.J. and van Bel, A.J.E. (1992b) 'Quantification of symplastic continuity as visualised by plasmodesmograms: diagnostic value for phloem-loading pathways', *Planta* **187**, 359–366.

Botha, C.E.J., Cross, R.H.M., van Bel, A.J.E. and Peter, C.I. (2000) 'Phloem loading in the sucrose-export-defective (*SXD-1*) mutant maize is limited by callose deposition at plasmodesmata in bundle sheath–vascular parenchyma interface', *Protoplasma* **214**, 65–72.

Botha, C.E.J., Evert, R.F., Cross, R.H.M. and Marshall, D.J. (1982) 'The suberin lamella, an impermeable barrier in the bundle sheath cells of *Themeda triandra* Forsk', *Protoplasma* **112**, 1–8.

Botha, C.E.J., Hartley, B.J. and Cross, R.H.M. (1993) 'The ultrastructure and computer-enhanced digital image analysis of plasmodesmata at the Kranz mesophyll-bundle sheath interface of *Themeda triandra* var. *imberbis* (Retz) A Camus in conventionally fixed leaf blades', *Annals of Botany* **72**, 255–261.

Crawford, K. and Zambryski, P. (2000) 'Plasmodesmata: gatekeepers for cell-to-cell transport of developmental signals in plants', *Annual Review of Cell and Developmental Biology* **16**, 393–421.

Crawford, K.M. and Zambryski, P.C. (2001) 'Non-targeted and targeted protein movement through plasmodesmata in leaves in different developmental and physiological states', *Plant Physiology* **125**, 1802–1812.

Ding, B., Turgeon, R. and Parthasarathy, M.V. (1992) 'Substructure of freeze-substituted plasmodesmata', *Protoplasma* **169**, 28–41.

Ehlers, K. and Kollmann, R. (1999) Multimorphology and nomenclature of plasmodesmata in higher plants, in *Plasmodesmata. Structure, Function and Role in Cell Communication* (eds A.J.E. van Bel and W.J.P. van Kesteren), Springer, Berlin, pp. 150–172.

Ehlers, K. and Kollmann, R. (2001) 'Primary and secondary plasmodesmata: structure, origin and functioning', *Protoplasma* **216**, 1–30.

Epel, B.L. (1994) 'Plasmodesmata: composition, structure and trafficking', *Plant Molecular Biology* **26**, 1343–1356.

Evert, R.F., Botha, C.E.J. and Mierzwa, R.J. (1985) 'Free-space marker studies on the leaf of *Zea mays* L.', *Protoplasma* **126**, 62–73.

Evert, R.F., Eschrich, W. and Heyser, W. (1977) 'Distribution and structure of the plasmodesmata in mesophyll and bundle-sheath cells of *Zea mays* L.', *Planta* **136**, 77–89.

Evert, R.F., Russin, W.A. and Bosabalidis, A.M. (1996a) 'Anatomical and ultrastructural changes associated with sink-to-source transition in developing maize leaves', *International Journal of Plant Sciences* **157**, 247–261.

Evert, R.F., Russin, W.A. and Botha, C.E.J. (1996b) 'Distribution and frequency of plasmodesmata in relation to photoassimilate pathways and phloem loading in the barley leaf', *Planta* **198**, 572–579.

Fricke, W., Pritchard, E., Leigh, R.A. and Tomes, A.D. (1994) Cells of the upper and lower epidermis of barley (*Hordeum vulgare* L.) leaves exhibit distinct patterns of vacuolar solutes', *Plant Physiology* **104**, 1201–1208.

Haritatos, E., Ayre, B.G. and Turgeon, R. (2000) 'Identification of phloem involved in assimilate loading in leaves by the activity of the galactinol synthase promoter', *Plant Physiology* **123**, 929–937.

Hattersley, P.W. and Watson, L. (1975) 'Anatomical parameters for predicting photosynthetic pathways of grass leaves.The maximum lateral cell count and minimum cells distance count', *Phytomorphology* **25**(3), 325–333.

Lee, J.-Y., Yoo, B.-C. and Lucas, W.J. (2000) 'Parallels between nuclear-pore and plasmodesmal trafficking of information molecules', *Planta* **210**, 177–187.

López-Sáez, J.F., Giménez-Martín, G. and Risueño, M.C. (1966) 'Fine structure of plasmodesmata', *Protoplasma* **61**, 81–84.

Oparka, K.J., Roberts, A.G., Boevink, P. *et al.* (1999) 'Simple, but not branched, plasmodesmata allow the nonspecific trafficking of proteins in developing tobacco leaves', *Cell* **97**, 743–754.

Overall, R.L. (1999) Substructure of plasmodesmata, in *Plasmodesmata: Structure, Function, Role in Cell Communication* (eds A.J.E. van Bel and W.J.P. van Kesteren), Springer-Verlag, Berlin, pp. 129–130.

Overall, R.L. and Blackman, L.M. (1996) 'A model of the macromolecular structure of plasmodesmata', *Trends in Plant Science* **1**(9), 307–311.

Radford, J.E. and White, R.G. (1998) 'Localization of a myosin-like protein to plasmodesmata', *The Plant Journal* **14**(6), 743–750.

Radford, J.E. and White, R.G. (2001) 'Effects of tissue-preservation-induced callose synthesis on estimates of plasmodesma size exclusion limits', *Protoplasma* **216**, 47–55.

Robards, A.W. (1968a) 'Desmotubule – a plasmodesmatal substructure', *Nature* **218**, 784.

Robards, A.W. (1968b) 'A new interpretation of plasmodesmatal ultrastructure', *Planta* **82**, 200–210.

Robards, A.W. and Lucas, W.J. (1990) 'Plasmodesmata', *Annual Review of Plant Physiology* **41**, 369–419.

Robinson-Beers, K. and Evert, R.F. (1991) 'Ultrastructure of and plasmodesmatal frequency in mature leaves of sugarcane', *Planta* **184**, 291–306.

Russin, W.A., Evert, R.F., van der Veer, P.J., Sharkey, T.D. and Briggs, S.P. (1996) 'Modification of a specific class of plasmodesmata and loss of sucrose export ability in the *sucrose export defective1* maize mutant', *The Plant Cell* **8**, 645–658.

Schulz, A. (1995) 'Plasmodesmal widening accompanies the short-term increase in symplasmic phloem unloading in pea root tips under osmotic stress', *Protoplasma* **188**, 22–37.

Schulz, A. (1999) Physiological control of plasmodesmal gating, in *Plasmodesmata. Structure, Function, Role in Cell Communication* (eds A.J.E. van Bel and W.J.P. van Kesteren), Springer, Berlin, pp. 173–204.

Soros, C. and Dengler, N. (1998) 'Quantitative leaf anatomy of $C_3$ and $C_4$ Cyperaceae and comparisons with the Poaceae', *International Journal of Plant Sciences* **159**, 480–491.

Turgeon, R., Medville, R. and Nixon, K.C. (2001) 'The evolution of minor vein phloem and phloem loading', *American Journal of Botany* **88**(8), 1331–1339.

White, R.G., Badelt, K., Overall, R.L. and Vesk, M. (1994) 'Actin associated with plasmodesmata', *Protoplasma* **180**, 169–184.

# 5 Plasmodesmata and plant morphogenesis

Ken Kobayashi, Insoon Kim, Euna Cho and
Patricia Zambryski

## 5.1 Introduction

The developmental strategy that most distinguishes plants from animals is that their cell fate is generally determined by positional information. In this regard, plant cells are extremely sensitive to localized cell-to-cell signals. To enable such signaling, plants have evolved unique intercellular connections called plasmodesmata (PD). Historically, PD were seen as facilitating cell-to-cell traffic of low-molecular-weight growth regulators and nutrients. Moreover, PD were seen simply as static 'pores' in the thick cell walls that surround individual plant protoplasts. This concept of PD has changed dramatically in the last 10 years. Now, native PD are no longer considered static, but instead fluctuate in aperture in different cell types during development and in response to the environment. In addition, PD have demonstrated roles in mediating transport of endogenous macromolecules, exemplified by several transcriptional factors, and other proteins and RNAs, such as green fluorescent protein (GFP), mRNA and gene silencing signals. Thus, PD are critical gatekeepers of informational signal exchange between cells. Such exchange is essential to regulate developmental programs that lead to morphogenesis of tissues and organs.

## 5.2 Role of PD and symplasmic domain boundaries during plant development

During plant development, plant cells must undergo differentiation, exemplified by meristematic cells, which switch from an undifferentiated state to specific identities with varied fates. During this process, different sets of genes and regulators are expressed, and in some cases, a series of regulatory genes exhibit overlapping expression. In addition, plant cells are responsive to cues from the surrounding environment, and can adopt alternate fates depending on diffusible signals that arrive in their vicinity. Subsequent differentiation is a prerequisite for morphogenesis, to develop organized tissue structure and shape that leads to the formation of organs.

In plants, there are many cases where cells become symplasmically disconnected before they undergo differentiation and acquire specific identities (Table 5.1). Symplasmic domains/fields refer to cells or tissues that are isolated from the cells around them. [For simplicity we refer to domains and fields synonymously, although one report distinguishes fields as regions with dynamic boundaries, versus domains as regions where boundaries contain cells with permanently closed PD (Gisel *et al.*, 1999).] The isolation can be permanent and complete, such as guard cells and root hair cells (Palevitz and Hepler, 1985; Duckett *et al.*, 1994), or transient for a particular time frame. For example, leaf buds reduce transport during dormancy and then regain relative continuity at the end of the dormancy phase (Rinne and Van Der Schoot, 1998). Thus, regulating PD permeability provides a simple means to control the quantity and quality of communication between individual cells.

Figure 5.1 graphically illustrates a tissue where all cells are either homogenously connected by open PD (no symplasmic domain) or where the central region has become isolated as a symplasmic domain. The possible scenarios that result from such isolation are that cells in the domain can undergo growth and/or differentiation that then leads to specific morphogenetic events. Generally, formation of symplasmic domains is mediated by down-regulation or loss of functional PD. For example, during callus formation in *Solanum nigrum* and *Molinia caerulea*

**Table 5.1** Symplasmic domains in plants

| Cell type/tissue | Organism | Functional state | References |
|---|---|---|---|
| Cotton fiber cells | Cotton | Temporal isolation during fiber cell elongation | Imlau *et al.*, 1999; Ruan *et al.*, 2001 |
| Shoot apex | *Arabidopsis thaliana* | Isolation of meristem inner central zone. Temporal isolation of meristem at transition to flowering | Gisel *et al.*, 1999, 2002 |
| Shoot apical meristem | Birch (*Betula pubescens*) | Short photoperiod induces breakdown of two concentric symplasmic fields in meristem tunica into symplasmic isolated cells | Rinne and Van Der Schoot, 1998 |
| Root hair cells | *Arabidopsis thaliana* | Symplasmic isolation | Duckett *et al.*, 1994 |
| Stomatal guard cells | Onion (*Allium cepa*) | Isolation and degradation of PD | Palevitz and Hepler, 1985 |
| Companion cells/sieve elements | Tobacco (*Nicotiana tabacum*), *Arabidopsis thaliana* | Isolation from non-phloem cells in source tissue | Imlau, *et al.*, 1999; Oparka, *et al.*, 1999 |

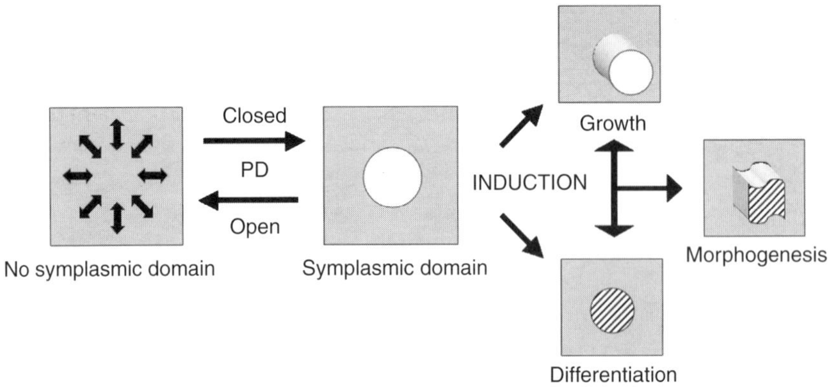


**Figure 5.1** Symplasmic domain formation precedes developmental events. Symplasmic domains are formed either by the disappearance/absence of PD, or by down-regulation of SEL (both cases indicated as CLOSED). Maintenance of these domains may be necessary for induction and coordination of growth, differentiation and morphogenetic events. This figure represents events that occur during simple differentiation of callus tissue, or more complex differentiation, e.g. at the shoot apex during floral primordial initiation. The shaded box represents a tissue that will undergo the morphogenic event. The double-ended arrows indicate homogeneous symplasmic communications within the tissue. The white circle represents the newly formed symplasmic-domain-containing cells, which are isolated from the surrounding tissue. Growth is represented by a three-dimensional cylinder. Differentiation is represented by a stripe-filled circle to indicate the different identities/properties of cells acquired within the symplasmic domain. Finally, morphogenesis is represented by a three-dimensional polygon filled with stripes to indicate that this event is a consequence of both differentiation and growth.

protoplasts, different symplasmic domains are formed coincident with morphogenesis of callus regions into globular proembryonic structures and subsequently to somatic embryos (Ehlers *et al.*, 1999). The boundaries of these domains contain PD occluded by dense material. Those occluded PD mark the limits of symplasmic domains, suggesting that demarcations of domains are necessary for morphogenesis. The formation of symplasmic domains is critical to regulate several different plant developmental programs (Table 5.1). Thus, the regulation of PD function provides a means whereby specific plant cells can be isolated from surrounding cells and initiate independent developmental programs, establishing cell and tissue boundaries within particular tissues and organs.

## 5.3 Generic changes in structure and function of PD during different developmental processes

In later sections, we will review how PD function correlates with developmental changes during vegetative, reproductive and embryonic development. Here, we first review some principles that apply to PD function in general.

### *5.3.1 PD type and frequency*

PD participate directly or indirectly in the coordination of developmental processes in plants by undergoing changes in structure, number and connectivity in the course of morphogenetic events.

PD are formed under two conditions. Primary PD arise during cytokinesis (Jones, 1976; Hepler, 1982) whereas secondary PD are formed through existing cell walls to increase the symplasmic connection between cells, or to connect cells not clonally related (Ding *et al.*, 1992, 1993; Evert *et al.*, 1996). Both primary and secondary PD are initially simple in structure, i.e. a single pore traversing the cell wall; but later, during tissue development, they can form more complex structures consisting of branched PD (see Chapter 1). Formation of secondary PD is critical to achieve intercellular communication in three-dimensions and may be under unique regulation, distinct from primary PD regulation.

An enormous loss of simple PD occurs during the sink-to-source transition in the tobacco leaf (Roberts *et al.*, 2001). In part, this may be a consequence of the decrease in shared-cell-wall length that occurs as the leaf undergoes cell enlargement and intercellular space expansion, leading to separation of cells. Figure 5.2 graphically displays an estimation of the relative frequency of PD along the length of a single leaf (counted as number of PD per micrometre of shared cell wall in different shared cell interfaces). Most PD in the sink region are simple and decrease in number during leaf expansion and the transition to the source state. At the same time, the number of branched PD increases (Figure 5.2). It is unknown whether these branched PD arise from existing simple PD or are formed *de novo* across preexisting cell walls.

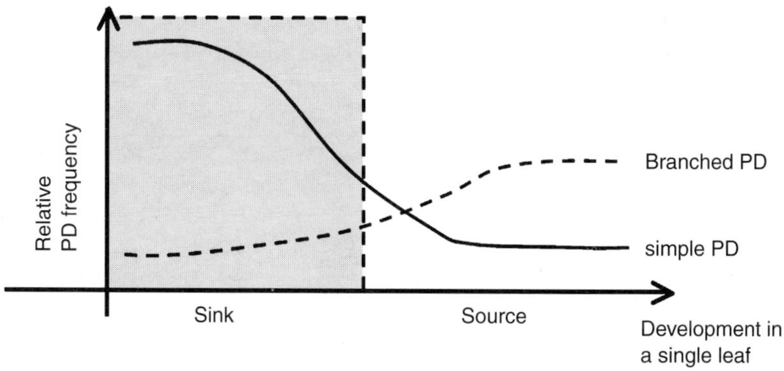


**Figure 5.2** Change in frequency of simple and branched PD in a single leaf undergoing the sink-to-source transition. The horizontal axis represents sink-to-source leaf development, which occurs from the base of the leaf toward the leaf tip. The (base) sink region of the leaf is shaded. The values corresponding to each curve are approximate estimations based on the frequency of PD in different cell interfaces, including anticlinal and periclinal cell walls between epidermal/epidermal cells, epidermal/palisade mesophyl cells, palisade mesophyl/spongy mesophyl cells, palisade/palisade mesophyl cells and spongy/spongy mesophyl cells. Adapted from Roberts *et al.*, (2001).

In the shoot apical meristem (SAM), PD frequency increases across all SAM cell layers during floral induction of *Sinapis alba*, including the central and peripheral zones (Ormenese *et al.*, 2000). Before floral induction, PD frequency in the SAM is around 0.2 PD per micrometre of wall. During induction, PD frequency increases up to 2.5-fold after 36 h. These PD are probably of secondary origin because they are formed in existing walls, such as periclinal walls between L1 and L2, and between L2 and L3. Interestingly, this increase in PD frequency is transient, since at 48 h after induction, PD frequency decreases and returns to the original values of the pre-floral induction state. Potentially, there is transient activation of secondary PD formation followed by their dilution due to continuous mitotic activity and cell expansion.

### *5.3.2 PD are dynamic*

Developmental changes in PD activity were first studied in leaves of tobacco plants from several species (Oparka and Turgeon, 1999; Crawford and Zambryski, 2000; Itaya *et al.*, 2000; Roberts *et al.*, 2001). Biolistic transformation of constructs leading to the transient expression of GFP allows one to monitor cell-to-cell transport in hundreds of epidermal cells from a single leaf, and reveals that the leaf consists of a heterogeneous population of cells with regard to PD aperture. This quantitative assay for intercellular movement of different GFP-tagged macromolecules furthermore revealed the dynamic nature of PD, whereby a single PD likely fluctuates in connectivity to its neighbors in temporally and spatially restricted manners. Thus, the leaf consists of a mosaic of cells having diverse intercellular trafficking potentials depending on leaf age and growth conditions (Figures 5.3 and 5.4). In all cases, leaf age affects PD dilation, as more cells exhibit protein movement in younger leaves (region A) than in older leaves (region B) (Figure 5.3). This developmental restriction is most dramatically illustrated if cells are transfected with DNA encoding

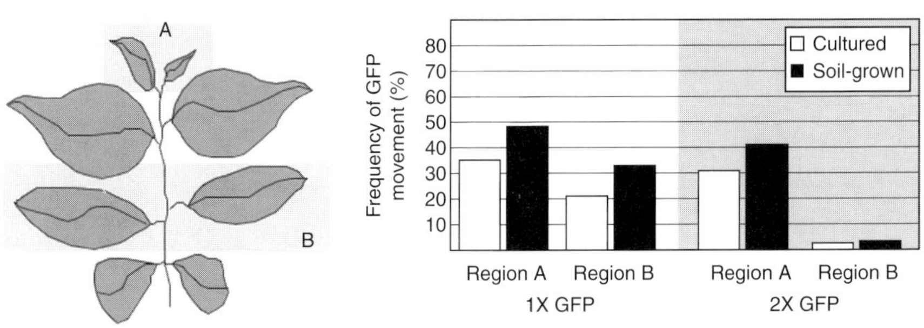


**Figure 5.3** Dynamic nature of PD in leaves. In this study, the two smallest visible leaves on the plant were scored as region A, whereas the older leaves near the base of the plant were scored as region B. The leaves of region A are sinks and the leaves of region B are in transition to source. The bar graph compares 1X GFP (white area) and 2X GFP (grey area) movement in region A leaves versus region B leaves in cultured (white bars) versus soil-grown (black bars) tobacco plants. Adapted from Crawford and Zambryski (2001).

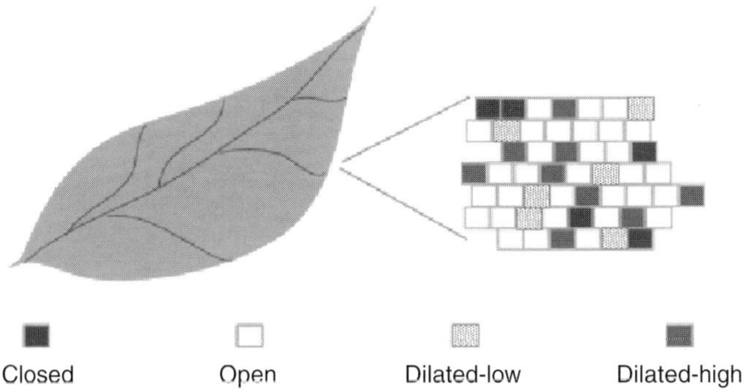


**Figure 5.4** A single leaf consists of a mosaic of cells with varied PD apertures. Each box represents an individual cell and the shading indicates one of the possible four PD dynamic states: closed, open, dilated-low or dilated-high. Some transfected cells do not exhibit GFP movement; thus, their PD are completely closed or only open to molecules smaller than GFP. Closed PD would not allow any transport, and could be permanently (as in stomata) or temporarily sealed (as has been reported in the shoot apical meristem). Open PD would allow for the exchange of nutrients and small dyes (i.e. sugars, HPTS). Other cells of the leaf have dilated PD of varying apertures that allow for macromolecular trafficking through PD. One population of cells allows GFP diffusion, implying their PD are dilated to a sufficient degree to allow this 27-kDa molecule to transit. A smaller population of transfected cells have PD that are dilated to a higher degree, as they allow 2X GFP (54 kD) transit. These results suggest that the leaf is a mosaic where cells exist with PD in varying states of distention, and that dilated PD do not have a single-set aperture. Modified from Crawford and Zambryski (2001).

double GFP (2X GFP; 54 kDa). In sink leaves, between 34 and 67% of the cells transport 1X GFP (27 kDa), and between 30 and 46% of the cells transport 2X GFP. In older leaves, transport is reduced; between 21 and 46%, and between 2 and 9% of cells transport 1X and 2X GFP, respectively. Thus, leaf age dramatically affects the ability of cells to exchange proteins nonselectively. Furthermore, the number of cells allowing 1X GFP or 2X GFP movement was sensitive to growth conditions, as greenhouse (soil) grown plants exhibited higher diffusion rates than axenic culture-grown plants.

These studies revealed there are at least three states of PD in the leaf: closed, open and dilated (Figure 5.4). Closed PD have no cytoplasmic connection between cells. In open and dilated PDs the cytoplasm between adjacent cells is connected, allowing passages of micro- and macromolecules. Obviously, macromolecular size and aperture of PD physically govern cell-to-cell traffic. There is likely no single size of dilated PD, but rather a continuum of apertures. For example, some PD (dilated low PD) allow transport of relatively small proteins such as 1X GFP and other PD (dilated high PD) allow transport of large proteins such as 2X GFP.

### *5.3.3 PD are passageways for nontargeted and targeted transport*

Nontargeted and targeted macromolecular movements were distinguished following biolistic bombardment of DNA constructs to express GFP protein and its

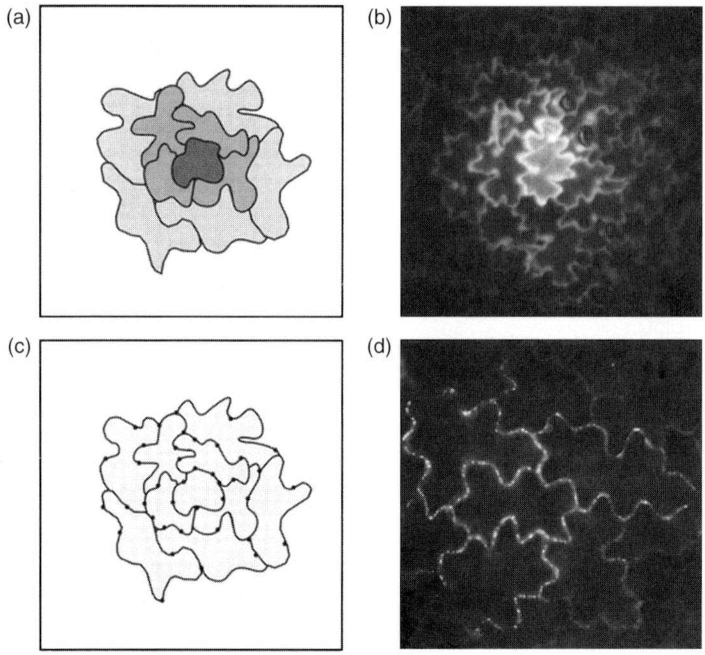


**Figure 5.5** Targeted and non-targeted movement via PD. Non-targeted versus targeted movement of GFP is compared diagrammatically in (a) and (c); (b) and (d) show their respective patterns in living cells. DNA encoding GFP [(a) and (b)] or GFP fused to viral movement protein (MP) [(c) and (d)] was introduced biolistically into single cells. The initial transfected cells are the black central cell in (a), and brightest central cell in (b); for (c) and (d), the initial cell is centrally located, but can no longer be visualized due to rapid movement of GFP. Following expression, two distinct patterns of GFP fluorescence are seen. (a) and (b) reflected 'non-targeted' movement where GFP fluorescence appears as a gradient from brightest in the initial cell to progressively less and less in adjacent cells; in (a) the brightest cell is represented as black and the adjacent cells are progressively lighter gray. In contrast, GFP–MP 'targeted' movement results in localization to 'punctae' [black dots in (c), or fluorescent dots in (d)] at the edges of cells, corresponding to PD channels. From Figure 3 of Zambryski and Crawford (2000); reproduced with permission.

derivatives (Figure 5.5; Crawford and Zambryski, 2001). Following transfection of single cells, one then monitors the movement of GFP to surrounding cells. In the case of nontargeted movement, GFP moves out of the initially transfected cell by diffusion, resulting in a gradient of diminishing fluorescence in tobacco epidermal cells. In contrast, targeted movement is characterized by localization of transported protein in cell walls in a punctate pattern (probably in PD) without the formation of a fluorescent gradient between cells. Moreover, the targeted fluorescent signal is quickly equalized between cells. To date, the best illustration of targeted movement is observed by fusing GFP to a viral movement protein (MP) such as the MP of tobacco mosaic virus (TMV), which is known to interact specifically with PD (Deom *et al.*, 1987; Wolf *et al.*, 1989). It is noteworthy that nontargeted protein movement is dramatically affected by protein size and the size exclusion limit (SEL) of PD, whereas targeted movement is able to overcome the size restriction of PD. While 2X GFP (54 kDa) diffuses less in comparison to 1X GFP (27 kDa), TMV-MP-GFP (57 kDa), which has similar molecular weight to 2X GFP, exhibits extremely efficient movement to actively access adjacent cells. In addition, the targeted transport of TMV-MP-GFP was unaffected by the conditions observed to limit nontargeted movement such as leaf age or physiology (Crawford and Zambryski, 2001). These data suggest that these two types of movement differ fundamentally in their mechanisms: nontargeted movement is a passive diffusional event, while targeted movement is an active event likely accompanying PD SEL modification. Targeted transport is remarkable in its capacity to skillfully manipulate PD status to promote protein transit.

That endogenous proteins have the capacity to move cell to cell by diffusion, presents a potential quandary to cell homeostasis. How does the cell prevent the loss of critical components? While GFP, a non-native protein, can diffuse cell to cell, native endogenous proteins have evolved particular cellular addresses in their innate function. Endogenous cell components are specifically anchored or sequestered within the cells via specific protein sorting/targeting amino acid sequences (Crawford and Zambryski, 2000). For example, additions of amino acid sequences to allow soluble GFP to anchor to actin filaments or to target and be retained into ER prevent its diffusion. However, GFP carrying a nuclear localization signal (NLS) retains the ability to diffuse between cells. Clearly, nuclear localization does not inhibit movement of GFP into neighboring cells as the nucleus is a soluble compartment connected to the cytoplasm. However, 2X GFP fused to the same NLS signal targeted to the nucleus but was unable to escape the nucleus or transit through PD. Thus, nuclear-localized proteins are capable of cell-to-cell transit in a size-dependent manner.

## 5.4 Changes in PD conductivity during vegetative development

Although we have already mentioned data relevant to PD functions during vegetative development, as cited for general PD functions, we highlight studies with GFP and its fusions in leaves and veins. Additional descriptions of the regulation of PD conductance in general, and in the vasculature system in particular, can be found in other chapters (Chapters 7, 8 and 13). As mentioned above, proteins of at least 50 kDa traffic through cells of the leaf; the extent of such movement is highly dependent on leaf age (Oparka *et al.*, 1999; Crawford and Zambryski, 2001), and PD structure (simple versus branched), which in turn correlates with its SEL (high versus low, respectively) (Oparka *et al.*, 1999; Roberts *et al.*, 2001).

Other features relating to PD conductance during the vegetative phase are worthy of brief mention. Nontargeted trafficking of GFP has directionality within organs and between cell layers. For example, in cucumber, GFP can move from epidermal to mesophyll cells in the hypocotyl, but not in the reverse direction

(Itaya *et al.*, 2000). This finding suggests that the regulation of PD conductance is sensitive to other parameters besides PD architecture and its SEL status. Potentially, specific signal molecules regulate PD to function in a directional manner.

The vascular system, where PD connect the veins to surrounding tissues, provides an important conduit for macromolecule dispersal. When GFP is expressed in companion cells (CC) immediately abutting the vascular system, it generally moves toward regions of new growth and morphogenesis, such as young leaves, upper shoot and newly arising floral organs (Imlau *et al.*, 1999). Again, young sink veins have PD with higher SEL allowing soluble GFP to unload from the phloem to surrounding tissues, but in mature source veins, GFP movement is restricted to the phloem (Oparka *et al.*, 1999).

A maize mutant, *sucrose export deficient 1* (*sxd1*) (Russin *et al.*, 1996; Provencher *et al.*, 2001) has defective PD which do not extend from cell to cell, but instead terminate within the bundle sheath–vascular parenchyma cell wall. This lack of conductivity causes a reduction of photosynthate delivery to the phloem, and provokes chlorosis in *sxd1* mutant leaves. Interestingly, SXD1 encodes a nuclear encoded chloroplast-localized protein, that is homologous to VTE1-encoding tocopherol cyclase from *Arabidopsis*. Although it is not immediately obvious how to link the cell-specific PD occlusion phenotype with synthesis of tocopherols, these data open the possibility that tocopherols or their breakdown products function as signal transduction molecules regulating PD structure. Alternatively, the reduced antioxidant levels within cells may induce plugging of PD pores.

## 5.5 Changes in PD conductivity during floral induction

Plants are fundamentally different from their animal counterparts in their reproductive strategy. In animals, reproductive organs are set aside early during embryogenesis. In contrast, plants mature to adults before producing reproductive organs. During vegetative plant development, roots and leaves are produced continuously from meristematic regions. The shoot meristem, at the plant apex, contains a dome of undifferentiated cells that give rise to new leaf primordia off its flanks. Late in life, in response to developmental and environmental signals, the adult plant reprograms this apical meristem to stop producing leaves and instead produce floral organs. This developmental switch results in a dramatic reorganization of the meristem; instead of giving rise to single leaf primordia, the meristem now produces floral meristems, which in turn produce multiple primordia corresponding to the four floral organ types.

The important resource of developmentally competent cells in the apical meristem is expected to be highly sensitive to signals that traffic cell to cell. Indeed, temporal and spatial changes in cell to cell transport occur at the shoot apex as a function of developmental age (Gisel *et al.*, 1999, 2002; Figure 5.6). An assay was developed, whereby a low-molecular-weight fluorescent tracer (HPTS; 8-hydroxypyrene

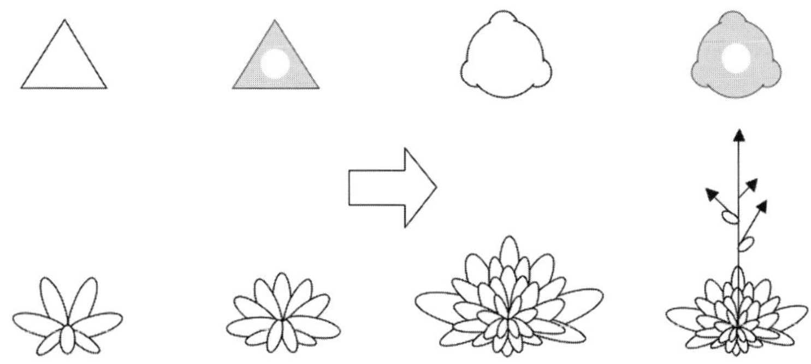


**Figure 5.6** Developmental transitions in tracer traffic in meristems during the floral transition. The shaded areas in the meristems indicate the presence of HPTS loaded from vasculature. Vegetative meristems have triangular points to indicate 'pointy' leaf primordia. Inflorescence meristems have rounded areas to indicate the more 'rounded' floral meristems. Note the existence of isolated symplasmic fields in the central zones of both meristems that contain loaded tracer. The large arrow indicates the transition from the vegetative to the reproductive state. Adapted from Gisel *et al.*, (1999).

1,3,6 trisulfonic acid; 520 Da) was loaded into leaf petioles; the tracer then moves via the phloem to neighboring leaves, and to the meristem and its surrounding young primordia. The meristem is then viewed by confocal microscopy. This method ensures that the target tissue of interest, the meristem, is far from the site of experimental perturbation during tracer loading. Furthermore, this method allows one to visualize the pattern of tracer movement to the meristem in living tissues. Apices of vegetative plants reveal tracer in the outer cell layers of the meristem and in very young primordia. Remarkably, fluorescence signal in the entire SAM region decreases prior to the onset of flowering. Then, after floral morphogenesis is under way, floral apices again traffic tracer (Figure 5.6). Additional in-depth analyses of cell-to-cell traffic at the shoot apex during this developmental switch were performed under six different photoperiodic treatments (flowering in *Arabidopsis* is highly dependent on day length), and in four late or early flowering mutants [*phyB-6* (Reed *et al.*, 1994), *ft-1* (Koornneef and Van Der Veen, 1991), *co-2* (Putterill *et al.*, 1995), *tfl-1* (Shannon and Meeks-Wagner, 1991)]. In all cases, the shoot apex reduced incoming cell-to-cell transport just prior to floral commitment.

These oscillating states of connectivity may correspond to the traffic of important developmental signals such as those regulating the induction of flowering. These symplasmic fields highlight that beyond the transcriptional activation of genes involved in morphogenesis, cell-to-cell signaling may regulate developmental transitions. Thus, regulation of apical symplasmic trafficking represents an additional parameter of the multifactorial control of the developmental transition to flowering, and supports a model where reduction of transport of a floral repressor(s) contributes to the induction of flowering. As the apical meristem undergoes profound changes in architecture and concomitant signaling and expression pathways, it may be advantageous to be sequestered from intercellular signal input during the establishment of the reproductive program.

Besides temporal fluctuations in conductivity, the meristem exhibits a precise spatial control of intercellular transport capacity. Although the outer tunica layers can transport tracer, the inner central zone (ICZ) does not (Figure 5.6). This distinction may reflect that the tunica cells arise from anticlinal divisions resulting in continual renewal of the outer surface area of the meristematic dome, while the central cells undergo division in all planes (Medford, 1992). This focus of transport to only the outer layers may allow developmentally important signals to move to outlying primordial via symplasmically connected tunica cells. The central zone may maintain its symplasmic isolation to ensure that it remains a pure source of fully competent undifferentiated cells.

In summary, PD change their conductivity during the transition from vegetative to floral development. Futhermore, as mentioned earlier for *S. alba*, PD frequency also fluctuates during the floral transition. Clearly, PD significantly influence this crucial morphogenetic program.

## 5.6 Changes in PD conductivity during embryogenesis

Until recently, regulation of PD trafficking has been studied predominantly in leaves (Ding *et al.*, 1996; Itaya *et al.*, 1997, 2000; Oparka *et al.*, 1999; Crawford and Zambryski 2000, 2001), roots (Oparka and Prior, 1992; Oparka *et al.*, 1995) and root apical meristem and SAM (Jackson *et al.*, 1994; Lucas *et al.*, 1995; Rinne and Van Der Schoot, 1998; Gisel *et al.*, 1999, 2002; Kim *et al.*, 2002b). In all cases, the regulation of cell-to-cell transport during development has important implications in the establishment of specific morphogenetic programs. Because embryogenesis is an essential developmental stage consisting of specific morphogenetic events, one might expect that there is also regulation of PD function during this essential time frame. Indeed, during *Arabidopsis* embryogenesis, PD SEL is reduced significantly at the torpedo stage (Kim *et al.*, 2002a).

In higher plants, all cells of the embryo are initially connected into a single symplasmic unit as monitored by ultrastructural observation of PD between embryo cells (Mansfield and Briarty, 1991). The recent development of a functional assay for PD connectivity during embryogenesis allows one to assess cell-to-cell transport in depth in living embryos (Kim *et al.*, 2002a). First, *Arabidopsis* embryos are released from their seed coats. This removal creates small nonlethal breaks in cell walls and plasma membranes and allows entry points for symplasmic tracers. Embryos then are incubated with either small (520 Da HPTS) or large

(10 kDa FITC(F)-dextrans) fluorescent tracers, and symplasmic movement is observed by epifluorescence and confocal microscopy. Figure 5.7 reveals that HPTS traffics throughout the entire embryo during all stages of embryogenesis tested, while F-dextran transport stops at the mid-torpedo stage. These functional analyses indicate that the embryo consists of a single symplasmic domain, yet there is a developmental down-regulation of PD SEL in later stages of embryogenesis. Once again, the same theme recurs. PD have higher SEL in earlier developmental times. Besides this temporal theme, there is a consistent structural aspect to PD function during development. As already mentioned, simple PD have larger SEL and are more likely to transport macromolecules during leaf development (Oparka and Turgeon, 1999; Crawford and Zambryski, 2000). Embryonic PD resemble primary simple PD in their structure (Mansfield and Briarty, 1991). This is not too surprising as embryo cells are continuously dividing, and thus producing more and more simple PD during cell division. Perhaps the short cell-cycle times

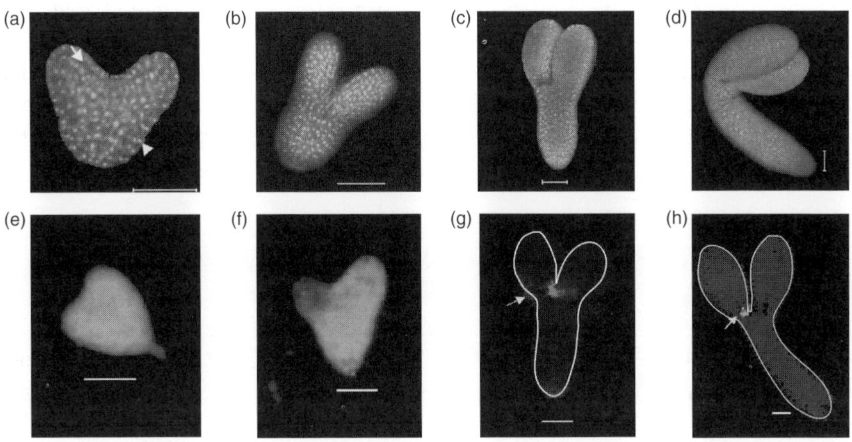


**Figure 5.7** Characterization of cell-to-cell transport in *Arabidopsis* wild-type embryos. Embryos at different stages of development are loaded with either HPTS (a–d) or 10-kDa FITC (F)-dextran (e–h). All cells in embryos allow the movement of HPTS, indicating that the embryo constitutes a single symplasmic domain, from early heart (a), late heart (b), early torpedo (c), to mid torpedo (d) stages of embryo development. Cellular localization of HPTS shows tracer in the cytoplasm as well as the nuclei (arrowhead in a). In contrast, 10-kDa F-dextrans move only in early heart (e) and mid heart (f) embryos. 80% of early (g) and 100% of mid-torpedo (h) embryos do not allow the movement of F-dextran. Instead, a small number of cells are loaded at the region where cotyledons join the hypocotyl [arrows in (g) and (h)]; such loading occurs due to breakage of cell walls and plasma membranes during release of embryo from their seed coats. The images in (a–d) are optical sections captured by confocal laser scanning microscopy. Images in (e–h) were obtained by epifluorescence microscopy, and are therefore less highly resolved. The images in (a–f) reveal that the shapes of the embryos as green fluorescence (from HPTS or F-dextran) can be copied into grey scale. The images of the embryos for (g) and (h) cannot be seen except where a small amount of tracer has entered the embryos (arrows). As the original red autofluorescence from chlorophyll in the images in (g) and (h) does not copy into grey scale in a manner that can be distinguished from the green fluorescence of F-dextran, the outline of the embryos is shown as a white line. Scale bars = 50 μm. Adapted from Kim *et al.*, (2002a).

do not provide opportunity for modification of primary to secondary PD, or for the *de novo* synthesis of secondary PD in the side walls of cells. Nevertheless, the existence of primary PD again correlates with increased cell-to-cell transport. Furthermore, no obvious structural changes can be detected in down-regulated PD observed at the mid-torpedo stage (Kim *et al.*, submitted). The possible mechanism involved in this down-regulation should be different than that observed for the transient lack of cell-to-cell transport in the shoot apex during the floral transition, as small tracers like 3-kDa F-dextran and HPTS diffuse freely through early as well as late stages of embryogenesis. Thus, the mechanism of decrease in transport is unlikely to be due to the physically observable plugging of PD pores, e.g. via callose deposition (Botha and Cross, 2000).

In another approach, different sizes of GFP protein including 1X GFP and 2X GFP were expressed specifically in the shoot apex, with the aim to determine the diffusion pattern of endogenously expressed macromolecules during embryogenesis (Kim *et al.*, 2005). Again, down-regulation of GFP diffusion was consistent with previously observed data with exogenously added tracer. While 1X GFP diffusion occurred throughout the whole embryo and throughout all stages of embryogenesis tested, including mid torpedo stages, 2X GFP diffusion was significantly reduced by the mid-torpedo stage (Figure 5.8).

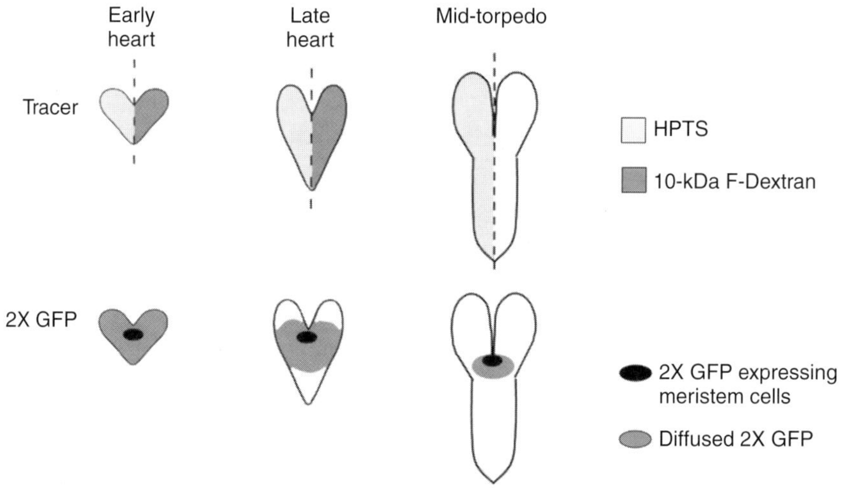


**Figure 5.8** Down-regulation of plasmodesmata during embryogenesis. Cell-to-cell movement of exogenously added tracers HPTS and F-dextran (upper panel). Left halves of embryos summarize results from an assay for HPTS tracer movement. The right halves of each embryo illustrate whether or not 10-kDa F-dextran moves cell-to-cell in similarly staged embryos, and diagrams the developmental down-regulation of PD SEL after mid-torpedo stage [adapted from Kim *et al.*, (2002a)]. Similarly, when 2X GFP (54 kDa) is expressed endogenously, specifically in shoot apical tissues, it diffuses freely in early heart embryos, but its diffusion starts to be restricted at the late-heart stages and at the mid-torpedo stage almost no diffusion is observed (Kim *et al.*, 2005).

It is interesting that the molecular weight of macromolecular tracers used for the movement assays differs significantly between the exogenous added tracer (10-kDa F-dextran) and the endogenous expressed tracer (such as 27-kDa GFP). At first glance, one might assume that these differences likely reflect dissimilarities in the effective hydrodynamic size between the globular carbohydrate F-dextran and the more barrel-shaped GFP. However, GFP has a larger Stokes radius (2.82 nm) than 10-kDa dextran (2.3 nm) (reviewed in Heinlein and Epel, 2004). Perhaps, as F-dextran has no precise secondary or tertiary structure, the predicted 'diameter' is only a statistical average of un-branched random coils, many of which might have larger 'diameters'. In contrast, the structures of proteins are precise and it is very likely that the barrel-shaped GFP (Ormo *et al.*, 1996) migrates through PD end to end rather than sideways; as such, GFP may have an effective narrower dimension than globular 10-kDa dextran. Thus, size and molecular mass alone do not accurately predict PD SEL. In addition, other factors such as hydrophobicity and molecular charge of cargo macromolecules (Heinlein and Epel, 2004) likely play significant roles in their transportability via PD.

What is the biological significance of this PD down-regulation during embryogenesis? Perhaps, this down-regulation in PD aperture is a prerequisite to symplasmic domain formation, e.g. leading to formation of cotyledons, provascular tissues and embryonic roots. The solid line in Figure 5.9 simplistically diagrams a single down-regulation in PD aperture at the mid-torpedo stage, identified by the tracer loading assay with F-dextran. Potentially, this graph could be broken into several step-downs (see dotted line) if we more accurately assess PD function during embryogenesis. This latter sort of analysis requires the identification of specific

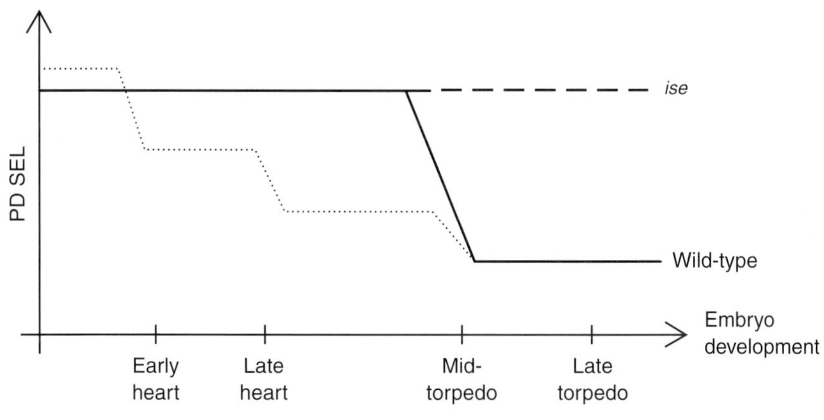


**Figure 5.9** Developmental transitions in cell-to-cell transport of macromolecules during *Arabidopsis* embryogenesis. The reduction of transport between early-torpedo and mid-torpedo stages is observed in wild-type embryos (solid curve), but it is absent in embryo defective mutants with altered PD such as *ise* (dashed line). Hypothetical transitions in PD aperture are indicated by the dotted line. Note the known reduction of PD (Kim *et al.*, 2002a) occurs between early- to mid-torpedo stages. This PD reduction still allows the passage of Dextran 3 kDa and HPTS of 520 Da, indicating that PD are not closed.

macromolecules that have highly regulated and unique regions of intercellular transport that can be utilized as probes to assess cell-to-cell transport during different stages of embryogenesis.

Another powerful approach to elucidate PD function during this critical stage of plant development is genetics, i.e. to identify mutants that affect PD structure or function. Alterations of PD function likely will have severe defects in growth and not produce viable plants. Thus, PD defects are expected to manifest first as defects during embryo development. However, as embryo-lethal lines can be propagated as heterozygotes that segregate wild-type and embryo-defective phenotypes in their seedpods, the plant viability issue is no longer an impeding factor for genetic analyses. Moreover, the determination of developmental transitions in PD function during embryogenesis provides entry points to identify mutants with altered cell-to-cell transport. Indeed, screening embryo-defective lines individually for their ability to transport fluorescent tracers identified 15 lines altered in their ability to traffic tracers during late embryogenesis (Kim *et al.*, 2002a). These mutants exhibited increased SELs (*ise*), by continuing to allow the transport of 10-kDa F-dextran beyond the mid-torpedo stage of embryogenesis (Figure 5.9). Interestingly, these embryos apparently develop normally until they reach the stage at which PD down-regulation is observed – at the mid-torpedo stage. Hence, there is a strong correlation between the regulation of PD SEL during embryogenesis and normal morphogenetic events.

One important question immediately arises. Either PD alterations result in the embryo defective phenotype we observe in *ise* mutants, or embryo defective mutations cause downstream effects leading to alterations in PD function. Considering that only 0.3% (15 out of 5000) of embryo defective lines analysed showed an altered PD phenotype, *ise* mutants are not generally associated with altered development at the torpedo stage. These data strongly support the view that alteration of proper PD function during embryogenesis indeed plays a critical role (Figure 5.10).

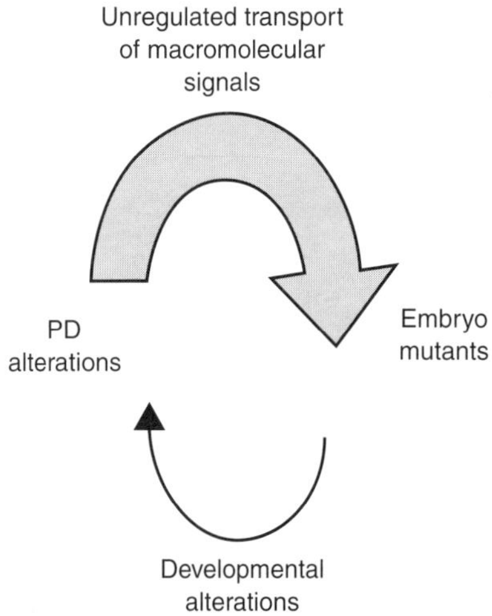


**Figure 5.10** Schematic illustration of two possible models showing the role of PD function during embryogenesis. As discussed in the text, the relatively low frequency of mutants with altered cell-to-cell transport, among total embryo defective mutants, suggests that altered PD cause the embryo defective phenotype and not vice versa.

Current studies in our laboratory aim to clone and characterize the genes affected by *ise* mutations. To date, we have map base-cloned the EMS (ethyl methyl sulfonate) induced *ise1* locus and determined that the *ISE1* gene encodes a DEAD-box RNA Helicase (Kim *et al.*, submitted). The lesion in the *ise1* mutant was determined to be a G to A transition, resulting in a glycine to glutamine alteration at position 228, corresponding to the position of a highly conserved 'GG loop' (Johnson and McKay, 1999) in the predicted structure of the ISE1 protein; an identical lesion causes a lethal phenotype in yeast (Schmid and Linder, 1991). ISE1 localizes to puncta in plant cell walls, and co-localizes with tobacco mosaic virus movement protein (TMV MP), known to localize to discrete punctate PD sites (Figure 5.5). Ultrastructural studies do not reveal an alteration of PD structure in *ise1* mutants. Thus, ISE1 may have a regulatory versus a structural role in PD function. Helicase domains of several plant viruses are essential for intercellular movement of viruses (Carrington *et al.*, 1998; Erickson *et al.*, 1999; Hirashima and Watanabe, 2001). Thus, ISE1 likely represents an analogous endogenous protein essential for macromolecular transport via PD in plant cells. The localization of an RNA helicase to PD has important implications for PD function. Either PD have an RNA component that can regulate PD aperture in conjunction with helicase activity, or helicase located to PD plays a role in RNA transport required for plant growth and development. Future studies will address the exact mechanism of PD regulation by helicase activity. Furthermore, it would be interesting to identify those macromolecular signals that are over-transported in *ise* mutants due to the lack of SEL reduction.

## 5.7 Importance of intercellular communication: roles of transported macromolecules in plant development

Local and systemic transport of specific proteins and RNAs play important roles during plant development (see Chapter 6). Briefly, we mention some landmark studies elucidating the role of macromolecular signals as a consequence of their transport during morphogenetic events. Elegant grafting experiments using two different types of tomato plants (Kim *et al.*, 2001) showed that long-distance transport of the Mouse ears (Me) transcriptional factor mRNA from root stock to scion can induce the dominant Me phenotype in newly formed younger leaves in shoots of the scion. This experiment strongly suggests that the long-distance transportable mRNA and/or protein can modify leaf morphology. Another example of cell fate determination by a mobile macromolecular signal is SHORT-ROOT (SHR) mRNA and protein, which are expressed in the central region of root tip, specifically in stele cells. SHR protein can move only to one adjacent cell file, inducing the expression of SCARECROW (SCR) genes, and differentiation of endodermal cells. Thus, the movement of a regulatory protein (SHR), and not its mRNA, determines the cell fates of adjacent cells (Nakajima *et al.*, 2001). Very recently, NCAPP1 (non-cell autonomous pathway protein 1) was identified from PD-enriched cell-wall extracts (Lee *et al.*, 2003). NCAPP1 is a highly basic protein with an ER trans-membrane domain that localizes to the cell periphery. When NCAPP1 was silenced or when a dominant negative form of NCAPP1 was present, transport through PD was abolished, provoking altered plant development. These data suggest that NCAPP1-mediated transport plays an important role during plant morphogenesis.

Flower development is also regulated by the cell-to-cell transport of regulatory proteins (Sessions *et al.*, 2000; Kim *et al.*, 2003; Wu *et al.*, 2003). In particular, transcription factor movement may play important roles in determining the timing and formation of reproductive organ development in plants. A description of transcriptional factor movement can be found in Chapter 6.

Note that these developmental macromolecules can either move between cells in a passive diffusion-like manner, forming a macromolecular gradient, or alternatively in an active targeted manner. The LEAFY transcription factor exemplifies the first class of mobile transcriptional factors, which can diffuse through different meristem cell layers in a nontargeted fashion (Sessions *et al.*, 2000; Wu *et al.*, 2003). Potentially, KN1 (Jackson *et al.*, 1994) and SHR (Nakajima *et al.*, 2001) exemplify targeted movement as they only move directionally to adjacent cell layers. However, additional studies are needed to clarify and provide further support for whether transcriptional factor transport can be classified as nontargeted as against targeted. Table 5.2 summarizes evidence for mobile endogenous plant proteins and RNAs.

Recently, miRNAs (Lagos-Quintana *et al.*, 2001, 2002; Lee and Ambros, 2001; Llave *et al.*, 2002; Mourelatos *et al.*, 2002; Park *et al.*, 2002; Reinhart *et al.*, 2002) have been identified as significant players with regulatory functions during plant morphogenesis (Carrington and Ambros, 2003; Palatnik *et al.*, 2003). Due to their small size, miRNAs are definitely within the realm of molecules capable of transport through PD to regulate development, analogous to protein regulatory molecules. In fact, miRNAs resemble another abundant class of small RNAs – gene silencing RNAs (siRNAs, reviewed in Chapter 11) – that are proven cargo for PD transit during plant defense responses. In support, ARGONAUTE1 (AGO1) (part of the RISC complex involved in RNA silencing; see Chapter 11) regulates miRNA distribution in developing leaves (Kidner and Martienssen, 2004). *ago1* mutant plants have impaired miRNA production, which results in the altered distribution of two target mRNAs (*PHABULOSA* and *PHAVULOTA*), causing defects in polar leaf growth. Although there is no direct evidence, the authors suggest that miRNAs act as mobile signaling molecules to direct adaxial and abaxial leaf cell polarization. In this scenario, miRNAs would act non-cell autonomously. As miRNAs complementary base pair with target mRNAs, they provide a precise mechanism for regulating mRNA function. Such an activity would be more efficient if it were cell autonomous. Future studies will undoubtedly reveal whether non-cell autonomous scenarios apply to miRNA function during plant morphogenesis. If miRNAs are found to move cell-to-cell, PD may play yet another role in plant development by regulating their distribution.

**Table 5.2** Transported macromolecular signals through PD

| Tissue/ developmental state | Tracers | Endogenous[a] | Virus/MP | RNA | References |
|---|---|---|---|---|---|
| Leaf/root | GFP1X, GFP2X | SHR, CPC, NACPP1 | + | N.d | Nakajima *et al.*, 2001; Wada *et al.*, 2002 |
| Phloem | GFP1X | CmPP16, RPP13-1, PP2, CmNACP | + | Thioredoxin h, mRNA, SUT1 mRNA, CmNACP, mRNA, siRNA | Mezitt and Lucas, 1996; Kuhn *et al.*, 1997; Ishiwatari *et al.*, 1998; Sasaki *et al.*, 1998; Vaucheret, *et al.*, 1998; Xoconostle-Cázares *et al.*, 1999; Golecki *et al.*, 1999; Ruiz-Medrano *et al.*, 1999 |
| Reproductive | GFP1X | KN1, FLO, LFY, GLO, DEF, STM, KNAT1 | + | N.d | Lucas *et al.*, 1995; Mezitt and Lucas, 1996; Kragler *et al.*, 1998; Sessions *et al.*, 2000; Kim, 2003 |
| Embryo | 10-kDa-Dex, GFP1X, GFP2X | N.d | N.d | N.d | Kim *et al.*, 2002a |

[a]Endogenously expressed proteins are either native or fused to GFP. N.d., not defined.

## 5.8 Concluding remarks

In the first part of this chapter, we described how PD change their basic structure (simple versus branched) to modulate their aperture (SEL). We also mentioned that the potential for cell-to-cell signaling and transport is further regulated by occlusion or *de novo* formation of PD. These changes are observed by microscopic or functional assays. We are beginning to understand the molecular components that regulate PD function, such as actin, myosin, centrin and calcium-binding calreticulin (see Chapters 1 and 4). However, it is remarkable that there is no relevant data on the nature of the intra- and intercellular signals that provoke modifications of

PD structural components, or that regulate the dynamic aperture of PD, such as PD formation or PD degradation. Such signals must be correlated with basic cellular events, such as mitosis, as well as more complex developmental programs.

There are additional relevant questions that surface. Why do PD undergo structural modification during plant development? In all studies reported to date, the structural changes that occur during the conversion of simple PD to branched PD correlates with reduction in their conductivity (Imlau *et al.*, 1999; Oparka *et al.*, 1999). Thus, maturation correlates with branched PD and lower SEL. What are the signals that provoke PD modification? PD modification may occur prior to observable changes in developmental programming; e.g. branched PD form prior to the sink–source transition in some tissues. Potentially, signaling pathways and specific signaling components may be identified by a genetic screen for mutants that no longer form branched PD. Such mutants may completely lack branched PD, or lack such PD in particular tissues.

The formation of *de novo* secondary PD is essential to increase the connectivity between cells. Another obvious question is: what is the signal(s) that trigger secondary PD formation in specific cell types? What signals determine the sites in the cell periphery where PD will form? Ultrastructural studies suggest that secondary PD can initiate in two adjoining cells leading to their joining in the middle of the cell wall (Ehlers and Kollmann, 1996). How do adjacent cells then coordinate such precise growth of two independent PD initials? Clearly, there are numerous critical questions regarding how PD form or not. The identification of the signals involved and the underlying signal transduction, perhaps via a novel genetic approach, may generate new discoveries regarding how plants produce new intercellular connections, and consequently increase their intra- and inter-symplasmic communication to coordinate new developmental events.

In the second part of this chapter, we summarized PD function and regulation during three developmental stages in plants: leaf maturation, the transition from vegetative to reproductive development in the SAM and during embryogenesis. All studies emphasize the correlation between the regulation of PD function and developmental changes. Even though there are limited data regarding the changes of PD conductivity during plant morphogenic events, recent data describing the down-regulation of PD during embryogenesis underscore that PD are essential gatekeepers for cell-to-cell transport of informational signals during plant development. Hopefully, future work on the characterization of existing and newly discovered embryo defective lines with altered PD function will provide important clues regarding PD function.

PD research is at a juncture. Their dynamic and critical roles have been established. Yet, there are many questions regarding how PD functions are regulated during plant development. Fortunately, new macromolecular signals are being identified by genomic approaches at a rapid pace in numerous laboratories, and other laboratories are pursuing novel genetic strategies to identify genes that regulate or form structural components of PD. Such studies will provide complementary information about structural PD components and its regulators. Future studies will continue to reveal the intricacies of the mechanisms involved in PD function during complex developmental programs. Forthcoming new information regarding PD will impact enormously on research relevant to understanding the complexitics of plant development. Finally, although PD are unique to plants, such fundamental knowledge likely will impact our understanding of intercellular communication in non-plant organisms as well.

## Acknowledgements

The entire PD group of our laboratory are listed as co-authors. We gratefully acknowledge former lab members, especially Katrina Crawford, Andreas Gisel and Fred Hempel, for their excellent research that contributed to our understanding of PD function, included in this review. Research on PD in our laboratory is supported by a research grant from the National Institutes of Health (GM 45244).

## References

Botha, C. and Cross, R. (2000) 'Towards reconciliation of structure with function in plasmodesmata-who is the gatekeeper?', *Micron* **31**, 713–721.

Carrington, J.C. and Ambros, V. (2003) 'Role of microRNAs in plant and animal development', *Science* **301**, 336–338.

Carrington, J.C., Jensen, P.E. and Schaad, M.C. (1998) 'Genetic evidence for an essential role for potyvirus CI protein in cell-to-cell movement', *Plant Journal* **14**, 393–400.

Crawford, K.M. and Zambryski, P.C. (2000) 'Subcellular localization determines the availability of non-targeted proteins to plasmodesmatal transport', *Current Biology* **10**, 1032–1040.

Crawford, K.M. and Zambryski, P.C. (2001) 'Non-targeted and targeted protein movement through plasmodesmata in leaves in different developmental and physiological states', *Plant Physiology* **125**, 1802–1812.

Deom, C.M., Shaw, M.J. and Beachy, R.N. (1987) 'The 30-Kilodalton gene product of tobacco mosaic virus potentiates virus movement', *Science* **327**, 389–394.

Ding, B., Haudenshield, J.S., Hull, R.J., Wolf, S., Beachy, R.N. and Lucas, W.J. (1992) 'Secondary plasmodesmata are specific sites of localization of the tobacco mosaic virus movement protein in transgenic tobacco plants', *The Plant Cell* **4**, 915–928.

Ding, B., Haudenshield, J.S., Willmitzer, L. and Lucas, W.J. (1993) 'Correlation between arrested secondary plasmodesmal development and onset of accelerated leaf senescence in yeast acid invertase transgenic tobacco plants', *Plant Journal* **4**, 179–189.

Ding, B., Kwon, M.-O. and Warnberg, L. (1996) 'Evidence that actin filaments are involved in controlling the permeability of plasmodesmata in tobacco mesophyll', *Plant Journal* **10**, 157–164.

Duckett, C.M., Oparka, K.J., Prior, D.A.M., Dolan, L. and Roberts, K. (1994) 'Dye-coupling in the root epidermis of *Arabidopsis* is progressively reduced during development', *Development* **120**, 3247–3255.

Ehlers, K. and Kollmann, R. (1996) 'Formation of branched plasmodesmata in regenerating *Solanum nigrum* protoplasts', *Planta* **199**, 126–138.

Ehlers, K., Binding, H. and Kollmann, R. (1999) 'The formation of symplasmic domains by plugging of plasmodesmata: A general event in plant morphogenesis?' *Protoplasma* **209**, 181–192.

Erickson, F.L., Holzberg, S., Calderon-Urrea, A. *et al.* (1999) 'The helicase domain of the TMV replicase proteins induces the N-mediated defense response in tobacco', *Plant Journal* **18**, 67–75.

Evert, R.F., Russin, W.A. and Bosabalidis, A.M. (1996) 'Anatomical and ultrastructural changes associated with sink-to-source transition in developing maize leaves', *International Journal of Plant Sciences* **157**, 247–261.

Gisel, A., Barella, S., Hempel, F.D. and Zambryski, P.C. (1999) 'Temporal and spatial regulation of symplastic trafficking during development in *Arabidopsis thaliana* apices', *Development* **126**, 1879–1889.

Gisel, A., Hempel, F.D., Barella, S. and Zambryski, P.C. (2002) 'Leaf-to-shoot apex movement of symplastic tracer is restricted coincident with flowering in *Arabidopsis', Proceedings of the National Academy of Sciences USA* **99**, 1713–1717.

Golecki, B., Schulz, A. and Thompson, G.A. (1999) 'Translocation of structural P proteins in the phloem', *The Plant Cell* **11**, 127–140.

Heinlein, M. and Epel, B.L. (2004) 'Macromolecular signaling through plasmodesmata', *International Review of Cytology* **235**, 93–164.

Hepler, P. (1982) 'Endoplasmic reticulum in the formation of the cell plate and plasmodesmata', *Protoplasma* **111**, 121–133.

Hirashima, K. and Watanabe, Y. (2001) 'Tobamovirus replicase coding region is involved in cell-to-cell movement', *Journal of Virology* **75**, 8831–8836.

Imlau, A., Truernit, E. and Sauer, N. (1999) 'Cell-to-cell and long-distance trafficking of the green fluorescent protein in the phloem and symplastic unloading of the protein into sink tissues', *The Plant Cell* **11**, 309–322.

Ishiwatari, Y., Fujiwara, T., McFarland, K.C. *et al.* (1998) 'Rice phloem thioredoxin h has the capacity to mediate its own cell-to-cell transport through plasmodesmata', *Planta* **205**, 12–22.

Itaya, A., Hickman, H., Bao, Y., Nelson, R. and Ding, B. (1997) 'Cell-to-cell trafficking of cucumber mosaic virus movement protein, green fluorescent protein fusion produced by biolistic gene bombardment in tobacco', *Plant Journal* **12**, 1223–1230.

Itaya, A., Liang, G.Q., Woo, Y.M., Nelson, R.S. and Ding, B. (2000) 'Nonspecific intercellular protein trafficking probed by green-fluorescent protein in plants', *Protoplasma* **213**, 165–175.

Jackson, D., Veit, B. and Hake, S. (1994) 'Expression of maize KNOTTED1 related homeobox genes in the shoot apical meristem predicts patterns of morphogenesis in the vegetative shoot', *Development* **120**, 405–413.

Johnson, E.R. and McKay, D.B. (1999) 'Crystallographic structure of the amino terminal domain of yeast initiation factor 4A, a representative DEAD-box RNA helicase', *RNA* **5**, 1526–1534.

Jones, M. (1976) The origin and development of plasmodesmata, in *Intercellular Communication in Plants: Studies on Plasmodesmata* (eds B.E.S. Gunning and A.W. Robards), Springer-Verlag, Berlin, pp. 81–105.

Kidner, C.A. and Martienssen, R.A. (2004) 'Spatially restricted microRNA directs leaf polarity through ARGONAUTE1', *Nature* **428**, 81–84.

Kim, I., Cho, E., Crawford, K.M., Hempel, F.D. and Zambryski, P.C. (2005) 'Cell-to-cell movement of GFP during embryogenesis and early seedling development in *Arabidopsis', Proceedings of the National Academy of Sciences USA* **102**, 2227–2231.

Kim, I., Hempel, F.D., Sha, K., Pfluger, J. and Zambryski, P.C. (2002a) 'Identification of a developmental transition in plasmodesmatal function during embryogenesis in *Arabidopsis thaliana*', *Development* **129**, 1261–1272.

Kim, I., Mindrinos, M., Otegui, M. and Zambryski, P.C. *Arabidopsis* ISE1, DEAD-box RNA helicase, regulates plasmodesmatal functions during embryogenesis, (Submitted).

Kim, J.Y., Yuan, Z. and Jackson, D. (2003) 'Developmental regulation and significance of KNOX protein trafficking in *Arabidopsis*', *Development* **130**, 4351–4362.

Kim, J.Y., Yuan, Z., Cilia, M., Khalfan-Jagani, Z. and Jackson, D. (2002b) 'Intercellular trafficking of a KNOTTED1 green fluorescent protein fusion in the leaf and shoot meristem of *Arabidopsis*', *Proceedings of the National Academy of Sciences USA* **99**, 4103–4108.

Kim, M., Canio, W., Kessler, S. and Sinha, N. (2001) 'Developmental changes due to long-distance movement of a homeobox fusion transcript in tomato', *Science* **293**, 287–289.

Koornneef, M.H.C. and Van Der Veen, J.H. (1991) 'A genetic and physiological analysis of late flowering mutants in *Arabidopsis thaliana*', *Molecular and General Genetics* **229**, 57–66.

Kragler, F., Monzer, J., Shash, K., Xoconostle-Cázares, B. and Lucas, W.J. (1998) 'Cell-to-cell transport of proteins., requirement for unfolding and characterization of binding to a putative plasmodesmal receptor', *Plant Journal* **15**, 367–381.

Kuhn, C., Franceschi, V.R., Schulz, A., Lemoine, R. and Frommer, W.B. (1997) 'Macromolecular trafficking indicated by localization and turnover of sucrose transporters in enucleate sieve elements', *Science* **275**, 1298–1300.

Lagos-Quintana, M., Rauhut, R., Lendeckel, W. and Tuschl, T. (2001) 'Identification of novel genes coding for small expressed RNAs', *Science* **294**, 853–858.

Lagos-Quintana, M., Rauhut, R., Yalcin, A., Meyer, J., Lendeckel, W and Tuschl, T. (2002) 'Identification of tissue-specific microRNAs from mouse', *Current Biology* **12**, 735–739.

Lee, J.Y., Yoo, B.C., Rojas, M.R., Gomez-Ospina, N., Staehelin, L.A. and Lucas, W.J. (2003) 'Selective trafficking of non-cell-autonomous proteins mediated by NtNCAPP1', *Science* **299**, 392–396.

Lee, R.C. and Ambros, V. (2001) 'An extensive class of small RNAs in *Caenorhabditis elegans*', *Science* **294**, 862–864.

Llave, C., Kasschau, K.D., Rector, M.A. and Carrington, J.C. (2002) 'Endogenous and silencing-associated small RNAs in plants', *The Plant Cell* **14**, 1605–1619.

Lucas, W.J., Bouche-Pillon, S., Jackson, D.P. *et al.* (1995) 'Selective trafficking of KNOTTED1 homeodomain protein and its mRNA through plasmodesmata', *Science* **270**, 1980–1983.

Mansfield, S.G. and Briarty, L.G. (1991) 'Early embryogenesis in *Arabidopsis thaliana*, II. The developing embryo', *Canadian Journal of Botany* **69**, 461–476.

Medford, J.I. (1992) 'Vegetative apical meristems', *The Plant Cell* **4**, 1029–1039.

Mezitt, L.A. and Lucas, W.J. (1996) 'Plasmodesmal cell-to-cell transport of proteins and nucleic acids', *Plant Molecular Biology* **32**, 251–273.

Mourelatos, Z., Dostie, J., Paushkin, S. *et al.* (2002) 'miRNPs, a novel class of ribonucleoproteins containing numerous microRNAs', *Genes and Development* **16**, 720–728.

Nakajima, K., Sena, G., Nawy, T. and Benfey, P.N. (2001) 'Intercellular movement of the putative transcription factor SHR in root patterning', *Nature* **413**, 307–311.

Oparka, K.J. and Prior, D.A.M. (1992) 'Direct evidence for pressure-generated closure of plasmodesmata', *Plant Journal* **2**, 741–750.

Oparka, K.J. and Turgeon, R. (1999) 'Sieve elements and companion cells-traffic control centers of the phloem', *The Plant Cell* **11**, 739–750.

Oparka, K.J., Prior, D.A.M. and Wright, K.M. (1995) 'Symplastic communication between primary and developing lateral roots of *Arabidopsis thaliana*', *Journal of Experimental Botany* **46**, 187–197.

Oparka, K.J., Roberts, A.G., Boevink, P. *et al.* (1999) 'Simple, but not branched, plasmodesmata allow the nonspecific trafficking of proteins in developing tobacco leaves', *Cell* **97**, 743–754.

Ormenese, S., Havelange, A., Deltour, R. and Bernier, G. (2000) 'The frequency of plasmodesmata increases early in the whole shoot apical meristem of *Sinapis alba* L, during floral transition', *Planta* **211**, 370–375.

Ormo, M., Cubitt, A.B., Kallio, K., Gross, L.A., Tsien, R.Y. and Remington, S.J. (1996) 'Crystal structure of the *Aequorea victoria* green fluorescent protein', *Science* **273**, 1392–1395.

Palatnik, J.F., Allen, E., Wu, X. *et al.* (2003) 'Control of leaf morphogenesis by microRNAs', *Nature* **425**, 257–263.

Palevitz, B.A. and Hepler, P.K. (1985) 'Changes in dye coupling of stomatal cells of *Allium* and *Commelina* demonstrated by microinjection of Lucifer yellow', *Planta* **164**, 473–479.

Park, W., Li, J., Song, R., Messing, J. and Chen, X. (2002) 'CARPEL FACTORY, a Dicer homolog, and HEN1, a novel protein, act in microRNA metabolism in *Arabidopsis thaliana*', *Current Biology* **12**, 1484–1495.

Provencher, L.M., Miao, L., Sinha, N. and Lucas, W.J. (2001) 'Sucrose export defective1 encodes a novel protein implicated in chloroplast-to-nucleus signaling', *The Plant Cell* **13**, 1127–1141.

Putterill, J., Robson, F., Lee, K., Simon, R. and Coupland, G. (1995) 'The CONSTANS gene of *Arabidopsis* promotes flowering and encodes a protein showing similarities to zinc finger transcription factors', *Cell* **80**, 847–857.

Reed, J.W., Nagatani, A., Elich, T.D., Fagan, M. and Chory, J. (1994) 'Phytochrome A and Phytochrome B have overlapping but distinct functions in *Arabidopsis* development', *Plant Physiology* **104**, 1139–1149.

Reinhart, B.J., Rhoades, M.W., Bartel, B. and Bartel, D.P. (2002) 'MicroRNAs in plants', *Genes and Development* **16**, 1616–1626.

Rinne, P.L.H. and Van Der Schoot, C. (1998) 'Symplasmic fields in the tunica of the shoot apical meristem coordinate morphogenetic events', *Development* **125**, 1477–1485.

Roberts, I.M., Boevink, P., Roberts, A.G., Sauer, N., Reichel, C. and Oparka, K.J. (2001) 'Dynamic changes in the frequency and architecture of plasmodesmata during the sink–source transition in tobacco leaves', *Protoplasma* **218**, 31–44.

Ruan, Y.L., Llewellyn, D.J. and Furbank, R.T. (2001) 'The control of single-celled cotton fiber elongation by developmentally reversible gating of plasmodesmata and coordinated expression of sucrose and K+ transporters and expansin', *Plant Cell* **13**, 47–60.

Ruiz-Medrano, R., Xoconostle-Cázares, B. and Lucas, W.J. (1999) 'Phloem long-distance transport of CmNACP mRNA, implications for supracellular regulation in plants', *Development* **126**, 4405–4419.

Russin, W.A., Evert, R.F., Vanderveer, P.J., Sharkey, T.D. and Briggs, S.P. (1996) 'Modification of a specific class of plasmodesmata and loss of sucrose export ability in the *sucrose export defective 1* mazie mutant', *Plant Cell* **8**, 645–658.

Sasaki, T., Chino, M., Hayashi, H. and Fujiwara, T. (1998) 'Detection of several mRNA species in rice phloem sap', *Plant and Cell Physiology* **39**, 895–897.

Schmid, S.R. and Linder, P. (1991) 'Translation initiation factor 4A from *Saccharomyces cerevisiae*, analysis of residues conserved in the D-E-A-D family of RNA helicases', *Molecular and Cellular Biology* **11**, 3463–3471.

Sessions, A., Yanofsky, M.F. and Weigel, D. (2000) 'Cell–cell signaling and movement by the floral transcription factors LEAFY and APETALA1', *Science* **289**, 779–782.

Shannon, S. and Meeks-Wagner, D. (1991) 'A mutation in the *Arabidopsis* TFL1 gene affects inflorescence meristem development', *Plant Cell* **3**, 877–892.

Vaucheret, H., Beclin, C., Elmayan, T. *et al.* (1998) 'Transgene-induced gene silencing in plants', *Plant Journal* **16**, 651–659.

Wada, T., Kurata, T., Tominaga, R. *et al.* (2002) 'Role of a positive regulator of root hair development, CAPRICE, in *Arabidopsis* root epidermal cell differentiation', *Development* **129**, 5409–5419.

Wolf, S., Deom, C.M., Beachy, R.N. and Lucas, W.J. (1989) 'Movement protein of tobacco mosaic virus modifies plasmodesmata size exclusion limit', *Science* **246**, 377–379.

Wu, X., Dinnemy, J.R., Crawford, K.M. *et al.* (2003) 'Modes of intercellular transcription factor movement in the *Arabidopsis* apex', *Development* **130**, 3735–3745.

Xoconostle-Cázares, B., Xiang, Y., Ruiz-Medrano, R. *et al.* (1999) 'Plant paralog to viral movement protein that potentiates transport of mRNA into the phloem', *Science* **283**, 94–98.

Zambryski, P. and Crawford, K. (2000) Plasmodesmata: gatekeepers for cell-to-cell transport of developmental signals in plants. *Annual Review of cell and Developmental Biology*, **16**, 393–421.

# 6 Transcription factor movement through plasmodesmata

David Jackson

## 6.1 Introduction

Intercellular communication is essential for the establishment and maintenance of cell fates in all multicellular organisms. Despite the highly regular cell lineages observed in some model plants, such as *Arabidopsis*, clonal analyses and ablation experiments indicate that cell fate is determined primarily by position and not by lineage (Van den Berg *et al.*, 1995; Hake and Char, 1997; Dolan *et al.*, 1998; Kidner *et al.*, 2000). Diverse modes of intercellular signaling transmit positional information. In animals, the predominant mechanism is through secreted ligands that signal to cell surface receptors on neighboring or distant cells. Secreted ligands and their corresponding receptors are also common in plants. However, since the discovery of plasmodesmata (PD) in plants over a hundred years ago, speculation about their role in signaling during development and in the coordination of physiology has abounded. Though widely accepted that PDs provide cytoplasmic continuity and allow passive cell-to-cell movement of small molecules, it was not until the late 1980s that clues arose about their potential role in the transport of viral proteins and RNAs (Wolf *et al.*, 1989, 1991; Ding *et al.*, 1992; Waigmann *et al.*, 1994). This discovery prompted speculation that endogenous plant proteins and nucleic acids might also make use of PDs for short- and long-range signaling (Lucas *et al.*, 1993). In 1995, the first endogenous plant protein with the ability to traffic cell to cell was reported (Lucas *et al.*, 1995). This protein, KNOTTED1 (KN1), and its homologs are homeodomain transcription factors (TFs) that regulate leaf and shoot meristem development in a variety of plants (Reiser *et al.*, 2000). KN1 has the remarkable ability to traffic itself and its mRNA from cell to cell. Subsequently, a handful of other developmental TFs have also been found to traffic, and, in specific cases, it is clear that trafficking serves an integral function in regulating plant cell fate (reviewed in Mezitt and Lucas, 1996; Jackson and Hake, 1997; Kragler *et al.*, 1998a; Hake, 2001; Haywood *et al.*, 2002; Wu *et al.*, 2002; Oparka, 2004). In this chapter, I will review the evidence for intercellular trafficking of TF proteins and their mRNAs, and discuss the possible mechanisms by which these macromolecules are able to move from cell to cell.

## 6.2 Plant TFs

As in other organisms, plants dedicate a significant proportion of their genome to encoding TFs. In *Arabidopsis*, for example, they account for more than 1500 genes, or around 6% of the total genes identified to date (Riechmann *et al.*, 2000). These proteins act as essential regulators of physiological, environmental and developmental responses, by activating downstream target genes through interactions with regulatory DNA elements. TFs are therefore sometimes referred to as master regulators, and they are widely studied for this reason. Each TF can be composed of different domains, typically a nucleic acid binding domain, which interacts with DNA usually with high sequence specificity, protein–protein interaction domain(s) for interaction with other TFs and activation or repressor domain(s) that govern the transcriptional outcome of the DNA binding interaction. Classification of TFs is usually based on their DNA binding domain. For example, in *Arabidopsis*, the AP2/EREBP (APETALA2/ethylene responsive element binding protein), MYB and bHLH (basic helix-loop-helix) families each represent around 9% of the total TFs. Other large families include the NAC, C2H2 zinc finger, homeobox, MADS, bZIP, WRKY, GARP, CCAAT and GRAS domain families (Riechmann *et al.*, 2000). Specific members of several of these TF classes traffic from cell to cell, presumably through PDs (Perbal *et al.*, 1996; Sessions *et al.*, 2000; Nakajima *et al.*, 2001; Wada *et al.*, 2002; Kim *et al.*, 2003). It would be interesting to determine whether the ability to traffic has been gained independently in the different TF lineages; however, to date, no systematic study of TF trafficking has been carried out.

## 6.3 Evidence for cell-to-cell trafficking of TFs

Various lines of evidence support the hypothesis of cell-to-cell trafficking of TFs. In general, the primary evidence comes from one of two sources. First, localization studies using either combinations of *in situ* mRNA hybridization and immunolocalization (e.g. Lucas *et al.*, 1995; Perbal *et al.*, 1996; Sessions *et al.*, 2000; Jackson, 2002) or tissue-specific promoter expression of green fluorescent protein (GFP) fusions (Nakajima *et al.*, 2001; Kim *et al.*, 2002b; Wada *et al.*, 2002; Wu *et al.*, 2003) has revealed the presence of TF protein outside of the domain in which its mRNA can be detected. Second, mosaic analysis has shown that some TFs act outside of the cells in which they are expressed (Hake and Freeling, 1986; Sinha and Hake, 1990; Hake and Char, 1997; Foster *et al.*, 1999; Marcotrigiano, 2001). In this case, the TF is said to act nonautonomously. This in itself is not evidence for cell-to-cell movement, because nonautonomy could also be caused by movement of a signal encoded by a downstream target gene, such as a secreted peptide or a hormone. However, in combination with localization studies, mosaic analyses provide rigorous evidence for functionally significant cell-to-cell TF movement. In the following sections, I will discuss examples where these kinds of evidence have been used to support the hypothesis that TFs act as intercellular signals by directly trafficking from cell to cell.

## 6.4 Plant TFs that traffic cell to cell

### *6.4.1 Homeobox proteins – KNOTTED1 and homologs*

KNOTTED1 was the first example of a TF found to traffic cell to cell (Lucas *et al.*, 1995). This gene encodes a homeodomain protein (Vollbrecht *et al.*, 1991), and got its name from dominant gain of function alleles that lead to abnormal outgrowths ('knots') over the lateral veins of maize leaves (Figures 6.1a, b). These outgrowths initiate from ectopic cell divisions in all cell layers of the leaf, including the epidermis (Hake and Freeling, 1986; Sinha and Hake, 1990). Knots are also characterized by changes in cell fate, as their cells resemble those of the basal part of the leaf (the sheath) even though they are found in the distal or blade part of the leaf (Sinha *et al.*, 1993a). Evidence that the dominant *Kn1* allele acts non-cell autonomously came from mosaic analyses performed by Hake and Freeling in the mid 1980s (Hake and Freeling, 1986). They used X-ray irradiation to create clones of cells that had lost the dominant *Kn1* allele, and these clones were marked using a linked albino mutant phenotype. Analysis of mosaics that expressed *Kn1* only in specific tissue layers showed that although the phenotype is caused by cell divisions in all cell layers, the presence of the dominant mutant allele in only the inner mesophyll and vascular cells was sufficient to generate knots (Hake and Freeling, 1986).

The conclusion of the mosaic studies was that *Kn1* acts nonautonomously to affect the fate and division of cells in which it is not expressed. Because *Kn1* sectors could be observed at cellular resolution, it was possible to determine with precision the range of its nonautonomy. *Kn1* could signal over a distance of several cell layers from vascular to epidermal cells. However, the range of signaling in a lateral dimension, across the plane of the leaf, was limited as completely albino sectors adjacent to green sectors did not produce knots (Hake and Freeling, 1986).

After the *kn1* gene was isolated, the basis for the dominant mutations became clear. *kn1* is not normally expressed in maize leaf primordia, but in the shoot apical meristem (SAM), the collection of stem cells at the apex of the shoot that is responsible for the indeterminate growth of the shoot system (Smith *et al.*, 1992; Jackson *et al.*, 1994) (Figure 6.1). Loss of function mutations in *kn1*, and in an *Arabidopsis* homolog *SHOOTMERISTEMLESS* (*STM*) cause a loss of indeterminacy in the SAM and premature termination of shoot development after the first leaf/leaves are initiated (Long *et al.*, 1996; Vollbrecht *et al.*, 2000). Therefore, *kn1* and *STM* are required for the initiation or maintenance of the SAM, perhaps by promoting stem-cell divisions and repressing differentiation (Sinha *et al.*, 1993b; Reiser *et al.*, 2000). In the dominant *Kn1* mutations, the gene is abnormally expressed in leaf primordia, where it causes aberrant growth of knots by prolonging the phase of cell proliferation and delaying the acquisition of a mature cell fate (Smith *et al.*, 1992).

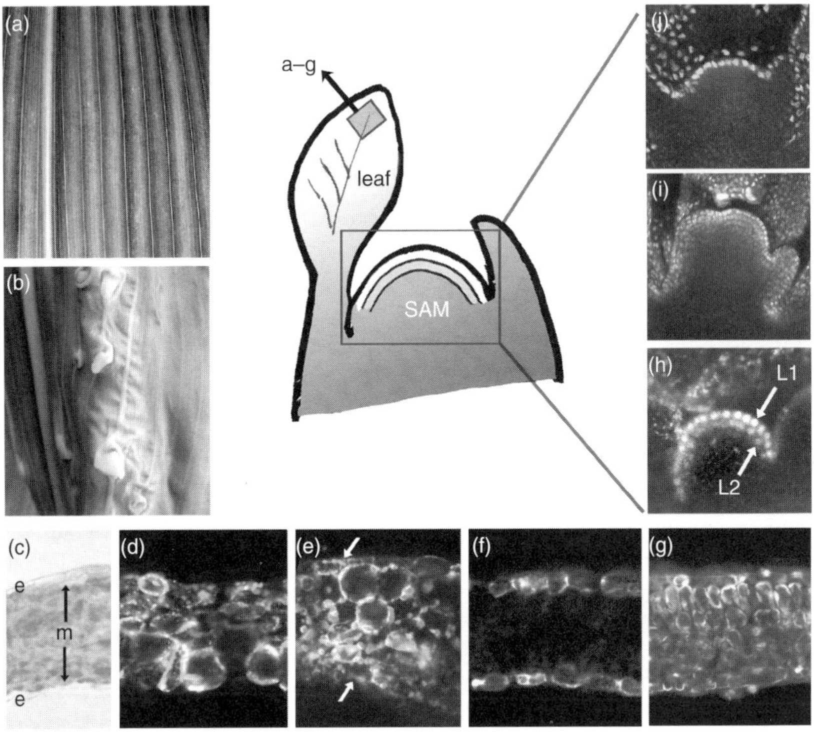


**Figure 6.1** Transcription factor movement in the shoot. The central image is a sketch of a shoot apex, showing the layered cellular arrangement of the shoot apical meristem (SAM) and lateral leaf primordia (leaf). Images in (a)–(g) are of leaves: (a) part of a normal maize leaf with parallel lateral veins; (b) *KNOTTED1* mutant leaf with outgrowths or knots over the lateral veins; (c) cross-section of part of an *Arabidopsis* leaf with epidermis (e) and mesophyll (m) tissues marked; (d)–(e) directional control of KN1 trafficking in the leaf. In (d), fluorescence from a cell-autonomous GFP driven by a mesophyll-specific promoter is seen as bright signal only in mesophyll tissue. In (e), GFP–KN1 under the same promoter shows trafficking to epidermal cells, arrowed. (f) When expressed in the epidermis, GFP–KN1 does not traffic to underlying mesophyll cells. In (g), a control GFP–MP fusion shows trafficking from epidermal to mesophyll cells. (h)–(j) Transcription factor trafficking in the SAM. All constructs are expressed using the *ML1* promoter. In (h), trafficking of GFP–STM from the L1 to the L2 layer is observed as bright signal in nuclei, arrowed. In (i), trafficking of GFP–LFY from L1 to L2 and L3 is shown. In (j), the cell-autonomous AP1–GFP fusion does not traffic out of L1 cells and is localized to nuclei. Images in (c)–(h) are from Kim *et al.* (2003), (i)–(j) are from Wu *et al.* (2003), drawing from Cilia and Jackson (2004).

The subsequent use of antibodies and *in situ* hybridization probes in localization experiments led to the unexpected finding that KN1 protein was present outside the visible domain of KN1 mRNA expression in the *Kn1* mutant leaf primordia (Smith *et al.*, 1992; Jackson *et al.*, 1994). This observation prompted speculation that the KN1 protein itself could be the cell-to-cell signal that acts to promote aberrant cell proliferation and cell fate in *Kn1* leaves. This hypothesis was partially tested by microinjection experiments (Lucas *et al.*, 1995). KN1 protein was expressed in *Escherichia coli*, purified and solubilized, then labeled with a fluorescent dye. The labeled KN1 protein was injected into mesophyll cells of maize or tobacco leaves, and almost immediately it moved to adjacent cells, showing that KN1 has the ability to traffic cell to cell, presumably through PDs. Control microinjections using fluorescently tagged 10- or 20-kD dextrans showed that these molecules could not traffic. However, in co-injection experiments, KN1 promoted the movement of dextrans, presumably by 'gating' or transiently increasing the size exclusion limit (SEL) of PDs in a manner analogous to the effect of viral movement proteins (MPs) (Wolf *et al.*, 1989; Waigmann *et al.*, 1994). Alanine scanning mutagenesis of KN1 also revealed a motif at the N-terminus of the homeodomain that was required for KN1 movement or for PD gating. Collectively, these observations suggest that KN1 moves through PDs by an active mechanism (Lucas *et al.*, 1995).

Although the localization experiments suggested that KN1 protein, and not its mRNA, was the mobile signal, Lucas *et al.* (1995) asked if KN1, like viral MPs, could also traffic RNA. Surprisingly, KN1 mRNA was indeed mobile, but only when injected along with KN1 protein. In contrast to the low-sequence specificity for RNA movement by viral MPs, KN1 mRNA trafficking was highly specific, as only the sense strand and not the antisense strand of KN1 RNA, nor a viral RNA, could be trafficked by KN1 protein (Lucas *et al.*, 1995). These results suggest that KN1 transports its own mRNA through PDs, perhaps to amplify intercellular signaling by KN1. The interaction between KN1 and its mRNA is also supported by other reports of interactions between homeodomain proteins and specific mRNAs (Dubnau and Struhl, 1996).

The microinjection data provide compelling evidence for cell-to-cell movement of KN1 and its mRNA; however, they do not directly address whether KN1 acts as a developmental signal in the cells into which it traffics. They also do not address whether KN1 traffics between different cell layers of the leaf or of the SAM, due to technical limitations of which tissues can be injected and imaged. To address these important points, a GFP fusion of KN1 was expressed in specific tissue or cell types in *Arabidopsis* (Kim *et al.*, 2002b). As an important control, function of the GFP–KN1 fusion was demonstrated by its overexpression phenotype, and ability to complement strong *stm* mutants (Long *et al.*, 1996; Kim *et al.*, 2002b; Kim *et al.*, 2003). Following micro-projectile bombardment or tissue-specific expression, the GFP–KN1 fusion protein moved cell to cell. Importantly, the fusion could move from peri-vascular cells to the epidermis, across several layers of mesophyll cells. This observation confirms that cell-to-cell movement is sufficient to explain the nonautonomy of *KN1* mutants in maize. The GFP–KN1 fusion localized to nuclei and to punctae in the cell wall, could be PD, though this should be confirmed by transmission electron microscopy (Kim *et al.*, 2002b).

Although GFP–KN1 moved in the leaf from peri-vascular or mesophyll cells to the epidermis, it was surprisingly unable to move in the opposite direction – from epidermis to mesophyll cells (Figure 6.1c–g, Kim *et al.*, 2003). This observation suggests that either the PDs between mesophyll and epidermis are directional, or that KN1 is modified in the epidermis or mesophyll to inhibit or allow its movement, respectively. The PDs do not appear to be directional, at least for other proteins, because an MP–GFP fusion could traffic from epidermal to mesophyll cells as expected (Figure 6.1g; Kim *et al.*, 2003).

In contrast to the situation in the leaf, KN1 and its *Arabidopsis* homologs KNAT1/BREVIPEDICILLUS and STM were able to traffic from the epidermal (L1) layer to underlying L2 and L3 cell layers in the SAM (Figure 6.1h; Kim *et al.*, 2002b). This movement appears to be functionally significant, since it allows the rescue of shoot meristem formation in strong *stm* mutants. In contrast, the expression of a cell autonomous β-glucuronidase–KN1 fusion in the L1 alone is insufficient to rescue meristem formation, indicating that movement is necessary for KN1 function in the context of L1-specific expression (Kim *et al.*, 2003).

Why do KN1 and STM traffic in the meristem? In *Arabidopsis*, the reason is not so obvious, because STM mRNA normally accumulates in all of the three cell layers of the SAM (Long *et al.*, 1996), and the *STM* promoter is active in all three layers (A. Fernandez and K. Barton, unpublished). However, in maize, which has only a single periclinal cell layer, the L1, overlying the L2 cells, *kn1* mRNA is undetectable in the L1, whereas the protein is clearly present there (Smith *et al.*, 1992; Jackson *et al.*, 1994). So, in this case, trafficking of KN1 from L2 to L1 could serve to coordinate the growth of the different cell layers in the SAM. It is not clear, however, whether the lack of KN1 mRNA expression in the L1 is functionally significant. An alternative hypothesis is that the promoter elements required for L1 expression were lost fortuitously, but because KN1 protein could move from L2 to L1, this loss had no obvious phenotypic consequence. In both *Arabidopsis* and maize, trafficking of KN1/STM might play a role in general coordination of cell fate in the SAM, to maintain the stem cell niche and perhaps to produce gradients of protein concentration that serve to maintain boundaries between the SAM and lateral organ primordia (Jackson, 2002). Further details of the possible mechanisms of KN1 trafficking will be discussed in Section 6.6.

### *6.4.2 MADS box proteins*

The MADS box proteins were first identified in plants as homeotic regulators of floral organ development, and subsequently have been found to regulate diverse aspects of plant development, including flowering time, root development and fruit dehiscence (Becker and Theissen, 2003). One of the founding members of this family, *DEFICIENS* (*DEF*) and its close homolog *GLOBOSA* (*GLO*) are *Antirrhinum* genes that regulate petal and stamen identity and make up the B function in the classic ‘ABC model’ of floral organ development (Sommer *et al.*, 1990; Coen and Meyerowitz, 1991; Trobner *et al.*, 1992). These two proteins are orthologs of the *Arabidopsis* APETALA3 and PISTILLATA proteins (Becker and Theissen, 2003), and form obligate heterodimers (Zachgo *et al.*, 1995) as well as larger transcriptional complexes (Egea-Cortines *et al.*, 1999).

As is common in *Antirrhinum*, the *def* and *glo* mutations were transposon induced, and excision of the transposon leads to reversion of gene function in clonal sectors (Carpenter and Coen, 1990; Perbal *et al.*, 1996). Since cells in the meristem tend to divide within the plane of their layer, clones usually occupy discrete cell layers of the meristem or floral organ primordia (Satina *et al.*, 1940; Figure 6.1). These kinds of chimeras are called 'periclinal', and can be used to assess the extent of signaling between cell layers during development. In the case of the floral homeotic genes, these interactions are complex, with different outcomes depending on whether one considers cellular level phenotypes such as cell fate, or more global phenotypes like organ growth and shape. Here, I will concentrate on the acquisition of cell fate. Expression of *DEF* or *GLO* in the inner L2 and L3 layers alone was sufficient to rescue petal cell identity in the overlying L1 or epidermal layer, even though this layer was mutant for *DEF* or *GLO* (Perbal *et al.*, 1996). Immunolocalization showed the presence of DEF or GLO proteins in the epidermal layer in these chimeras, indicating that the nonautonomous regulation of petal cell fate was correlated with cell-to-cell trafficking. However, in epidermal chimeras, where DEF was expressed only in the epidermal layer, no movement of DEF protein to underlying mesophyll cells could be detected, and the cell identity of the mesophyll was sepalloid, similar to the *def* mutants. In other words, DEF acts autonomously to control cell fate when expressed in the L1, but nonautonomously when expressed in L2 and L3 (Perbal *et al.*, 1996; Jackson and Hake, 1997). This polar regulation of nonautonomy appears to be due to the directional control of DEF protein movement, and is reminiscent of KN1, which also traffics from mesophyll to epidermal cells but not in the reverse direction (Figure 6.1c–g).

Transposon excision can also generate sectoral chimeras, where only a part of a periclinal layer has reverted to wild type. These chimeras can address the autonomy of gene function in a lateral dimension across a tissue layer, and in the case of *DEF*, these types of sector boundaries were sharply defined, indicating little nonautonomy (and implying little or no protein movement) across a tissue layer (Perbal *et al.*, 1996).

Similar experiments have been performed with *AP3* and *PI*, the *Arabidopsis* orthologs of *DEF* and *GLO*, respectively. In contrast to the situation in *Antirrhinum*, epidermal chimeras of *PI* act nonautonomously to control petal and stamen development; however, the distribution of PI protein was not tested in those chimeras (Bouhidel and Irish, 1996). AP3 also acts nonautonomously to regulate organ shape, and in stamens (but not in petals) to control epidermal cell identity. However, in this case, no movement of AP3 protein could be detected (Jenik and Irish, 2001). It is therefore likely that AP3 nonautonomy is asserted through movement of downstream signals rather than of the TF itself, although it is formally possible that a small amount of AP3 protein below the limit of sensitivity of the immunolocalization technique traffics and is responsible for its nonautonomy.

Are the differences between *Arabidopsis* and *Antirrhinum* in signaling between cell layers, and in cell-to-cell movement of MADS TF proteins, species specific or protein specific? This question was addressed by expressing DEF or GLO using an epidermal specific promoter in either *Antirrhinum* or *Arabidopsis* (Efremova *et al.*, 2001). DEF and GLO acted mostly autonomously to control cell fate in *Antirrhinum*; however, in *Arabidopsis*, they showed a much greater degree of nonautonomy. Whether this was due to protein movement was not tested, but it seems unlikely, as DEF or AP3 movement out of the epidermis could not be detected in previous experiments. Therefore, it appears that species-specific differences in interlayer signaling, rather than differences in the orthologous proteins, account for the different degrees of autonomy observed in *Arabidopsis* and *Antirrhinum*. In conclusion, MADS TF trafficking appears to be important to communicate developmental signaling between cell layers, but this is probably only one of several intercellular signaling mechanisms that these proteins use to orchestrate floral organ development.

### *6.4.3 FLORICAULA and LEAFY*

FLORICAULA (FLO) and LEAFY (LFY) are members of a plant-specific TF family (Coen *et al.*, 1990; Weigel *et al.*, 1992). Most plants have only a single member of this family, and it is required for the development of floral identity, acting through downstream floral homeotic genes (Parcy *et al.*, 1998). *flo/lfy* mutants undergo the inflorescence transition, but produce only bract or leaf-like organs arranged in a floral phyllotaxy.

*FLO* was the first of these genes to be isolated, by transposon tagging in *Antirrhinum* (Coen *et al.*, 1990). As in the case of *DEF* and *GLO*, mosaic revertant sectors of *FLO* were identified, in this case as branches bearing flowers (Carpenter and Coen, 1995). These flowers were abnormal, but easily spotted among a field of green leafy inflorescences. Three types of revertant sectors could be identified, according to the appearance of the flowers. The first class had almost normal flowers, with occasional extra sepals; the second class had flowers with an intermediate phenotype of narrower petals and sepals and a divided corolla; and the third class had extreme phenotypes, such as abnormal flowers with distorted organs and homeotic conversions (Carpenter and Coen, 1995).

*In situ* hybridizations using a *FLO* probe revealed that the three classes of sectors corresponded to periclinal chimeras where *FLO* was expressed in only the L1, the L2 or the L3 layer of the inflorescence shoot (Hantke *et al.*, 1995). The L1 chimeras had the near normal phenotype, the L2 chimeras were intermediate and the L3 chimeras had the extreme phenotype. The L2 chimeras were also confirmed by transmission of the revertant wild-type *FLO* allele to the progeny, since the gametes are derived from L2 cells. The expression of downstream target genes *DEF* and *PLENA* (*PLE*), B and C function homeotic genes, respectively, was also monitored by *in situ* hybridization in the different chimeric plants. In all of them, *DEF* and *PLE* were expressed in all layers, despite *FLO* being expressed in only a single layer. In some cases, however, the *DEF* or *PLE* expression domain was reduced, or the onset of expression was delayed, and these differences probably account for the observed homeotic phenotypes (Hantke *et al.*, 1995). Nonetheless, the chimera analysis clearly indicates that *FLO* has the capacity to signal between cell layers in the floral meristem to activate downstream transcriptional target genes.

Evidence for the nature of the layer-to-layer signal came from experiments where *LFY*, the *Arabidopsis* ortholog of *FLO*, was expressed specifically in the L1 layer using the tissue-specific *ML1* promoter (Sessions *et al.*, 2000). In a *lfy* mutant background, this expression was sufficient to fully rescue the mutant phenotype. *LFY* is normally expressed in all three layers of young floral primordia, but as expected in the ML1 transgenics, its mRNA accumulates only in the L1 layer. However, LFY protein is detected in all layers of the rescued floral meristems, indicating that LFY protein moves from the L1 to L2 and L3, and suggesting that this movement is required for its function (Sessions *et al.*, 2000). Interestingly, another TF APETALA1 (AP1), a member of the MADS domain family of TFs that has similar functions as LFY, behaves predominantly cell-autonomously when expressed in the L1, though in this study the distribution of AP1 protein was not measured (Sessions *et al.*, 2000). These findings indicate selectivity of nonautonomous regulation, and imply selective cell-to-cell movement of TFs, in line with the selective regulation of MADS domain and KN1 protein movement discussed earlier.

GFP–LFY fusion proteins were used to address the mechanism of LFY movement in floral meristems. These were first shown to functionally complement *lfy* mutants when expressed using either the *LFY* or *ML1* promoters (Wu *et al.*, 2003). Distribution of the fusion proteins expressed from the *ML1* promoter was imaged in floral meristems, and as expected both N- and C-terminal LFY–GFP fusions were detected in cell layers underlying the L1, indicating cell-to-cell movement. Movement of the N-terminal GFP–LFY fusion protein could be seen over a range of 3–4 cell layers, whereas the C-terminal LFY–GFP fusion moved up to 10 cell layers. This difference in movement correlates with subcellular distribution, as the C-terminal fusion protein shows a higher relative accumulation in the cytoplasm, and so may have more ready access to PDs (Wu *et al.*, 2002, 2003).

The distribution of GFP–LFY in floral meristems was compared to that of a GFP–GFP fusion protein and a viral MP–GFP fusion protein, all expressed from the L1-specific promoter. As expected, the MP–GFP fusion moved extensively in the apex, but somewhat surprisingly so did the GFP–GFP fusion, which has a molecular weight of 54 kDa. This result suggests that the SEL for nontargeted movement in the floral meristem is somewhat larger than in the mature leaf, where GFP–GFP is usually cell-autonomous (Crawford and Zambryski, 2001; Kim *et al.*, 2002b). However, not all fusions of this size could move in the floral meristem, for example, a functional AP1–GFP fusion did not move, despite being practically the same molecular weight as GFP–GFP at 55 kDa (Wu *et al.*, 2003). This lack of AP1–GFP movement is consistent with its cell-autonomous behavior (Sessions *et al.*, 2000), and also correlates with its accumulation predominantly in nuclei, where it is presumably limited in its ability to gain access to PDs. Further support for the idea that subcellular localization could limit cell-to-cell movement in the meristem was obtained from the addition of a nuclear localization sequence to the GFP–GFP fusion. This addition greatly hindered its movement, so that it was only detected in L1 nuclei and at a low level in the L2 when expressed from the *ML1* promoter (Wu *et al.*, 2003).

These results highlight an important consideration in regulation of PD movement, which is that if a protein is targeted with very high efficiency to a particular sub-cellular compartment, it is unlikely to be able to gain access to PD for movement (Crawford and Zambryski, 2000). However, the simple imaging of GFP fusions is probably not going to be sufficient to distinguish the targeting efficiency. For example, a protein could have high affinities for both nuclear targeting and for PD transport, and in this case would appear nuclear in both source and movement-targeted cells.

To address whether LFY protein might have any motif or domain that actively targets it to PDs, three overlapping deletions of the protein were expressed as GFP fusions using the *ML1* promoter (Wu *et al.*, 2003). All three fusions were able to move, although one of them formed aggregates and was probably unable to fold properly. These results suggest that either LFY does not contain a PD-targeting sequence, or that it contains redundant targeting sequences in different parts of the protein. If the first hypothesis is true, then LFY may move by a passive diffusion-based mechanism. Can this hypothesis be extended to other proteins? Although it is true that some proteins such as AP1 and AP3 did not show evidence for movement, it is possible that they are unable to move either because they are targeted with high affinity to the nucleus or to multi-protein complexes that are larger than the PD SEL. Clearly, more studies are required to establish whether the passive mode of trafficking suggested for LFY can be extended to a global model of protein trafficking in the meristem by a default passive-diffusion mechanism.

### 6.4.4 *Shortroot*

The *Arabidopsis* root is an outstanding model to understand the role of patterning systems in development, largely due to its simple radial symmetry of cell types that are easily imaged. Like the shoot, the root is also built from an organized meristem, in this case the root apical meristem. The major root tissues are the inner vascular tissues within the stele, surrounded by a single layer of endodermal cells, a layer of cortical cells and finally the epidermal cells making up the outermost layer (Figure 6.2a). This pattern is created through regular divisions of root meristem initial cells followed by cell fate specification in a position-dependent manner (Van den Berg *et al.*, 1995; Kidner *et al.*, 2000). A number of mutants affecting specific aspects of the root radial pattern have been isolated. In one of these, *shortroot* (*shr*), the cortex/endodermis initial daughter cells fail to divide, so the mutant root contains a single layer of cells that resembles cortex in place of the usual endodermis and cortical layers (Helariutta *et al.*, 2000). Therefore, *SHR* is required not only for asymmetric divisions of the cortex/endodermis initial daughter cells, but also for the specification of endodermal fate (Helariutta *et al.*, 2000).

*SHR* encodes a putative TF of the GRAS family, related to *SCARECROW* (*SCR*), which is also required for the division of the cortex–endodermal initial daughter cell and is probably a transcriptional target of *SHR* (Di Laurenzio *et al.*, 1996; Helariutta *et al.*, 2000). Surprisingly, *SHR* mRNA is not detected in the

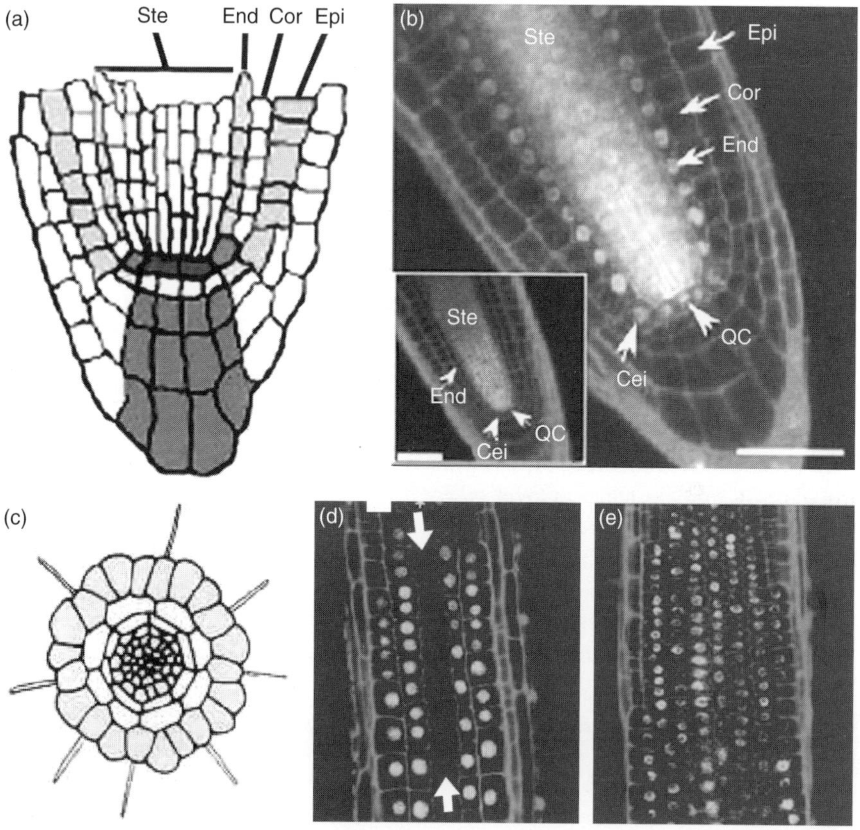


**Figure 6.2** Transcription factor movement in the root. A diagrammatic longitudinal section of the *Arabidopsis* root is shown in (a), with stele (Ste), endodermis (End), cortex (Cor) and epidermal (Epi) cell files labeled. In (b), the localization of SHR–GFP in stele and in endodermal cells is shown as bright staining; note that endodermal cells show localization predominantly in nuclei. Inset in (b) is an image of a cell-autonomous GFP expressed from the *SHR* promoter; note that the promoter is active only in the stele. QC = quiescent center, Cei = cortical–endodermal initial. (a,b) reproduced, with permission from Nature Publishing Group (http://www.nature.com/), Nakajima *et al.* (2001). In (c), a diagrammatic cross-section of the root shows alternating cell files with root hair projections. In (d), the expression of a cell-autonomous nuclear GFP from the *CAPRICE* (*CPC*) promoter shows expression only in non-hair cells; note that the file of hair cells (arrowed) has no expression. In (e), CPC–GFP expressed from the *CPC* promoter reveals protein localization in all cell files, presumably as a result of CPC–GFP trafficking into hair cells. (c–e) reproduced, with permission from The Company of Biologists Ltd., Wada *et al.* (2002).

cortex/endodermal initial cell, or in its daughter cells, where it functions. Instead, it is present in the internally adjacent cells of the stele (Helariutta *et al.*, 2000). The reason for this anomaly became obvious from experiments where a GFP fusion of SHR was expressed using the *SHR* promoter (*pSHR*). Expression of this GFP fusion could fully rescue the *shr* mutant phenotype, and microscopy of the *pSHR::SHR–GFP* roots revealed SHR–GFP fluorescence not only in the stele, where *SHR* mRNA is present, but also in the adjacent cell layer that includes the quiescent center, the cortex/endodermal initial and initial daughter cells and the endodermis (Figure 6.2b) (Nakajima *et al.*, 2001). Therefore, SHR–GFP traffics from the stele to a single adjacent layer of cells, where it functions to promote asymmetric cell division and endodermal cell fate specification.

To investigate further the underlying mechanism and range of movement, SHR was expressed ectopically using the *SCR* promoter (*pSCR*). This promoter drives expression specifically in the cell layer adjacent to the stele, in other words in those cells into which SHR normally traffics. Intriguingly, the transgenic plants developed several additional layers of cells outside of the endodermis, and each of these extra layers expressed endodermal cell fate markers (Nakajima *et al.*, 2001). The extra cell layers appear to be produced by an autocatalytic relay mechanism that depends on the fact that *SCR* is a downstream target of *SHR*. In the transgenic roots, SHR protein moves out of the layer of cells where it is being expressed under the control of pSCR, into the adjacent layer, where it activates the *SCR* promoter and thereby reactivates its own expression. This cycle reiterates to generate multiple additional cell layers, each of which acquire endodermal fate, due to the presence of SHR (Nakajima *et al.*, 2001).

To ask if the regulation of SHR movement was dependent on tissue-specific factors, *SHR–GFP* was also expressed using its own promoter in the *pSCR::SHR* transgenic plants. In these plants, SHR–GFP is still restricted in its movement to only a single cell layer surrounding the stele, despite the fact that the mutant seedlings form multiple endodermal layers. This observation suggests that the mechanism for restriction of SHR movement is not dependent on tissue-specific regulation at the endodermis–cortex boundary, since in this case, the cell layer outside of the endodermis resembles another endodermal layer. In other words, it is unlikely that specific PDs at the endodermis–cortex boundary are involved in the normal restriction of SHR movement to a single layer (Sena *et al.*, 2004).

SHR trafficking provides a robust mechanism for radial patterning of the root. Its expression in the stele and subsequent movement to the adjacent layer ensures the correct division of the cortex/endodermal initial cell through the action of the *SCR* target gene. SHR movement, therefore, provides the essential positional information to ensure that the endodermis layer is faithfully positioned adjacent to the stele. This is perhaps the clearest example to date of TF movement that is absolutely required for function.

Studies of SHR movement also raise important questions about the specificity and range of TF movement. For example, what governs the ability of SHR protein, but not the closely related SCR protein, to move? Also, is SHR movement regulated in a directional manner, as described for some of the TFs that traffic during shoot development? It will be of particular interest to understand how SHR trafficking stops after movement into a single cell layer outside the stele. One simple possibility is that further movement is restricted because of subcellular localization, since in the stele, the SHR–GFP fusion protein accumulates both in the cytoplasm and in the nuclei, whereas in the adjacent cell layer it is found only in nuclei (Figure 6.2b). Therefore, a highly efficient nuclear import mechanism in endodermal cells might limit the accessibility of SHR to PDs and prevent it from moving to the next layer of cells. Alternatively, SHR might be modified as it traffics from the stele to the neighboring cell layer in a way that prevents it from further movement.

To address some of these questions, SHR–GFP was expressed in other root tissues using specific promoters (Sena *et al.*, 2004). When expressed in the phloem companion cells within the stele, SHR–GFP did not move out of these cells, despite the fact that this cell type is particularly active in both passive and active PD transport. The fusion protein accumulated mostly in nuclei in the companion cells, so inaccessibility to PDs might explain this lack of movement. Similarly, when expressed in epidermal cells, SHR–GFP accumulates in nuclei and does not move to underlying cells. In this transgenic line, the epidermal cells expressing SHR developed some endodermal characters, indicating that movement is not a prerequisite for the ability of SHR to control cell fate (Sena *et al.*, 2004). The competence of cells outside of the endodermis (e.g. epidermal cells in this case) to respond to SHR expression also highlights the importance of restricting SHR protein movement into only the endodermal layer. If SHR was able to move freely into other cell layers, then the consequence would be catastrophic because cell fates would be mis-specified.

In summary, the intercellular movement of SHR is tightly regulated, as it only moves out of the stele, and then only to the single adjacent cell layer. Expression in companion cells or epidermal cells does not result in movement, at least of a SHR–GFP fusion (Sena *et al.*, 2004). What might be the mechanism of this tissue-specific regulation? One obvious factor, as also proposed for LFY, is subcellular localization, because in the stele cells, SHR–GFP accumulates in the cytoplasm and nucleus, whereas it becomes predominantly nuclear in the adjacent endodermal cells. Tissue-specific expression in other root cell types also leads to nuclear accumulation and no movement. Therefore, it is possible that a stele-specific factor is required to keep SHR in the cytoplasm and allow access to the PD translocation pathway. The subcellular localization hypothesis is supported by an unexpected result that in an *scr* mutant background, epidermally expressed SHR–GFP accumulates in the cytoplasm and nucleus, and is able to move into the adjacent underlying cell layer. Therefore, cytoplasmic localization appears to be necessary for SHR movement; however, it is not yet clear if it is sufficient. It may not be, because the product of one SHR mutant allele accumulates in the cytoplasm but does not move (Gallagher *et al.*, 2004), and this observation argues that SHR moves by a specific, regulated mechanism.

### *6.4.5 Caprice*

Roots produce cylindrical outgrowths of epidermal cells, called root hairs, in order to increase their surface area for nutrient and water uptake. In *Arabidopsis*, the root hairs are initiated from alternating vertical files of cells in the epidermis (Figure 6.2c), and this patterning is under genetic control. *GLABRA2* (*GL2*) is a negative regulator of root hair development, because in *gl2* loss-of-function mutants, all epidermal cells make hairs (Masucci *et al.*, 1996). In contrast, *CAPRICE* (*CPC*) is a positive regulator, as *cpc* mutants make only occasional hairs (Wada *et al.*, 1997). *gl2* is epistatic to *cpc*, since the double mutants make ectopic root hairs just like *gl2*, indicating that *CPC* probably acts upstream of *GL2* in the root hair patterning program (Wada *et al.*, 1997).

As predicted for a negative regulator, *GL2* is expressed mainly in the hairless cells of the epidermis. In *cpc* mutants, the *GL2* expression domain expands to include the hair cells also, leading to almost hairless roots. The simplest model to explain these interactions is that *CPC* should be expressed in the hair cells where it can repress *GL2* transcription. However, *CPC in situ* hybridizations and promoter analysis found that *CPC* is actually transcribed and its mRNA accumulates mainly in hairless cells, like *GL2* (Wada *et al.*, 2002). To study the distribution of CPC protein, the *CPC* promoter (*pCPC*) was used to drive expression of GFP or of a CPC–GFP fusion. Although the GFP expression was mostly in hairless cells (Figure 6.2d), reflecting the activity of *pCPC*, the CPC–GFP fusion protein accumulated in hair as well as hairless cells, suggesting that it traffics into the hair cells (Figure 6.2e; Wada *et al.*, 2002). A model to explain these observations is that CPC traffics from hairless to hair cells where it represses GL2 expression, resulting in hair outgrowth. Clearly, this model is an oversimplification, and one unanswered question is why does CPC not repress GL2 in the hairless cells where it is produced? Perhaps the relative levels in different cell types, or a modification of CPC during its cell-to-cell trafficking, are involved in the differential effects of CPC on GL2 expression in the hair and hairless cell types.

CPC is a very small protein of only 94 amino acids. However, it appears to move by a targeted mechanism, because the larger GFP protein does not move between epidermal cells, and CPC confers a gain of trafficking activity to GFP. It is also interesting to note that the mobile CPC–GFP fusion is predominantly nuclear, so in this case, subcellular localization does not appear to limit trafficking ability, though the range of movement that can be observed is only one cell.

### *6.4.6 Other candidate cell-to-cell trafficking TFs*

So far, I have discussed examples of TFs for which there is convincing experimental evidence for cell-to-cell movement. Of course, a number of developmental TFs are expressed and/or behave cell-autonomously, e.g. AP1 (Sessions *et al.*, 2000), AP3 (Jenik and Irish, 2001) and SCR (Di Laurenzio *et al.*, 1996). Therefore, movement is clearly a property only of specific TFs. However, how many of the hundreds of plant TFs might traffic is not known. Our level of understanding is partly limited by the low number of specific antibodies available for plant TFs. However, the continuing use of GFP fusions to study protein distribution will no doubt continue to reveal new candidate trafficking proteins, when used in combination with a cell-autonomous GFP or GUS reporter to assay promoter specificity.

Other nonautonomously acting TFs have been identified by mosaic analyses, and, of course, these are good candidates for trafficking proteins. These include the maize KN1-related homeodomain proteins GNARLY1 (Foster *et al.*, 1999) and ROUGHSHEATH1 (Becraft and Freeling, 1994), and LIGULELESS2, a basic leucine zipper protein required for ligule formation in maize (Harper and Freeling, 1996; Walsh *et al.*, 1998). Interestingly, LIGULELESS1, another protein required for ligule initiation, also encodes a TF but it functions cell-autonomously (Becraft *et al.*, 1990; Moreno *et al.*, 1997). Nonautonomous effects have also been reported for PHANTASTICA, a MYB-related protein (Waites *et al.*, 1998), and GRAMINIFOLIA, a YABBY TF (Golz *et al.*, 2004). These proteins control the dorsiventral polarity of shoot lateral organs. Nonautonomy of TF activity does not, of course, imply that the TF itself traffics. An excellent example of this is WUSCHEL (WUS), a homeodomain protein that functions nonautonomously to regulate stem cell maintenance in shoot and floral meristems (Laux *et al.*, 1996; Mayer *et al.*, 1998). WUS also acts nonautonomously during ovule development, but localization studies indicate that WUS protein does not traffic, and suggest that a downstream WUS target acts as the mobile signal (Gross-Hardt *et al.*, 2002). Another example is FUSCA3, a member of the plant specific B3 TF family that regulates developmental timing during shoot development. FUS3 controls leaf development nonautonomously, but the protein itself does not traffic, and in this case, the mobile signal is likely to be a hormone (Gazzarini *et al.*, 2004).

## 6.5 Long-distance signaling by TFs, TF mRNAs and miRNAs

The examples of TF movement and developmental signaling discussed so far occur over a very small range of one to a few cell diameters, or up to a few tens of microns. Some TFs however have nonautonomous effects that can operate over much longer distances. This is especially true of TFs involved in floral initiation, in cases where the gene is expressed and functions in the leaf, but its phenotypic consequence is at the plant apex where the meristem is induced to flower. For example, the maize *INDETERMINATE* gene is expressed in leaves and its expression induces flowering at the shoot apex (Colasanti *et al.*, 1998). ID encodes a zinc-finger DNA binding protein (Kozaki *et al.*, 2004). In *Arabidopsis*, *CONSTANS* regulates the photoperiod-response flowering pathway and also encodes a nuclear zinc-finger protein though its sequence is not closely related to ID (Putterill *et al.*, 1995). Grafting experiments indicate that CO also acts non-cell-autonomously, and in fact can induce flowering when expressed only in phloem companion cells (An *et al.*, 2004). However, the analysis of plants expressing GFP-CO indicates that CO protein itself does not move, or at least its movement cannot be detected at the level of sensitivity of this system (An *et al.*, 2004). Therefore, *CO* is likely to be involved in the synthesis or transport of a downstream signal that travels to the apex to induce flowering. *ID*- and *CO*-dependent long-range signals are therefore candidates for the elusive flowering hormone florigen (Zeevaart, 1962).

Although there is good evidence that the CO protein is nonmobile, another possibility is that TF mRNAs act as long-range signals. Sequencing of cDNAs prepared from phloem sap revealed the presence of several TFs mRNAs, and grafting experiments indicate that these are mobile over long distances (Ruiz-Medrano *et al.*, 1999). These phloem-mobile TF or transcription-associated mRNAs include members of the NAC, STM, WRKY and RING families, and it will be interesting to determine the functional significance of movement of these transcripts.

One candidate for a functionally significant phloem mobile TF mRNA is the *KN1*-related *MOUSE EARS* (*Me*) mRNA in tomato (Kim *et al.*, 2001). The *Me* mutation results in enhanced leaf lobing, and is caused by a chromosomal rearrangement that fused the *LeT6 KN1*-related homeobox gene to the 5′ coding and promoter region of *PYROPHOSPHATE-DEPENDENT PHOSPHOFRUCTOKINASE* (*PFP*), which encodes a metabolic enzyme. This fusion results in overexpression of *LeT6*, leading to the altered leaf morphology.

In grafting experiments where wild-type shoots were grafted onto *Me* stocks, remarkably, the *Me* leaf phenotype was graft transmissible; in other words, the wild-type scion developed the typical *Me* morphology. An indication of what the graft transmissible signal may be came from *in situ* polymerase chain reaction experiments, where *Me* transcripts were detected in the wild-type scion. These observations suggest that the altered morphology of the grafted scions is caused by the phloem-mobile *Me* RNA (Kim *et al.*, 2001). However, a causal relationship between presence of the *Me* mRNA and the altered morphology is difficult to prove, and an alternative possibility is that some other signal, such as a cytokinin or giberellin, acts as the nonautonomous signal. These hormones are mis-expressed in KNOX overexpressing plants, and may also be phloem mobile (Ori *et al.*, 1999; Sakamoto *et al.*, 2001; Hay *et al.*, 2002).

Although awaiting direct proof, the possibility that phloem-mobile TF mRNAs regulate plant development is extremely exciting, and suggests a unique RNA-based systemic signaling pathway (Jorgensen *et al.*, 1998). Another possible mechanism of long-range TF signaling and regulation is by microRNAs. These short regulatory RNAs target several developmental TFs, and were recently found to be present and mobile in the phloem (Juarez *et al.*, 2004; Yoo *et al.*, 2004). By regulating mRNA stability or translation, miRNAs are extremely powerful modulators of gene expression. Despite their small size of around 21–25 nucleotides in length, trafficking of miRNAs through PD appears to be regulated, as a small RNA binding protein that is required for cell-to-cell transport of miRNAs in microinjection assays was recently identified (Yoo *et al.*, 2004). This protein is present in phloem sap, so may act as a carrier to transport miRNAs systemically throughout the plant. This is a fascinating discovery and it will be interesting to figure out how widespread this phenomenon is, as well as which specific processes may be regulated by systemic miRNA trafficking.

## 6.6 Mechanisms of TF movement

Two possible mechanisms exist for protein transport through PDs. The first is nontargeted movement, which probably occurs by passive diffusion. Nontargeted movement is governed simply by the size of the protein, which should be lower than the PD SEL, as well as its accessibility to PDs. A good example of a protein that moves in a nontargeted fashion is the GFP (Imlau *et al.*, 1999; Oparka *et al.*, 1999; Itaya *et al.*, 2000; Crawford and Zambryski, 2001; Kim *et al.*, 2002b). An example of a TF that appears to move by a nontargeted mechanism is LFY, because no trafficking signal sequence could be found in this protein, and because the range of LFY movement in the meristem correlates with the relative amount of protein in the cytoplasm (Wu *et al.*, 2003). How widespread is the nontargeted movement of TFs? The fact that many of them are cell-autonomous and do not traffic would argue against a general nontargeted mechanism. However, this lack of movement could still be consistent with a nontargeted mechanism if those proteins do not move simply because they are present in complexes larger than the PD SEL, or because they localize to the nucleus or another site away from PDs (Crawford and Zambryski, 2000; Wu *et al.*, 2002). According to the nontargeted movement model, the default state is for a protein to move, unless it is actively retained in the cell.

The second mechanism of movement is through an active and targeted process. There are several ways in which such a mechanism could operate, but perhaps the simplest to imagine would be through a receptor, at or close to PDs, that a TF has to interact with in order to traffic to the neighboring cell. Evidence for receptor-mediated transport of TFs has come from studies on KN1 trafficking. For example, the ability of KN1 to increase the SEL, and the observation that a specific mutation in KN1 can block its ability to traffic or to increase the PD SEL both argue for an active mechanism of transport (Lucas *et al.*, 1995). Biochemical studies also indicate the involvement of a chaperone activity and a putative receptor in KN1 trafficking (Kragler *et al.*, 1998b). Importantly, competition experiments using gold-tagged KN1 suggest that KN1 shares a common PD receptor with a viral MP (Kragler *et al.*, 1998b). Specific interaction between KN1 and PDs is also highlighted by the discovery of a peptide antagonist of KN1 that can interfere with its ability to increase the SEL or to transport its mRNA (Kragler *et al.*, 2000).

To date, only a single candidate PD receptor, Non Cell Autonomous Pathway Protein1 (NCAPP1) has been isolated (Jackson and Kim, 2003; Lee *et al.*, 2003). Although this protein was purified by virtue of its interaction with a plant homolog of a viral MP, it may also regulate the trafficking of developmental TFs. Silencing of NCAPP1, or expression of a dominant negative version, resulted in a range of developmental phenotypes, some of which resembled LFY overexpression. These results suggest either that interference with NCAPP1 expression changes the PD SEL for passive transport of LFY, or that NCAPP1 may be involved in active transport of LFY.

Although the available evidence for cell-to-cell transport of TFs suggests that they move through PDs, one should not forget that TF trafficking between animal cells has also been reported. In this case, specific and regulated transport of homeodomain proteins occurs by an unconventional secretion and uptake mechanism (Maizel *et al.*, 2002). The ability of KN1 to increase PD SEL, and the localization of GFP–KN1 to punctate sites in the cell wall argue against such a mechanism of transport for KN1, but the possibility should be considered.

Clearly, we are just beginning to understand the mechanisms of TF trafficking. Efforts currently under way in several labs to find PD-targeting sequences, as well as continued efforts in biochemical isolation of PD components, will hopefully fill in the mechanistic details in the near future.

## 6.7 Perspectives

The last decade has seen the birth of a new paradigm in plant biology, that of developmental signaling by cell-to-cell movement of TFs and their mRNAs. Although the first reports were greeted with healthy skepticism, this phenomenon is now indisputable. However, questions remain about how widespread TF trafficking is and its mechanism. The continued application of diverse cellular, genetic and genomic tools will hopefully answer these questions and shed further light on this fascinating new mechanism of developmental signaling.

## Acknowledgements

Research in the Jackson lab on PD is supported by the National Science Foundation Integrative Plant Biology program, and by CSHL institutional funds.

## References

An, H., Roussot, C., Suarez-Lopez, P. *et al.* (2004) 'CONSTANS acts in the phloem to regulate a systemic signal that induces photoperiodic flowering of *Arabidopsis*', *Development* **131**(15), 3615–3626.

Becker, A. and Theissen, G. (2003) 'The major clades of MADS-box genes and their role in the development and evolution of flowering plants', *Molecular Phylogenetics and Evolution* **29**(3), 464–489.

Becraft, P.W. and Freeling, M. (1994) 'Genetic analysis of Rough sheath 1 developmental mutants of maize', *Genetics* **136**(1), 295–311.

Becraft, P.W., Bongard-Pierce, D.K., Sylvester, A.W., Poethig, R.S. and Freeling, M. (1990) 'The liguleless-1 gene acts tissue specifically in maize leaf development', *Developmental Biology* **141**(1), 220–232.

Bouhidel, K. and Irish, V.F. (1996) 'Cellular interactions mediated by the homeotic PISTILLATA gene determine cell fate in the *Arabidopsis* flower', *Developmental Biology* **174**(1), 22–31.

Carpenter, R. and Coen, E.S. (1990) 'Floral homeotic mutations produced by transposon-mutagenesis in *Antirrhinum majus*', *Genes and Development* **4**(9), 1483–1493.

Carpenter, R. and Coen, E.S. (1995) 'Transposon induced chimeras show that floricaula, a meristem identity gene, acts non-autonomously between cell layers', *Development* **121**(1), 19–26.

Cilia, M.L. and Jackson, D. (2004) 'Plasmodesmata form and function', *Current Opinion in Cell Biology* **16**(5), 500–506.

Coen, E.S. and Meyerowitz, E.M. (1991) 'The war of the whorls: genetic interactions controlling flower development', *Nature* **353**(6339), 31–37.

Coen, E.S., Romero, J.M., Doyle, S., Elliott, R., Murphy, G. and Carpenter, R. (1990) 'Floricaula: a homeotic gene required for flower development in *Antirrhinum majus*', *Cell* **63**(6), 1311–1322.

Colasanti, J., Yuan, Z. and Sundaresan, V. (1998) 'The indeterminate gene encodes a zinc finger protein and regulates a leaf-generated signal required for the transition to flowering in maize', *Cell* **93**(4), 593–603.

Crawford, K.M. and Zambryski, P.C. (2000) 'Subcellular localization determines the availability of non-targeted proteins to plasmodesmatal transport', *Current Biology* **10**(17), 1032–1040.

Crawford, K.M. and Zambryski, P.C. (2001) 'Non-targeted and targeted protein movement through plasmodesmata in leaves in different developmental and physiological states', *Plant Physiology* **125**(4), 1802–1812.

Di Laurenzio, L., Wysocka-Diller, J., Malamy, J.E. *et al.* (1996) 'The SCARECROW gene regulates an asymmetric cell division that is essential for generating the radial organization of the *Arabidopsis* root', *Cell* **86**(3), 423–433.

Ding, B., Haudenshield, J.S., Hull, R.J., Wolf, S., Beachy, R.N. and Lucas, W.J. (1992) 'Secondary plasmodesmata are specific sites of localization of the tobacco mosaic virus movement protein in transgenic tobacco plants', *The Plant Cell* **4**(8), 915–928.

Dolan, L., Linstead, P., Kidner, C., Boudonck, K., Cao, X.F. and Berger, F. (1998) 'Cell fate in plants. Lessons from the *Arabidopsis* root', *Symposia of the Society for Experimental Biology* **51**, 11–17.

Dubnau, J. and Struhl, G. (1996) 'RNA recognition and translational regulation by a homeodomain protein', *Nature* **379**(6567), 694–699.

Efremova, N., Perbal, M.C., Yephremov, A., Hofmann, W.A., Saedler, H. and Schwarz-Sommer, Z. (2001) 'Epidermal control of floral organ identity by class B homeotic genes in *Antirrhinum* and *Arabidopsis*', *Development* **128**(14), 2661–2671.

Egea-Cortines, M., Saedler, H. and Sommer, H. (1999) 'Ternary complex formation between the MADS-box proteins SQUAMOSA, DEFICIENS and GLOBOSA is involved in the control of floral architecture in *Antirrhinum majus*', *EMBO Journal* **18**(19), 5370–5379.

Foster, T., Veit, B. and Hake, S. (1999) 'Mosaic analysis of the dominant mutant, Gnarley1-R, reveals distinct lateral and transverse signaling pathways during maize leaf development', *Development* **126**(2), 305–313.

Gallagher, K.L., Paquette, A.J., Nakajima, K. and Benfey, P.N. (2004) 'Mechanisms regulating SHORT-ROOT intercellular movement', *Current Biology*, **14**, 1847–1851.

Gazzarrini, S., Tsuchiya, Y., Lumba, S., Okamoto, M. and McCourt, P. (2004) 'The transcription factor FUSCA3 controls developmental timing in *Arabidopsis* through the hormones gibberellin and abscisic acid', *Developmental Cell*, **7**(3), 373–385.

Golz, J.F., Roccaro, M., Kuzoff, R. and Hudson, A. (2004) 'GRAMINIFOLIA promotes growth and polarity of *Antirrhinum* leaves', *Development* **131**(15), 3661–3670.

Gross-Hardt, R., Lenhard, M. and Laux, T. (2002) 'WUSCHEL signaling functions in inter-regional communication during *Arabidopsis* ovule development', *Genes and Development* **16**(9), 1129–1138.

Hake, S. (2001) 'Transcription factors on the move', *Trends in Genetics* **17**(1), 2–3.

Hake, S. and Char, B.R. (1997) 'Cell–cell interactions during plant development', *Genes and Development* **11**(9), 1087–1097.

Hake, S. and Freeling, M. (1986) 'Analysis of genetic mosaics shows that the extra epidermal cell divisions in *Knotted* mutant maize plants are induced by adjacent mesophyll cells', *Nature* **320**(6063), 621–623.

Hantke, S.S., Carpenter, R. and Coen, E.S. (1995) 'Expression of floricaula in single cell layers of periclinal chimeras activates downstream homeotic genes in all layers of floral meristems', *Development* **121**(1), 27–35.

Harper, L. and Freeling, M. (1996) 'Interactions of liguleless1 and liguleless2 function during ligule induction in maize', *Genetics* **144**(4), 1871–1882.

Hay, A., Kaur, H., Phillips, A., Hedden, P., Hake, S. and Tsiantis, M. (2002) 'The gibberellin pathway mediates KNOTTED1-type homeobox function in plants with different body plans', *Current Biology* **12**(18), 1557–1565.

Haywood, V., Kragler, F. and Lucas, W.J. (2002) 'Plasmodesmata: pathways for protein and ribonucleoprotein signaling', *The Plant Cell* **14**(Suppl.), 303–325.

Helariutta, Y., Fukaki, H., Wysocka-Diller, J. *et al.* (2000) 'The SHORT-ROOT gene controls radial patterning of the *Arabidopsis* root through radial signaling', *Cell* **101**(5), 555–567.

Imlau, A., Truernit, E. and Sauer, N. (1999) 'Cell-to-cell and long-distance trafficking of the green fluorescent protein in the phloem and symplastic unloading of the protein into sink tissues', *The Plant Cell* **11**(3), 309–322.

Itaya, A., Liang, G., Woo, Y.M., Nelson, R.S. and Ding, B. (2000) 'Nonspecific intercellular protein trafficking probed by green fluorescent protein', *Protoplasma* **213**, 165–175.

Jackson, D. (2002) 'Double labeling of KNOTTED1 mRNA and protein reveals multiple potential sites of protein trafficking in the shoot apex', *Plant Physiology* **129**(4), 1423–1429.

Jackson, D. and Hake, S. (1997) 'Morphogenesis on the move: cell-to-cell trafficking of plant regulatory proteins', *Current Opinions in Genetics and Development* **7**(4), 495–500.

Jackson, D. and Kim, J.Y. (2003) 'Intercellular signaling: an elusive player steps forth', *Current Biology* **13**(9), 349–350.

Jackson, D., Veit, B. and Hake, S. (1994) 'Expression of maize *KNOTTED1* related homeobox genes in the shoot apical meristem predicts patterns of morphogenesis in the vegetative shoot', *Development* **120**(2), 405–413.

Jenik, P.D. and Irish, V.F. (2001) 'The *Arabidopsis* floral homeotic gene APETALA3 differentially regulates intercellular signaling required for petal and stamen development', *Development* **128**(1), 13–23.

Jorgensen, R.A., Atkinson, R.G., Forster, R.L.S. and Lucas, W.J. (1998) 'An RNA-based information superhighway in plants', *Science* **279**(5356), 1486–1487.

Juarez, M.T., Kui, J.S., Thomas, J., Heller, B.A. and Timmermans, M.C. (2004) 'MicroRNA-mediated repression of rolled leaf1 specifies maize leaf polarity', *Nature* **428**(6978), 84–88.

Kidner, C., Sundaresan, V., Roberts, K. and Dolan, L. (2000) 'Clonal analysis of the *Arabidopsis* root confirms that position, not lineage, determines cell fate', *Planta* **211**(2), 191–199.

Kim, I., Hempel, F.D., Sha, K., Pfluger, J. and Zambryski, P.C. (2002a) 'Identification of a developmental transition in plasmodesmatal function during embryogenesis in *Arabidopsis thaliana', Development* **129**(5), 1261–1272.

Kim, J.Y., Yuan, Z. and Jackson, D. (2003) 'Developmental regulation and significance of KNOX protein trafficking in *Arabidopsis*', *Development* **130**(18), 4351–4362.

Kim, J.Y., Yuan, Z.A., Cilia, M., Khalfan-Jagani, Z. and Jackson, D. (2002b) 'Intercellular trafficking of a KNOTTED1 green fluorescent protein fusion in the leaf and shoot meristem of *Arabidopsis*', *Proceedings of the National Academy of Sciences USA* **99**(6), 4103–4108.

Kim, M., Canio, W., Kessler, S. and Sinha, N. (2001) 'Developmental changes due to long-distance movement of a homeobox fusion transcript in tomato', *Science* **293**(5528), 287–289.

Kozaki, A., Hake, S. and Colasanti, J. (2004) 'The maize ID1 flowering time regulator is a zinc finger protein with novel DNA binding properties', *Nucleic Acids Research* **32**(5), 1710–1720.

Kragler, F., Lucas, W.J. and Monzer, J. (1998a) 'Plasmodesmata: dynamics, domains and patterning', *Annals of Botany* **81**(1), 1–10.

Kragler, F., Monzer, J., Shash, K., Xoconostle-Cázares, B. and Lucas, W.J. (1998b) 'Cell-to-cell transport of proteins: requirement for unfolding and characterization of binding to a putative plasmodesmal receptor', *Plant Journal* **15**(3), 367–381.

Kragler, F., Monzer, J., Xoconostle-Cázares, B. and Lucas, W.J. (2000) 'Peptide antagonists of the plasmodesmal macromolecular trafficking pathway', *EMBO Journal* **19**(12), 2856–2868.

Laux, T., Mayer, K.F.X., Berger, J. and Jurgens, G. (1996) 'The WUSCHEL gene is required for shoot and floral meristem integrity in *Arabidopsis*', *Development* **122**(1), 87–96.

Lee, J.Y., Yoo, B.C., Rojas, M.R., Gomez-Ospina, N., Staehelin, L.A. and Lucas, W.J. (2003) 'Selective trafficking of non-cell-autonomous proteins mediated by NtNCAPP1', *Science* **299**(5605), 392–396.

Long, J.A., Moan, E.I., Medford, J.I. and Barton, M.K. (1996) 'A member of the KNOTTED class of homeodomain proteins encoded by the SHOOTMERISTEMLESS gene of *Arabidopsis*', *Nature* **379**, 66–69.

Lucas, W.J., Bouche-Pillon, S., Jackson, D.P. *et al.* (1995) 'Selective trafficking of KNOTTED1 homeodomain protein and its mRNA through plasmodesmata', *Science* **270**, 1980–1983.

Lucas, W.J., Ding, B. and Van der Schoot, C. (1993) 'Plasmodesmata and the supracellular nature of plants', *New Phytologist* **125**, 435–476.

Maizel, A., Tassetto, M., Filhol, O., Cochet, C., Prochiantz, A. and Joliot, A. (2002) 'Engrailed homeoprotein secretion is a regulated process', *Development* **129**(15), 3545–3553.

Marcotrigiano, M. (2001) 'Genetic mosaics and the analysis of leaf development', *International Journal of Plant Sciences* **162**(3), 513–525.

Masucci, J.D., Rerie, W.G., Foreman, D.R. *et al.* (1996) 'The homeobox gene GLABRA2 is required for position-dependent cell differentiation in the root epidermis of *Arabidopsis thaliana*', *Development* **122**(4), 1253–1260.

Mayer, K.F.X., Schoof, H., Haecker, A., Lenhard, M., Jurgens, G. and Laux, T. (1998) 'Role of WUSCHEL in regulating stem cell fate in the *Arabidopsis* shoot meristem', *Cell* **95**(6), 805–815.

Mezitt, L.A. and Lucas, W.J. (1996) 'Plasmodesmal cell-to-cell transport of proteins and nucleic acids', *Plant Molecular Biology* **32**(1), 251–273.

Moreno, M.A., Harper, L.C., Krueger, R.W., Dellaporta, S.L. and Freeling, M. (1997) 'Liguleless1 encodes a nuclear-localized protein required for induction of ligules and auricles during maize leaf organogenesis', *Genes and Development* **11**(5), 616–628.

Nakajima, K., Sena, G., Nawy, T. and Benfey, P.N. (2001) 'Intercellular movement of the putative transcription factor SHR in root patterning', *Nature* **413**(6853), 307–311.

Oparka, K.J. (2004) 'Getting the message across: how do plant cells exchange macromolecular complexes?', *Trends in Plant Science* **9**(1), 33–41.

Oparka, K.J., Roberts, A.G., Boevink, P. *et al.* (1999) 'Simple, but not branched, plasmodesmata allow the nonspecific trafficking of proteins in developing tobacco leaves', *Cell* **97**(6), 743–754.

Ori, N., Juarez, M.T., Jackson, D., Yamaguchi, J., Banowetz, G.M. and Hake, S. (1999) 'Leaf senescence is delayed in tobacco plants expressing the maize homeobox gene knotted1 under the control of a senescence-activated promoter', *The Plant Cell* **11**(6), 1073–1080.

Parcy, F., Nilsson, O., Busch, M.A., Lee, I. and Weigel, D. (1998) 'A genetic framework for floral patterning', *Nature* **395**(6702), 561–566.

Perbal, M.C., Haughn, G., Saedler, H. and Schwarz-Sommer, Z. (1996) 'Non-cell-autonomous function of the *Antirrhinum* floral homeotic proteins DEFICIENS and GLOBOSA is exerted by their polar cell-to-cell trafficking', *Development* **122**(11), 3433–3441.

Putterill, J., Robson, F., Lee, K., Simon, R. and Coupland, G. (1995) 'The CONSTANS gene of *Arabidopsis* promotes flowering and encodes a protein showing similarities to zinc finger transcription factors', *Cell* **80**(6), 847–857.

Reiser, L., Sanchez-Baracaldo, P. and Hake, S. (2000) 'Knots in the family tree: evolutionary relationships and functions of knox homeobox genes', *Plant Molecular Biology* **42**(1), 151–166.

Riechmann, J.L., Heard, J., Martin, G. *et al.* (2000) '*Arabidopsis* transcription factors: genome-wide comparative analysis among eukaryotes', *Science* **290**(5499), 2105–2110.

Ruiz-Medrano, R., Xoconostle-Cázares, B. and Lucas, W.J. (1999) 'Phloem long-distance transport of CmNACP mRNA: implications for supracellular regulation in plants', *Development* **126**(20), 4405–4419.

Sakamoto, T., Kamiya, N., Ueguchi-Tanaka, M., Iwahori, S. and Matsuoka, M. (2001) 'KNOX homeodomain protein directly suppresses the expression of a gibberellin biosynthetic gene in the tobacco shoot apical meristem', *Genes and Development* **15**(5), 581–590.

Satina, S., Blakeslee, A.F. and Avery, A.G. (1940) 'Demonstration of the three germ layers in the shoot apex of *Datura* by means of induced polyploidy in periclinal chimeras', *American Journal of Botany* **27**, 895–905.

Sena, G., Jung, J.W. and Benfey, P.N. (2004) 'A broad competence to respond to SHORT ROOT revealed by tissue-specific ectopic expression', *Development* **131**(12), 2817–2826.

Sessions, A., Yanofsky, M.F. and Weigel, D. (2000) 'Cell–cell signaling and movement by the floral transcription factors LEAFY and APETALA1', *Science* **289**(5480), 779–782.

Sinha, N. and Hake, S. (1990) 'Mutant characters of knotted maize leaves are determined in the innermost tissue layers', *Developmental Biology* **141**(1), 203–210.

Sinha, N., Hake, S. and Freeling, M. (1993a) 'Genetic and molecular analysis of leaf development', *Current Topics in Developmental Biology* **28**, 47–80.

Sinha, N.R., Williams, R.E. and Hake, S. (1993b) 'Overexpression of the maize homeo box gene, *KNOTTED-1*, causes a switch from determinate to indeterminate cell fates', *Genes and Development* **7**(5), 787–795.

Smith, L.G., Greene, B., Veit, B. and Hake, S. (1992) 'A dominant mutation in the maize homeobox gene, *Knotted-1*, causes its ectopic expression in leaf cells with altered fates', *Development* **116**(1), 21–30.

Sommer, H., Beltran, J.P., Huijser, P. *et al.* (1990) 'Deficiens, a homeotic gene involved in the control of flower morphogenesis in *Antirrhinum majus*: the protein shows homology to transcription factors', *EMBO Journal* **9**(3), 605–613.

Trobner, W., Ramirez, L., Motte, P. *et al.* (1992) 'GLOBOSA: a homeotic gene which interacts with DEFICIENS in the control of *Antirrhinum* floral organogenesis', *EMBO Journal* **11**(13), 4693–4704.

Van den Berg, C., Willemsen, V., Hage, W., Weisbeek, P. and Scheres, B. (1995) 'Cell fate in the *Arabidopsis* root meristem determined by directional signaling', *Nature* **378**(6552), 62–65.

Vollbrecht, E., Reiser, L. and Hake, S. (2000) 'Shoot meristem size is dependent on inbred background and presence of the maize homeobox gene, knotted1', *Development* **127**(14), 3161–3172.

Vollbrecht, E., Veit, B., Sinha, N. and Hake, S. (1991) 'The developmental gene Knotted-1 is a member of a maize homeobox gene family', *Nature* **350**(6315), 241–243.

Wada, T., Kurata, T., Tominaga, R. *et al.* (2002) 'Role of a positive regulator of root hair development, CAPRICE, in *Arabidopsis* root epidermal cell differentiation', *Development* **129**(23), 5409–5419.

Wada, T., Tachibana, T., Shimura, Y. and Okada, K. (1997) 'Epidermal cell differentiation in *Arabidopsis* determined by a Myb homolog, CPC', *Science* **277**(5329), 1113–1116.

Waigmann, E., Lucas, W.J., Citovsky, V. and Zambryski, P.C. (1994) 'Direct functional assay for tobacco mosaic virus cell-to-cell movement protein and identification of a domain involved in increasing plasmodesmal permeability', *Proceedings of the National Academy of Sciences USA* **91**, 1433–1437.

Waites, R., Selvadurai, H.R., Oliver, I.R. and Hudson, A. (1998) 'The PHANTASTICA gene encodes a MYB transcription factor involved in growth and dorsoventrality of lateral organs in *Antirrhinum*', *Cell* **93**(5), 779–789.

Walsh, J., Waters, C.A. and Freeling, M. (1998) 'The maize gene liguleless2 encodes a basic leucine zipper protein involved in the establishment of the leaf blade–sheath boundary', *Genes and Development* **12**(2), 208–218.

Weigel, D., Alvarez, J., Smyth, D.R., Yanofsky, M.F. and Meyerowitz, E.M. (1992) 'LEAFY controls floral meristem identity in *Arabidopsis*', *Cell* **69**(5), 843–859.

Wolf, S., Deom, C.M., Beachy, R. and Lucas, W.J. (1991) 'Plasmodesmatal function is probed using transgenic tobacco plants that express a virus movement protein', *The Plant Cell* **3**(6), 593–604.

Wolf, S., Deom, C.M., Beachy, R.N. and Lucas, W.J. (1989) 'Movement protein of tobacco mosaic virus modifies plasmodesmata size exclusion limit', *Science* **246**, 377–379.

Wu, X., Weigel, D. and Wigge, P.A. (2002) 'Signaling in plants by intercellular RNA and protein movement', *Genes and Development* **16**(2), 151–158.

Wu, X.L., Dinneny, J.R., Crawford, K.M. *et al.* (2003) 'Modes of intercellular transcription factor movement in the *Arabidopsis* apex', *Development* **130**(16), 3735–3745.

Yoo, B.C., Kragler, F., Varkonyi-Gasic, E. *et al.* (2004) 'A systemic small RNA signaling system in plants', *The Plant Cell* **16**(8), 1979–2000.

Zachgo, S., Silva Ede, A., Motte, P., Trobner, W., Saedler, H. and Schwarz-Sommer, Z. (1995) 'Functional analysis of the *Antirrhinum* floral homeotic DEFICIENS gene *in vivo* and *in vitro* by using a temperature-sensitive mutant', *Development* **121**(9), 2861–2875.

Zeevaart, J.A.D. (1962) 'Physiology of flowering', *Science* **13**, **7**723–7731.

# 7 Role of plasmodesmata in solute loading and unloading

Alexander Schulz

## 7.1 Introduction

In the adult plant, the symplasm is divided into different domains that reflect the different specialisations that plant organs and tissues achieve during their development. The domain borders are characterised by tightly controlled plasmodesmata that may have a smaller size exclusion limit (SEL; see Erwee and Goodwin, 1985) than those in cells within the domain. Structurally, plasmodesmata at domain borders may look similar to those within a domain.

As a general phenomenon, it might be taken that plant development is accompanied by more and more control of solute fluxes in the symplasm. Some cells, such as root hairs and leaf guard cells, become totally excluded from the symplasm by occlusion or by loss of functional plasmodesmata. In contrast, preferential pathways are established with high symplasmic connectivity. A high connectivity for small solutes can be formed by many constricted plasmodesmata in a given interface, or by a relatively few dilated plasmodesmata. Experimentally, it is not trivial to discriminate between these possibilities, since the fluorescent tracers available (fluorescein-coupled oligopeptides, dextrans or GFP molecules) interact differently with plasmodesmal proteins and are not natural constituents of plant cells (see Terry and Robards, 1987; Goodwin and Cantrill, 1999).

At the cell-to-cell level, membrane transport is a competing pathway for intercellular transport via plasmodesmata. The former is characterised by two membrane steps – release from one cell and uptake into another – and depends on the presence of solute-specific membrane channels and carriers. Because of its specificity and limitations in capacity, membrane transport is more regulated than plasmodesmatal transport, giving the cell a large degree of control of what to release and what to take up. For transport against a solute concentration gradient, the only direct option is given by plasma membrane transporters. Uphill transport depends upon plasma membrane proton pumps and is always energy consuming (Sondergaard *et al.*, 2004). The question of whether a given transport event between cells uses the apoplasmic or the symplasmic route cannot easily be solved for many cells. Indication for apoplasmic transport is, however, given where cells show an increase of their plasma membrane surface by wall invaginations (transfer cells), or when cells lack functional plasmodesmata totally (guard cells). Neither the localisation of transporter proteins in the plasma membrane nor a high frequency of plasmodesmata can be taken as evidence for their respective functionality.

For long-distance transport of solutes, the phloem is specialised into a well-connected symplasmic domain. Not only do the sieve elements (SEs) have a much reduced cytoplasm without nucleus and vacuoles, but they also have wide sieve pores. Sieve pores can be considered as specialised plasmodesmata with a 10–300-fold widened diameter (from 50 nm to 0.5 to 15 μm, depending on the plant species; Schulz, 1998). Together, reduced cytoplasm and sieve pores ensure a low-resistance pathway for solutes from the photosynthetic plant organs (sources) to the assimilate-consuming (or storing) organs (sinks).

While short-distance transport of solutes in the symplasm is diffusional and solute non-specific, long-distance transport is solute-specific, requires energy and depends on the accumulation of solutes in source organs and their release in sink organs. The significance of the phloem for signal transport and systemic silencing is presented in Chapters 8 and 10, respectively. This chapter will focus on the role of plasmodesmata in solute loading in collection phloem, in the lateral exchange of solutes along the transport phloem and in solute unloading in sink tissues.

Entry of assimilates into the sieve tube system is a decisive checkpoint for long-distance transport. Assimilates might arrive there via an entirely symplasmic route or leave the symplasm temporarily for an apoplasmic step. The generally accepted mechanism of phloem transport by osmotically generated pressure flow involves the participation of membrane transport, even when assimilates are not leaving the symplasm from the mesophyll into the sieve tubes. The high sugar concentration in source sieve tubes osmotically attracts water which can cross the plasma membrane of the sieve element/companion cell complex (SECCC) via aquaporin proteins. Permeability of the plasma membrane for water but not (or only weakly) for sugars allows the build-up of hydrostatic pressure in the sieve tube system (Patrick *et al.*, 2001).

The phloem translocates sugars, amino acids, nucleic acids, proteins and RNAs. This chapter will focus on sugars, since they are the major osmolytes of the phloem sap. Amino acids and potassium are also important phloem constituents and contribute to the osmotic potential. Osmolyte concentrations in the phloem identify source and sink tissues and decide over transport direction and rate. Fine tuning of amino acid and potassium concentration in relation to sugars can be considered to occur by specific transporter proteins in the SECCC plasma membrane and by xylem/phloem exchange (Patrick *et al.*, 2001). For transport of RNA and non-cell-autonomous proteins in the phloem, the reader is referred to Chapter 8, and to recent reviews (Thompson, 1999; Thompson and Schulz, 1999; Haywood *et al.*, 2002; Oparka, 2004; Yoo *et al.*, 2004).

## 7.2 Role of plasmodesmata in collection phloem

Solutes produced by photosynthesis reach the phloem by cell-to-cell transport from mesophyll to bundle sheath. Here they may or may not pass vascular parenchyma cells before they enter the SECCC. Transport up to this interface is termed pre-phloem transport; phloem loading denotes only the entry of the solutes into the SECCC.

After discussing the modes of phloem loading, their evolution and the putative default mode, the last part of this section will consider the significance of plasmodesmatal branching.

### 7.2.1 Pre-phloem transport

The architecture and arrangement of veins in source leaves are optimised to keep the distance between any mesophyll cell and the nearest minor vein minimal. Some pre-phloem pathways are only three–six mesophyll cells from the most distant cell to the bundle sheath (e.g. Haritatos *et al.*, 2000b). Electron-microscopic (EM)-studies of developing leaves, and the spread of tracers and viruses, indicated that the smallest order veins are specialised for loading and only export assimilates, while the midrib and larger veins (class I–III) change their role from being delivery phloem in small sink leaves to being transport phloem in mature source leaves (Ding *et al.*, 1988; Roberts *et al.*, 1997). This was confirmed by expression studies with phloem-specific reporter-gene constructs (SUC2, galactinol synthase) (Oparka *et al.*, 1999; Haritatos *et al.*, 2000a; Roberts *et al.*, 2001; Ayre *et al.*, 2003a).

As a rule for angiosperms, only the smallest vein class terminates freely in an areole. Interestingly, terminal minor veins rarely end with SEs, but mostly with xylem elements, and can often be without phloem at all (Horner *et al.*, 1994). This indicates that terminal veins are more important for xylem than for phloem transport. It can be concluded that pre-phloem transport is perpendicular towards the veins bordering a leaf areole and that SECCC become loaded via their lateral interfaces.

Experimental data and theoretical considerations indicate that transport of solutes from mesophyll to the bundle sheath is symplasmic. The presence of plasmodesmata between mesophyll cells and the bundle sheath was shown by electron microscopy in many plant species. The different structural forms of plasmodesmata of the pre-phloem pathway have been comprehensively discussed by Beebe and Russin (1999). Tracer transport studies supported the view that the plasmodesmata on this pathway are functional in symplasmic as well as apoplasmic phloem loaders (Martens and Schulz, 2003). Besides apoplasmic loading into the cotyledonary sieve tubes, *Ricinus* seedlings have, in addition, a symplasmic component in the pre-phloem transport, as shown by sucrose uptake and pool-exchange studies (Orlich *et al.*, 1998). A repetitive apoplasmic pre-phloem pathway from cell to cell would be energetically costly, since it would involve the hydrolysis of ATP for each uptake step. Bulk flow of assimilates from the respective source cell to the phloem exclusively in the apoplasm is improbable because of the opposite directions of water and assimilate transport. Moreover, suberised wall regions at the bundle sheath interface exclude this type of transport, for example, in grasses (Beebe and Russin, 1999; see Chapter 4) and in leaves having an endodermis with Casparian strips (e.g. Gymnosperm needles).

Even though the pre-phloem pathway is considered to be symplasmic, the high concentration of sucrose in the mesophyll (100–200 mM; Patrick *et al.*, 2001) certainly results in a considerable leakage. Leaked sucrose is retrieved by sucrose/$H^+$ symport (Bush, 1993). Photosynthesis in the mesophyll, vacuolar pools and retrieval along the pre-phloem pathway together implicate a downhill gradient towards the veins, thus making a concentration difference the most probable driving force for pre-phloem transport. Whether this transport is just by diffusion or by bulk flow is open to discussion. Diffusional transport would allow bidirectional plasmodesmal transport in the pre-phloem pathway, while bulk flow from mesophyll to the phloem would exclude transport in the opposite direction, except for the gated transport of protein complexes (Oparka, 2004). We have been trying to solve this question by following the movement of uncaged tracers in attached leaves of suitable plant species (Martens *et al.*, 2004).

It appears that not all plant species need to accumulate assimilates in the phloem for long-distance transport. Phloem loading can experimentally be tested by plasmolysis studies, and by autoradiography of leaves or leaf disks after application of radiolabelled $CO_2$ or sucrose. While studying willow leaves, Turgeon and Medville (1998) found that incipient plasmolysis occurred in mesophyll and SECCC at about the same concentration of the osmoticum (about 1.0 M sorbitol) while in other plant species (except for *Populus*, which is also a member of the Salicaceae), the mesophyll typically plasmolyses at about half of that concentration. Autoradiography did not reveal the minor vein network as in other plant species. Calculation of the sugar concentration in the mesophyll (and literature data of phloem exudate determined by the aphid stylet technique) revealed a higher sucrose concentration in the cytosol of mesophyll cells than in the phloem exudate (between 300 and 900 mM in the mesophyll and between 200 and 300 mM in the exudates; see Turgeon and Medville, 1998). Thus, the gradient of assimilates extends from the source mesophyll beyond the pre-phloem pathway, via the phloem, into the sink tissue, making willow a true representative of Münch's mass-flow mechanism (Figure 7.1a). Interestingly, Münch did not include a loading step at the phloem in his model (Münch, 1930).

### *7.2.2 The two modes of phloem loading*

Phloem loading can be defined as the energy-dependent accumulation of solutes in the SECCC of minor veins (Turgeon and Medville, 1998). Two modes have been described that depend on specialised plasmodesmata and membrane transporter proteins in the minor vein SECCC, respectively. Plant species in which sugars stay in the symplasm all the way from mesophyll to the SECCC are defined as symplasmic loaders, while species where the sugars are taken up by the SECCC from the apoplasm are defined as apoplasmic loaders. It should be noted that the abundance of plasmodesmata between the SECCC and neighbouring cells alone cannot be taken as evidence for symplasmic phloem loading. Neither is it possible to derive from the presence of membrane transporters in the plasma membrane of SECCC that the plants might be apoplasmic loaders. However, it is safe to state that plants having an isolated minor vein configuration (e.g. type 2; Gamalei, 1991) most probably are apoplasmic loaders. Even though an abundance of plasmodesmata and membrane transporters alone are

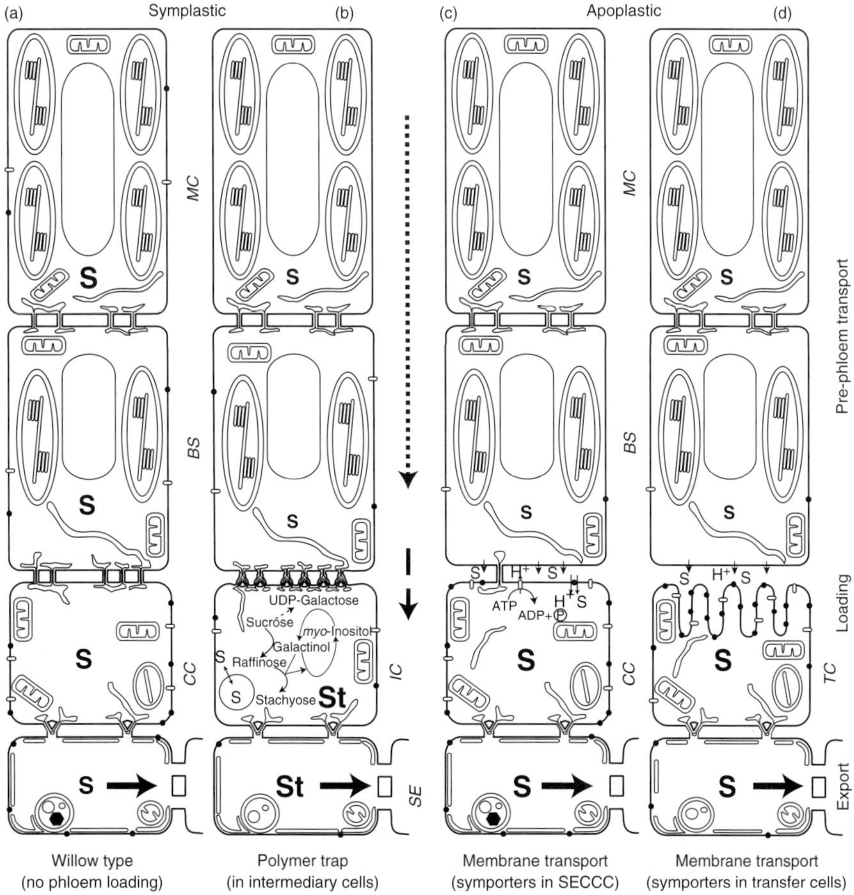


**Figure 7.1** Solute loading pathways. Four modes are established for the transport of photoassimilates from mesophyll into sieve elements. Pre-phloem transport (stippled arrow) up to the SECCC interface is symplastic in any case. Phloem loading (dashed arrow) is the enrichment of solutes in the SECCC, from which they are exported into sink organs (solid arrow). Relative solute concentrations are symbolised by the font size for **S** (sucrose) and **St** (stachyose). Membrane proteins involved in phloem loading are sucrose symporters (closed circles) and proton pumps (open ovals). (a) There is no phloem loading *sensu strictu* in willow where the highest sucrose concentration (**S**) is in the mesophyll cells (*MC*), leading to a downhill transport into the SECCC. (b) Symplastic loading involves the formation of a raffinose-family oligosaccharide (RFO; raffinose, stachyose or verbascose) in intermediary cells (*IC*) involving galactinol synthase. RFOs cannot move back, since the branched plasmodesmata between intermediary cells and bundle sheath (*BS*) cells are too small (Polymer trap; Turgeon, 1996). (c) Apoplastic loading in plants with a moderate connectivity between SECCC and surrounding cells shows sucrose/proton symporters in the plasma membrane of companion cell (*CC*) and/or sieve element (*SE*). Accumulation of sucrose in the SECCC is energised by proton pumping and demands ATP hydrolysis. (d) Apoplastic loading in plant species having transfer-cell type companion cells (*TC*). Symporters and ATPase are enriched in the infolded plasma membrane of the wall labyrinth opposite top bundle-sheath or phloem parenchyma cells. Virtually no plasmodesmata connect the SECCC with the surrounding tissue.

inappropriate, the structural features of minor-vein companion cells (CCs) do reflect the mode of loading in a number of plant species. For example, intermediary-type CCs and transfer-type CCs indicate symplasmic and apoplasmic loading, respectively (Figures 7.1b and d; Schulz, 1998).

#### *7.2.2.1 Symplasmic loading*

The term symplasmic loading implies an uphill transport across plasmodesmata. Considering that plasmodesmata, as a general rule, allow the free diffusional flow of solutes smaller than the SEL (Schulz, 1999), an uphill transport from one cytosolic compartment to that of the neighbouring cell is not conceivable. An elegant solution to this dilemma is offered in the polymer-trap hypothesis for plant species that translocate raffinose, stachyose and/or verbascose in the phloem (Figure 7.1b; Turgeon, 1996). According to this hypothesis, plants that transport raffinose-family oligosaccharides (RFO) in the phloem have intermediary-type CCs in the minor vein and have numerous branched plasmodesmata towards the bundle-sheath cells and many small vacuoles. It is characteristic that the plasmodesmata between intermediary cell and bundle sheath are more branched on the intermediary-cell side (see Beebe and Russin, 1999; Turgeon and Medville, 2004). In the intermediary cells, sucrose entering from the bundle sheath is converted into tri-, tetra- and pentasaccharides. The enzymes (e.g. galactinol synthase) necessary for this conversion are localised in the intermediary cells, and expressed when the leaf starts to export sugars (Haritatos *et al.*, 2000a). RFO are too large to diffuse back, since the branched plasmodesmata between bundle-sheath and intermediary cells have a small SEL, at least on the intermediary cell side (Turgeon, 1996).

It is not clear yet whether plant species that do not translocate RFO (or only minor amounts) can also be symplasmic loaders. Willow cannot be considered a symplasmic loader, since the accumulation of sugars occurs in the mesophyll and not at the SECCC (see earlier). Candidates for an as-yet-undefined symplasmic loading mode are gymnosperms. They have a high plasmodesmatal connectivity between bundle sheath cells and the different Strasburger cell types found in the needle. Thus, symplasmic bridges of living cells connect endodermis with Strasburger cells, which again are well connected to the exporting sieve cells by plasmodesma/sieve area contacts (Schulz, 1990; Kollmann and Glockmann, 1999). The apoplasmic pathway from mesophyll to sieve cells is interrupted by an endodermis. Within the stele, release of sugars into the apoplasmic space is improbable. The apoplasmic space consists of an extensive network of transfusion tracheids where water flow can be assumed to be in the direction opposite to assimilate transport. The mode of phloem loading in gymnosperms, which translocate sucrose, is not yet resolved. As a working hypothesis, Schulz (1998) proposed an important role for the endoplasmic reticulum (ER) in this process. The contacts between Strasburger cell and sieve cells are dome-shaped contact areas with an extended and branched plasmodesmatal network, unified in a median cavity and penetrated by the ER (Glockmann and Kollmann, 1996). On the sieve-element side, the area is covered by an extensive network of tubular ER membranes. The dilation of the desmotubules in these plasmodesmata is indicative of micro-osmotic differences between the plasmodesmal sleeve and the ER lumen. Schulz (1998) speculated that these differences indicate participation of the ER compartment in loading. Membrane transporters in the ER of Strasburger cells might differentially 'load' this compartment, and the release into the sieve cell might be via the enlarged surface of the ER complexes.

#### *7.2.2.2 Apoplasmic loading*

Apoplasmic phloem loading involves the activity of sugar membrane transporters in the SECCC that are typically proton symporters. This transport is energised by the activity of proton pumps and thus demands ATP hydrolysis. A number of symporters have been cloned and functionally characterised (for review, see Kühn, 2003). As pointed out earlier, lack or paucity of plasmodesmata between the SECCC and surrounding cells (type 2; Gamalei, 1991) is a clear indication of this loading type (Figure 7.1c). This type is even more evident where the CCs in minor veins differentiate into transfer cells. Wall labyrinths form that drastically increase the plasma membrane surface towards phloem parenchyma and bundle sheath cells (Figure 7.1d). However, apoplasmic loading was also shown to occur in other species without transfer-cell-type CCs that have a moderate number of plasmodesmata connecting them to neighbouring cells.

Experimentally, the mode of phloem loading can be identified by a three-step screening procedure (van Bel *et al.*, 1994). Apoplasmic loading is effectively inhibited by PCMBS, which out-competes sucrose at the apoplasmic uptake side of the sucrose symporters. As demonstrated in a considerable number of publications, PCMBS reduces the uptake of labelled sucrose into minor veins only in apoplasmic loaders; symplasmic loaders are insensitive to this treatment.

In the context of this chapter, the question is whether plasmodesmata between the SECCC are just non-functional relics from the cell division phase or whether they are functional? If the latter, the uphill transport of sucrose into the SECCC would always be counteracted by plasmodesmatal leakage of sucrose back into the pre-phloem pathway. Studies using caged fluorescein as a marker indicated that these plasmodesmata are indeed functional in *Solanum* species (Martens and Schulz, 2003). It is conceivable that sucrose backflow is negligible in relation to the large uptake activity of the numerous sucrose transporters in the SECCC, or that the plasmodesmata at this interface are closed by a pressure-valve mechanism when sugars in the SECCC have reached a critical concentration threshold value (see Oparka and Prior, 1992).

#### *7.2.2.3 The evolution of minor vein phloem loading*

A high frequency of plasmodesmata and the occurrence of sugar symporters at the outer boundary of SECCC are prerequisites for symplasmic and apoplasmic loading, respectively, but do not necessarily identify the mode of loading.

The plasmodesmatal frequency in minor-vein CCs was determined for a large number of plant families and grouped into three types (Gamalei, 1991). In Type 1 (open minor vein configuration), plasmodesmata are abundant, in Type 1-2a, relatively abundant, and in Type 2, infrequent (closed minor vein configuration). Species with intermediary cells fall into the Type 1 category and those with transfer cells into the Type 2 category (Type 2b). The loading mode of all others cannot be determined structurally.

The approach of finding an ecophysiological 'key' for the distribution of minor vein types among seed plants eventually turned out to be unsuccessful. Type 1 plant species were proposed to be more cold-sensitive than Type 2 which could account for their geographical distribution (van Bel and Gamalei, 1992). However, reports on ultrastructural changes in plants with open minor vein configuration could not be confirmed by a detailed study on broadleaf evergreen species with an open minor vein configuration under artificial cold exposure (Hoffmann-Thoma *et al.*, 2001).

Turgeon and coworkers (2001) have used the comprehensive material available on phloem structure and phloem loading and compared these parameters with modern, multiple nucleotide sequence-based interpretations of angiosperm evolution. They conclude that extensive plasmodesmatal connectivity between minor vein phloem and surrounding cells is ancestral in evolution. Reduction in plasmodesmatal frequency is a general evolutionary trend, which is punctuated by instances of the reverse. This is especially true in the case of intermediary cells that have arisen independently at least four times in derived lineages. The character of highly reduced plasmodesmatal frequency in minor vein phloem, common in crop plants, has several points of origin. Transfer cells, optimised for membrane uptake, have many independent points of origin, which are not always derived from lineages with reduced plasmodesmatal frequency (Turgeon *et al.*, 2001).

In conclusion, both structurally distinct loading types, characterised by specialised plasmodesmata and the transport of RFO on the one hand (Figure 7.1b), and a major surface area increase of the plasma membrane and reduction of plasmodesmata on the other hand (Figure 7.1d), are derived in evolution. This contrasts with the prevalent idea that symplasmic loading is ancient and apoplasmic loading is more advanced. It seems safe to generalise that the evolutionary optimisation of minor veins for loading led to adaptations of CCs rather than of SEs.

When considering the evolution of apoplasmic loading, it is evident that a reduction of the number of plasmodesmata, i.e. an increasing isolation of the SECCC, is advantageous, since it means a reduction of the loss of sugars via the symplasm back into the pre-phloem pathway. However, a total isolation of the SECCC in source leaves would exclude any long-distance signal transport from the phloem to the mesophyll. Comparative cost–benefit calculations of polymer-trap loading versus apoplasmic loading have not been undertaken so far. If conducted, the investments for the structural specialisation (plasmodesmatal branching, wall labyrinth, enlarged surface of the plasma membrane, expression of relevant enzymes and transporters) have to be included.

### 7.2.2.4 *The default mode of phloem loading*

As pointed out, the plasmodesmatal frequency between SECCC and surrounding cells can only be taken as an indication of the mode of phloem loading. There is a high probability of apoplasmic loading when the frequency is low, irrespective of whether the plant in question has transfer cells (Type 2b) or not (Type 2a; Gamalei, 1991). A large number of species fall in between the frequent and rare plasmodesmal contact types (Type 1-2a) and cannot readily be assigned to one loading mode, though important members of this type have been shown to be apoplasmic loaders (e.g. the Solanaceae; Kühn *et al.*, 1996). *Arabidopsis*, being a Type 1-2a according to Gamalei (1991), was shown to be an apoplasmic loader (Haritatos *et al.*, 2000b). A number of recent publications tried to uncover the mode of loading in those Type 1 plant species that do not translocate RFO. Even the tulip tree (*Liriodendron tulipifera*; Magnoliaceae) which is a basal member of the angiosperms, and has numerous plasmodesmata at the SECCC border, is an apoplasmic loader (Goggin *et al.*, 2001). Also, other Type 1 plant species such as *Clethra barbinervis* and *Liquidambar styraciflua* load sucrose apoplasmically (Turgeon and Medville, 2004).

Even plant species translocating RFO are not necessarily exclusive symplasmic loaders. Cloning and characterisation of a sucrose transporter from *Alonsea*, and sensitivity of loading to PCMBS, showed that in this species, sucrose is loaded apoplasmically, parallel to the symplasmic loading of RFO (Knop *et al.*, 2004).

Polyol-translocating plants have previously been considered to be symplasmic loaders because of their insensitivity to PCMBS (Flora and Madore, 1996). In a recent study on the sorbitol- and sucrose-translocating *Plantago*, two sorbitol symporters have been identified and localised to minor-vein CCs. Insensitivity to PCMBS of one, but not the other, of the two transporters to PCMBS is a caveat for the use of this inhibitor (Ramsberger-Gleixner *et al.*, 2004). The authors conclude that in *Plantago*, both sucrose and sorbitol are loaded apoplasmically.

Collectively, evidence emerging in recent years indicates that symplasmic loading is restricted to RFO-translocating plants. Speculations that symplasmic loading occurs in gymnosperms are as yet unfounded and demand a mechanistic explanation of how the translocate accumulates in the SEs. Thus, apoplasmic loading might be considered as the default mode of loading in angiosperm plants. Identification of the loading type in gymnosperms and in lower plants demands physiological investigations on the sugar concentration in source leaves/needles and the isolation of transporter proteins and their localisation.

For apoplasmically loading angiosperms, the role of the plasmodesmata between SECCC and surrounding cells is as yet unclear. A possible involvement in transport of signals and transcription factors from source to sink is a suggestive explanation. Here, a targeted form of transport is postulated that would allow trafficking even across those plasmodesmata closed by a pressure-valve mechanism (see Itaya *et al.*, 2002).

The importance of sucrose transporters for assimilate export in the minor veins was shown by antisense transformants of the SUT1 transporter in potato and by knockout mutants of the SUC2 transporter in *Arabidopsis*. In antisense plants, huge starch grains and oil vacuoles reflected accumulation of assimilates in all cells of the pre-phloem transport pathway, but not in the SECCC (Schulz *et al.*, 1998). Knockout mutants of *Arabidopsis* showed stunted growth, developed only small rosette leaves and had no inflorescences (Gottwald *et al.*, 2000). This phenotype could partially be rescued by the addition of sucrose to the growth medium. The rosette leaves that developed once again showed an accumulation of huge starch grains.

According to tuber development and labelling experiments, the antisense transformants and knockout mutants were, to a limited extent, still able to export some sugar (Kühn *et al.*, 1996; Gottwald *et al.*, 2000). Considering that the starch build up reflects an increase in soluble sugar in the mesophyll, this residual phloem transport might well be possible in plants containing a sufficient number of plasmodesmata between the mesophyll and the SECCC (Type 1 and 1-2a). Transport might be driven by a willow-type mechanism, where the highest osmolyte concentration is in the mesophyll and not in the SECCC (Figure 7.1a). Opening of the pressure-valve regulated plasmodesmata at the SECCC border would occur immediately when the sucrose concentration in the SECCC drops below the concentration of the neighbouring cells. In apoplasmic loaders, this option of phloem transport is masked by the activity of the sucrose transporters in the SECCC. It might, however, under distinct environmental conditions play a role in phloem transport. It is conceivable that the diurnal regulation of sucrose transporter transcription leads temporarily to a shortage in loading capacity in the morning and to an increase in mesophyll sugar concentration, accompanied by willow-type phloem transport. The study of genetically modified plants lacking a sufficient number of sucrose transporters in the minor veins seems to be able to unmask this option. With a detailed analysis of sugar flux rates and use of inhibitors, Orlich and coworkers (1998) showed that even in the 'model' plant for apoplasmic phloem loading, the *Ricinus* seedling, symplasmic mass flow driven by turgor pressure can load the SECCC.

#### *7.2.2.5 Symmetrical and asymmetrical branching of plasmodesmata in the loading pathway*

Plasmodesmata involved in pre-phloem transport and loading consist of complex plasmodesmata rather than simple, straight structures (Beebe and Russin, 1999; Oparka *et al.*, 1999; Roberts *et al.*, 2001). Complex plasmodesmata contain an extended median cavity and show branching on one or both sides of the cell-wall interface. Developmental EM-studies showed that secondary plasmodesmata, which develop *de novo* across existing walls to establish new symplasmic contact, are branched as a rule (Kollmann and Glockmann, 1985, 1990). Branching can, however, also be the result of the secondary modifications of primary plasmodesmata (Ehlers and Kollmann, 1996, 2001). This was shown for intermediary cells (Volk *et al.*, 1996), Strasburger cells (Glockmann and Kollmann, 1996; Kollmann and Glockmann, 1999) and occurs for the pore/plasmodesma units (PPUs) at the CC/SE interface (see later). Soon after the unequal division of the sieve-element/CC mother cell, the plasmodesmata differentiate into sieve pores on the sieve-element side, while they become branched on the CC side.

Why do plasmodesmata branch? As a working hypothesis, branching occurs wherever the demand for symplasmic contact increases during development, or when a reduction in symplasmic contact occurs during cell expansion that has to be compensated for (Kollmann and Glockmann, 1999; Ehlers and Kollmann, 2001). If so, then *asymmetric* branching implies that there is a difference in the permeability of the plasmodesmata on both sides of the wall. Careful measurements and calculations show that the secondary plasmodesmata that develop between epidermis and mesophyll in sectorial chimeras keep their species-specific diameter (Steinberg and Kollmann, 1994). The same appears logical for PPUs: the passage area of sieve pores is much larger than that of the plasmodesmata. Branching could, therefore, be understood as a developmental compensation for the narrowing or widening of plasmodesmal connections on one side of the common wall. This working hypothesis can also be applied to the contacts between bundle-sheath cells and intermediary cells. When the leaf is undergoing the sink/source transition, branching starts (Volk *et al.*, 1996). This is coincident with the necessity to avoid backflow of sugars. Here, sucrose (but not trisaccharides and tetrasaccharides) would be able to pass. Thus, narrowing of the cytoplasmic sleeve on the IC side may lead to a compensatory branching of the plasmodesmata.

The plasmodesmal neck region, thought to regulate the rate of intracellular transport (see e.g. Schulz, 1995), is the bottleneck in transport. When plasmodesmal widening is impossible (loading mode, domain border, wound response), this potential reduction in intercellular transport rate has to be compensated for by an increase in the number of plasmodesmatal branches. If this working hypothesis holds, it will be exciting to find out how plasmodesmal branching is regulated developmentally. An 'on-demand' connectivity could be tested by exposing plants to changed environmental conditions that invoke a higher demand in transport capacity.

## 7.3 Role of PPUs between SE and CCs

SEs originate from phloem mother cells by an unequal division, and together with the CCs become 'sister' cells, i.e. they are ontogenetically related. The division wall between SE and CC has numerous plasmodesmata/future PPUs that were shown by tracer studies to be open throughout development (van Bel and van Rijen, 1994). SE and CC are, however, not only a developmental but also a functional unit. There are only a few situations among angiosperms where CCs are lacking, as for example, within the root protophloem of dicotyledons. Here, the sieve tubes are intimately linked to neighbouring phloem parenchyma cells which may have taken over the function of the CC for the short period (hours or a few days) during which the protophloem is alive. Even in wound phloem, which regenerates around stele disruptions, CCs are obligatory sister cells to the wound-SEs, which show the typical contrast between the electron-dense CC and the electron-lucent SE and also the characteristic PPUs between both cell types (Schulz, 1990).

During differentiation, the selective autophagy that occurs in SEs (see Chapter 8) demarks the start of its functioning. Now, the integrity of the plasma membrane, the recycling of proteins and the supply of essential molecules are totally dependent on the accompanying CC (van Bel and Knoblauch, 2000). Moreover, the structural phloem proteins PP1 and PP2 are transcribed in CCs, and are found in large amounts in the SE (Thompson, 1999). Other phloem exudate-specific proteins share this fate (Thompson and Schulz, 1999). The exchange of molecules between SE and CCs is across the PPUs which consist of a sieve pore with some callose on the sieve-element side, and several branched plasmodesmata on the CC side.

The SEL of the PPUs is huge when compared with other differentiated cell types. Tracer studies indicated that molecules of at least 25 kDa can permeate between CC and SE. Monomeric GFP, expressed in the CCs, moves freely from source to sink (Imlau *et al.*, 1999). Among the endogenous phloem proteins, both the phloem lectin (PP2) with a molecular mass of 25 kDA, and even the phloem filament proteins with a monomer size of 88 kDa can permeate the PPUs (Golecki *et al.*, 1999). Dimeric GFP, however, cannot escape the CC (Itaya *et al.*, 2002). This indicates that PP1 trafficking, at least, is the result of targeted transport, i.e. result of a specific interaction of the transported protein with proteins in the PPU. From its molecular mass, PP2 might slip through the PPU without targeting, but this cannot be determined without the knowledge of the hydrodynamic radius of this protein.

Taking all the functions of CCs together, there are a number of different substances transported across the PPU, including assimilates, proteins and endogenous RNAs (for review see Oparka and Turgeon, 1999). Transport is mostly from CC into SE, but grafting experiments with antibodies specific for one partner only gave evidence that transport of phloem-specific proteins is bidirectional (Golecki *et al.*, 1999). This supports the Fisher model of protein turnover in the phloem (Fisher *et al.*, 1992). According to this model, proteins are both degraded and synthesised within the CC (see also Thompson and Schulz, 1999).

In order to avoid entry into the sieve-tube system, some proteins seem to have a companion-cell retention signal that prevents them from accessing the sieve tubes. For example, the SUC2 protein, which is expressed only in CCs, is eventually integrated into the CC plasma membrane (Stadler *et al.*, 1995; Knop *et al.*, 2004). This applies to all cytosolic enzymes of the CC, which only in artificial situations are present in phloem exudates (Lehmann, 1973; Ayre *et al.*, 2003b).

The transport pathway for those sucrose transporters that are localised to the sieve-tube plasma membrane (e.g. SUT1, SUT2 and SUT4; Kühn, 2003), and which are transferred from the CC into the SE, is not yet determined. It is, however, not very probable that these membrane proteins appear free in the cytosol of the CC. Insertion into the ER during translation may or may not be followed by processing in the Golgi apparatus and subsequent vesicle fusion with the plasma membrane of the CC. In the former case, the transporters could access the SE by membrane flow via the desmotubule of the PPUs and be distributed by the sieve-element reticulum (SER).

In the latter case, the transporter proteins would flow across the PPU in the plasma membrane (see also Oparka and Turgeon, 1999). Crucial is a free flow of membrane constituents across the PPU. For the endomembrane pathway, continuity of the ER between CC and SE was recently shown using an ER-specific dye and the technique of fluorescence redistribution after photobleaching (FRAP; Martens and Schulz, 2004). For the plasma membrane pathway, experimental evidence points to a barrier between the cells; lipids were not able to move into neighbouring cells across the plasmodesmata of suspension culture cells (Grabski *et al.*, 1993).

## 7.4 Role of plasmodesmata in the transport phloem

The mechanism of phloem transport – an osmotically generated pressure flow from source to sink – does not demand any lateral symplasmic connections in order to work; a water-tight pipe would be sufficient. However, the sieve tubes are living cells that have an intact plasma membrane. This means that there is water exchange between the phloem sap and the surrounding cells all the way from source to sink. The large concentration difference between the phloem sap and the surrounding cells leads to a considerable loss of sugars across the plasma membrane (Ayre *et al.*, 2003b). Sugar membrane-transporter proteins have been found all the way from source to sink and are localised in the SECCC (Kühn *et al.*, 1997, 2003; van Bel, 2003) which can account for the retrieval of sugars lost across the plasma membrane. It appears that RFO-translocating species have the advantage that the loss through the plasma membrane is much less than that of sucrose (Ayre *et al.*, 2003b).

The transport phloem was described to be symplasmically isolated in several plant species (Kempers *et al.*, 1998), but a few plasmodesmata between the SECCC and neighbouring cells are present in all plants investigated. It is not clear whether translocates are lost via plasmodesmata. Studies with the symplasmic tracer carboxyfluorescein indicate that the transport phloem is indeed functionally isolated (see Oparka *et al.*, 1994; Itaya *et al.*, 2002). However, the lateral escape of some tracer might go undetected. Physiological changes, and changes of the source/sink ratio, made lateral plasmodesmata operational (Patrick and Offler, 1996). Use of metabolic inhibitors led to lateral unloading of carboxyfluorescein (Wright and Oparka, 1997). Moreover, targeted transport out of the phloem of MP : GFP constructs expressed under a companion-cell specific transporter was demonstrated to be possible in all transport phloem regions (Itaya *et al.*, 2002).

Mechanistic explanations for the closure of plasmodesmata along the transport phloem are lacking. However, the pressure-valve mechanism postulated by Oparka and Prior (1992) is compatible with the observed virtual isolation of the sieve-tube system. As discussed above, plasmodesmata would be closed whenever there is a major difference in turgor pressure between neighbouring cells. Under normal transport conditions, the high concentration of sugars in the transporting sieve tubes in comparison to the adjacent cells will lead to closure of the pressure valve, and thus to the apparent isolation of the functional SECCC. Plasmolytic and pressure-probe studies have indeed indicated a steep turgor step between phloem and surrounding tissues (e.g. Meshcheryakov *et al.*, 1992; Turgeon and Medville, 1998). The pressure-valve hypothesis of transport phloem isolation implies that plasmodesmata could open temporarily under physiological conditions. A drop in water potential in surrounding tissue, or an increase in water potential in the sieve-tube system, would level out the pressure difference and release the plasmodesmata into the open state. It is even conceivable that these situations develop diurnally, for example when day/night cycles are not compensated for by mobilisation of starch in the mesophyll of source leaves. In any case, the elegant experiments by Itaya and coworkers (2002) indicate that the turgor-valve can be overcome by targeted transport.

Lateral isolation is part of the differentiation process of SECCC (van Bel and van Rijen, 1994). However, the isolation of the SECCC can be broken when developmental conditions demand it. This is the case for connection of lateral roots via connective cells, for wound phloem contact, nodule development and certain higher-plant parasites (*Orobanche*) or tumours (Schulz, 1986, 1998; Oparka *et al.*, 1995; Pradel *et al.*, 1999; Complainville *et al.*, 2003). Use of lateral plasmodesmata is moreover indicated wherever the axial assimilate flow has to change sieve tubes from primary to secondary phloem and *vice versa*. In a tree, for example the primary phloem of the leaf is connected to the secondary phloem in branches, stems and older root parts, but all assimilates have to leave the secondary phloem and move via the metaphloem, and finally the protophloem to the root tip. Lateral transfer from one phloem system to the other was analysed for the contact zone between wound phloem and regular phloem, and was shown to occur via lateral sieve pores and sieve areas in an extended overlapping area of secondary and primary phloem (Schulz, 1986).

## 7.5 Role of plasmodesmata in release phloem

Phloem unloading occurs in sink organs which are differentiated into permanent sinks (e.g. roots, shoot apices), temporary sinks (young leaves) and storage sinks (tubers, tap roots, fruits). *Sensu strictu*, unloading is the release of assimilates out of the SECCC only; the transport from there towards the target cells is defined as post-phloem transport. Permanent sinks show a one-way assimilate transport. The root grows by a combination of divisional activity of the apical meristem and of cell expansion, and the phloem pathway keeps pace with growth by the continuous elongation of protophloem sieve tubes shortly behind the apex. Temporary sinks, such as young leaves, import assimilates for a period like permanent sinks. However, from the onset of photosynthesis in the tip region, the direction of assimilate transport is inverted and the major veins (class I–III veins) become involved in bidirectional transport. Chlorophyll-less petals, however, behave like permanent sinks and remain a sink all of their lifetime. Storage sinks are also temporary sinks. Eventually, they separate from the mother plant and sprout/germinate. Sprouting/germination is supplied by the stored assimilates which means that during sprouting the sink organ becomes a source organ (e.g. in tubers).

### 7.5.1 General considerations

Using companion-specific expression of free GFP, recent publications have shown that unloading in most sink organs is primarily symplasmic, and that the plasmodesmata involved in the release of solutes from the phloem have a particularly high SEL. The pioneering paper of Imlau *et al.* (1999) showed that GFP, expressed in the CCs, entered the sieve tubes and was translocated into all plant organs and even crossed graft interfaces. Unloading and subsequent post-phloem transport was documented for the terminal phloem of petals, anthers, funiculi and root tips, and for the major veins in young rosette leaves of *Arabidopsis* (Imlau *et al.*, 1999). Transformed tobacco plants showed GFP unloading in young and transition leaves (Oparka *et al.*, 1999; Wright *et al.*, 2003). These results indicated the occurrence of symplasmic phloem unloading in all these sinks and confirmed earlier publications in which carboxyfluorescein was used as a tracer (Oparka *et al.*, 1994, 1995; Roberts *et al.*, 1997), but in addition show clearly that the plasmodesmata along the unloading/post-phloem pathway have a much higher SEL than those in other tissues.

Unloading of free GFP is non-targeted transport, and the plasmodesmata involved in this transport have an enlarged SEL corresponding to the Stokes radius of GFP, determined by fluorescence correlation spectroscopy to be about 2.82 nm (Terry *et al.*, 1995). Accordingly, this large SEL should be 'available' for all sugars and amino acids coming from the phloem stream whenever GFP unloading occurs. The common phloem tracer carboxyfluorescein has a Stokes radius of 0.61, and sucrose of 0.47 (Wang and Fisher, 1994). Following the estimations that Wang and Fisher (1994) did for the unloading pathway of developing wheat grains, a plasmodesmal pore size of about 6.5 nm can be expected for plasmodesmata, that allow GFP passage (GFP has a hydrodynamic diameter of 5.64). This is in good agreement with size-graded fluorescein-labelled dextrans shown to pass the vascular parenchyma and the post-phloem pathway towards the nucellus (4.5 and 7 nm, respectively; Wang and Fisher, 1994). It can be estimated that the plasmodesmata that allow passage of GFP are roughly 14 times more conductive for sucrose than ordinary plasmodesmata with a SEL of 0.8 kDa (Terry and Robards, 1987; see also discussion in Oparka and Santa Cruz, 2000). This is only a minimal value since the upper SEL of plasmodesmata on the unloading pathway is difficult to determine (Fisher and Cash-Clark, 2000). Dimeric GFP (54 kDa; Stokes radius unknown), expressed in CCs, did not pass the PPU and stayed in the CCs (Itaya *et al.*, 2002). However, information on the unloading properties of this tracer are unavailable. Endogenous proteins behave differently to GFP and are most probably subject to targeted transport through plasmodesmata, either by protein interaction with the PPU or by membrane integration (for review, see Oparka and Santa Cruz, 2000).

In contrast to membrane proteins, the cytosolic phloem proteins PP1 and PP2 (88 and 26 kDa, respectively) can pass PPUs in both directions. The proteins appear in large amounts in the sieve tube system, are graft-transmissible and were found to accumulate in the CCs of the graft partner after crossing the graft interface (Golecki *et al.*, 1999). Despite their ability to pass PPUs, and their considerable long-distance mobility, these proteins are limited to the SECCC system. There is only one report indicating their occurrence outside the SECCC – in phloem parenchyma cells of the root neck of pumpkin (Petersen *et al.*, 2001). An accumulation of phloem proteins might be expected in root tips because of unidirectional phloem transport. Unloading of these proteins was, however, not evident in immuno-localisations of root tips (Petersen and Schulz, unpublished data). This might be due to the detection limit of the method used, or to a rapid degradation of the proteins in cells neighbouring the protophloem sieve tubes. It appears that entry and exit of endogenous proteins into and out of the sieve tube system is a process that requires a targeting mechanism and/or involves retention signals (see also Oparka and Santa Cruz, 2000).

While the phloem lectin PP2 (26 kDa) is retained in the phloem, free GFP of approximately the same size (27 kDa) is unloaded from the phloem in most sink situations (Imlau *et al.*, 1999; Oparka *et al.*, 1999; Complainville *et al.*, 2003). It can be concluded that the bottleneck in phloem unloading is not located at the SECCC interface, but might well be acting later on the symplasmic transport pathway. In conclusion, the high SEL at this interface means that the sugar gradient from source to sink is extended beyond the protophloem boundary into the neighbouring cells.

### *7.5.2 Unloading in permanent sinks*

The model system for unloading in permanent sinks is a growing root tip. In contrast to the shoot apex, the root has a simple architecture and lacks any development of lateral organs in the growth zone. Collectively, physiological and structural evidence indicates that phloem unloading and post-phloem transport are purely symplasmic from the apical terminus of the phloem, consisting mainly of protophloem-sieve tubes (Figure 7.2a; Giaquinta *et al.*, 1983; Warmbrodt, 1985; Oparka *et al.*, 1994; Schulz, 1994, 1995; Patrick, 1997). In dicot roots, protophloem sieve tubes typically lack CCs. In contrast to most other phloem regions, these sieve tubes have numerous pore/plasmodesma contacts to phloem parenchyma and pericycle cells, the plasmodesmata of which may or may not be branched (Figures 3 and 4 in Schulz, 1995).

Treatment of the terminal 5 mm of root tips of intact pea seedlings with different mannitol and sucrose concentrations led to osmotic stress that was transiently compensated for by an increase in phloem import (Schulz, 1994). In electron microscopy, the plasmodesmata connecting the expanding cortex cells showed a significant increase in plasmodesmal diameter and a decrease in the frequency of plasmodesmata with neck constrictions after 1 h of treatment. The diameter of the desmotubules remained constant. The widening of the electron-translucent passage area of the plasmodesmata under osmotic stress (on average from ca. 110 to 260 $nm^2$) was correlated with the increase in phloem import (Schulz, 1995). The documented results are compatible with the view that the bottleneck in phloem unloading/post-phloem transport is the functional diameter of the plasmodesmata between the target cells in the symplasmic transport pathway. As long as phloem transport capacity is not saturated, an increase in the functional diameter of the bottleneck will have immediate effects on the phloem transport rate. The increase

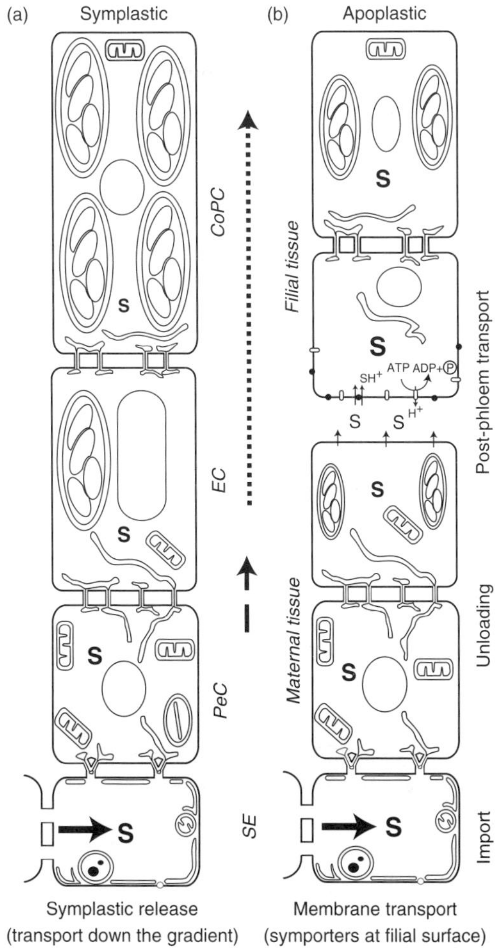


**Figure 7.2** Solute unloading pathways. Unloading of photoassimilates from the phloem follows one of two principal pathways. Phloem unloading (dashed arrow) is the symplastic release of solutes from the sieve elements or their immediate neighbour cells, as evidenced by GFP unloading in different sink types (Imlau *et al.*, 1999). This transport is followed by post-phloem transport (stippled arrow) to the receiver cells. Relative solute concentrations are symbolised by the font size for **S** (sucrose). Membrane proteins involved in the post-phloem transport are sucrose symporters (closed circles) and proton pumps (open ovals). (a) Post-phloem transport is symplastic down the gradient in many permanent sinks like root apices. Here, sieve elements (*SE*) are directly connected to pericycle cells (*PeC*) from which the solutes enter endodermis cells (*EC*) and cortex parenchyma cells (*CoPC*). (b) Post-phloem transport involves an apoplastic release in many storage sinks and in developing ovules, where the maternal and filial tissues are separated by an apoplastic interface. Uptake into the filial cells is by symporters, which are energised by proton pumping.

of the plasmodesmal diameter was transient. After a 3-h treatment, plasmodesmata turned back to control diameters (Schulz, 1995). For maize roots, solute mapping after a 24-h mannitol treatment showed that outer and inner cortex cells, as well as stelar parenchyma cells, accumulated twice the amounts of sugars in relation to controls (Tomos and Sharrock, 2001). This indicates that intracellular partitioning of assimilates, after extended treatment and/or the frequency of newly developing plasmodesmata on the post-phloem pathway, become adjusted in response to water stress within a few hours.

The mechanism of non-targeted changes in the functional diameter of plasmodesmata remains to be determined. Plasmodesmal widening was speculated to be regulated by a drop in cytosolic ATP-concentration and the dephosphorylation of plasmodesmal proteins (Schulz, 1999). A more general explanation was offered by Baluŝka and coworkers (2001), who localised the cytoskeletal protein myosin VIII and a calcium-sequestering ER-protein to cortex plasmodesmata. They suggested that myosin VIII and calreticulin might be molecular determinants of sink strength.

Carboxyfluorescein unloading in *Arabidopsis* was limited to the protophloem sieve tubes. The metaphloem, starting 1000 μm above the apical meristem, was isolated and did not show lateral spread of the tracer (Oparka *et al.*, 1994). These relations might be species-specific. An option for lateral unloading is given also in older root parts of barley and onion according to plasmodesmatal frequencies (Warmbrodt, 1985; Ma and Peterson, 2001). In 'plasmodesmograms' a continuous post-phloem pathway up to the exodermis was demonstrated. However, a paucity of plasmodesmata was obvious at two tangential interfaces; between pericycle and endodermis, as well as between exodermis and epidermis (Ma and Peterson, 2001). As pointed out before, the presence of plasmodesmata is no evidence for their functionality, and plasmodesmata functioning as pressure-valves would isolate the sieve-tube system, even if there were a large number of plasmodesmata (cf. Wright and Oparka, 1997).

In shoot apices, movement of symplasmic tracers turned out to be restricted coincident with flowering (Gisel *et al.*, 1999, 2002; see Chapter 5). It remains to be determined whether this switch is only effective for non-targeted transport of small molecules like gibberellins, cytokinins and sucrose, that are known to promote flowering, or whether the targeted transport of non-cell autonomous proteins and transcripts is also reduced in this period of reduced symplasmic communication.

In any case, developmental regulation of cell communication by changing plasmodesmal permeability is a fascinating topic for further research. As for the physiological regulation of plasmodesmatal permeability, discussed above, the challenge will be to uncover the mechanisms of developmental closure of plasmodesmata and to understand the signal pathways necessary to initiate closure and reopening of communication.

### *7.5.3 Unloading in temporary sinks*

Developing leaves show a fascinating pattern of phloem development during which the plasmodesmata change from simple/wide channels (allowing proteins up to

50 kDa to cross) to complex branched structures with a much more limited SEL (Oparka *et al.*, 1999). Though complicated, the picture emerging today is unified for different plant species.

A young sink leaf imports assimilates through the midrib and the class II and III vein system (Ding *et al.*, 1988). Unloading and post-phloem transport are symplasmic (Ding *et al.*, 1988) and involve only major veins. Xenobiotics like carboxyfluorescein or free GFP are swept with the phloem stream into the leaf symplasm and unload towards the mesophyll (Roberts *et al.*, 1997; Imlau *et al.*, 1999; Oparka *et al.*, 1999). This is not valid for some viruses and viroid RNA. Some viruses are phloem-limited and cannot escape the sieve-tube system. A recent study demonstrates that Potato spindle tuber viroid RNA evolved to potentiate an efficient trafficking from the bundle sheath into the mesophyll (Qi *et al.*, 2004). A few mutations transformed the viroid into a phloem-limited version. Interestingly, transport in the reverse direction (from mesophyll into the bundle sheath) did not depend on the source motif that allows the viroid to gate plasmodesmata (Qi *et al.*, 2004).

Maturation of the leaf spreads from the leaf tip basipetally and is accompanied by the start of photosynthetic activity. Genes specific for phloem loading, such as SUC2 (for apoplasmic loading) and galactinol synthase (for symplasmic phloem loading), are expressed in the minor veins, showing that these veins become active in phloem loading (Oparka *et al.*, 1999; Roberts *et al.*, 2001; Ayre *et al.*, 2003a).

Interestingly, the midrib (possibly also class II and III veins) and the petiole seem to be able to translocate bidirectionally in transition leaves. Assimilates produced in the leaf tip exit the leaf, while the basal leaf parts are still supplied by import from other leaves. There does not seem to be a direct symplasmic connection between the importing sieve tubes of class I–III veins and the differentiating sieve tubes in the minor veins, since viruses do not directly enter minor-vein sieve tubes (Roberts *et al.*, 1997).

The sink/source transition of leaves is, in addition, characterised by dynamic changes in the frequency and architecture of plasmodesmata (Roberts *et al.*, 2001). Viral movement proteins seem only to be able to bind to branched plasmodesmata, not to simple ones, as shown by MP–GFP targeting which follows the tip-to-base pattern (Oparka *et al.*, 1999; Roberts *et al.*, 2001). Vein maturation, as visualised by reporter-gene fusion, is related to the activity of the SUC2 promotor (Wright *et al.*, 2003). Functional studies indicate a symplasmic mode of unloading in sink leaves of barley (Haupt *et al.*, 2001), though protophloem and metaphloem sieve tubes in this plant appear to be symplasmically isolated in developing leaves (Evert and Russin, 1993). Plasmodesmata of intermediary cells of symplasmic loaders are secondarily modified and change from simple plasmodesmata to branched ones, concomitant with the onset of vascular differentiation (Volk *et al.*, 1996).

### *7.5.4 Unloading in storage sinks*

As in the case of leaves, storage sinks are temporary sink organs. The time period between filling the sink (sink condition) and sprouting/germination (source condition) is however much longer. In addition, there is a clear dormancy period in storage sinks between the end of organ filling and the start of sprouting/germination, while the sink/source transition of leaves is an integral part of a continuous development process. In comparison to permanent sinks and leaves, the structural and functional conditions in storage sinks are much more diverse, spanning vegetative organs such as tubers, tap roots, bulbs of biennial plants, ray and pith storage tissues of perennial plants and reproductive organs (i.e. fruits with seeds). Seed development includes transfer of assimilates across the maternal/filial interface with an obligatory apoplasmic step since plasmodesmata are lacking here (Figure 7.2b; but see Roberts *et al.*, 2003). The importance of phloem transport and unloading for fruits is reflected by the fact that the parallel xylem pathway is very often disrupted at later stages of fruit development (Van Ieperen *et al.*, 2003; Drazeta *et al.*, 2004). Both assimilates and water are then carried through the phloem.

Research in the last decade has focused on the sink condition of storage sinks, with emphasis on specific release/uptake mechanisms at the apoplasmic step (Wang and Fisher, 1994; Patrick and Offler, 2001; Borisjuk *et al.*, 2003; van Dongen *et al.*, 2003). Occurrence of transfer cells in cereal grains and legume seeds reflects the significance of membrane transport at this interface. Feed-back and feed-forward control integrate the assimilate flux rates across the apoplasm and the subsequent storage in endosperm cells or storage parenchyma of cotyledons (Patrick, 1997). Plasmodesmata between the sieve tubes and the apoplasmic interface seem to be specialised for transport by having a high permeability for tracers and reporter genes (Wang and Fisher, 1994; Imlau *et al.*, 1999; Fisher and Cash-Clark, 2000). Calculations of the available plasmodesmal channel size indicate that the symplasmic transport capacity is much higher than the observed transport rates into the wheat grain (Fisher and Cash-Clark, 2000). This emphasises again that the apoplasmic step at the maternal/filial interface controls assimilate transfer rates rather than the plasmodesmal permeability on the post-phloem pathway (Patrick and Offler, 2001).

The mode of phloem unloading to storage sink might well change during development. A detailed investigation of phloem unloading using carboxyfluorescein and radiolabelled assimilates as tracers showed that extension growth of the stolon was catered for by apoplasmic phloem unloading, while tuberisation is coincident with a switch to symplasmic unloading (Viola *et al.*, 2001). It might be a more general phenomenon that unloading in extending stems is apoplasmic. This is also valid in elongating bean stems, as shown by structural and functional studies (Wood *et al.*, 1997, 1998). Also, in roots the extension zone is isolated according to tracer unloading studies (Oparka *et al.*, 1994). It might well be that both parenchyma cells and SECCC in the metaphloem have an insufficient number of lateral plasmodesmal contacts to keep pace with the lateral unloading rate demanded. The switch from apoplasmic to symplasmic unloading was accompanied by a decrease of cell-wall bound invertase activity. It appears also that unloading in sugar cane becomes symplasmic in mature internodes since the suberisation of the vascular bundles excludes apoplasmic transport (Jacobsen *et al.*, 1992).

Kühn and coworkers (2003) localised the sucrose transporter SUT1 to SEs in potato tubers. Their somewhat surprising localisation in a sink organ, mainly showing symplasmic unloading, was discussed in relation to their involvement in unloading, when they work as an efflux carrier, or in sugar retrieval where they might control the osmotic potential (Kühn *et al.*, 2003). Interestingly, antisense transformants under the tuber-specific patatin promoter showed reduced fresh weight accumulation during early stages of tuber development, but eventually normal tuber yields. In the developing apple fruit, the virtual isolation of the sepal bundles and the presence of the plasma membrane ATPase and a putative monosaccharide transporter indicate that unloading is apoplasmic into the fruit flesh (Zhang *et al.*, 2004).

The reactivation of the storage organs/tissues after the dormant period include mobilisation of stored assimilates and uptake into the phloem. While mobilisation of proteins and starch is studied intensely, phloem uptake of solutes originating from storage tissue is a relatively neglected research area. Only in model systems like the *Ricinus* cotyledon, does detailed information on the contributions of apoplasmic and a parallel symplasmic loading pathway exist (Orlich *et al.*, 1998). In trees, storage of assimilates in the ray parenchyma system fuels the cambial activation and onset of phloem transport. Transport through the rays (= pre-phloem transport) is symplasmic (see Sauter and Kloth, 1986; van Bel, 1990) and might at least in gymnosperms end with a symplasmic sieve-element loading step via ray (Strasburger) cells (see Schulz, 1990). Even less is known about sprouting vegetative organs. While bulb scales (modified leaves) can be expected to show the same mode of phloem loading as ordinary leaves, it is unclear how the phloem of tap roots such as carrots or sugar beet structurally and functionally reorganise from the release mode of phloem unloading to subsequently function in sugar uptake and export.

## 7.6 Concluding remarks

Non-targeted transport of solutes through plasmodesmata precedes and follows long-distance transport, and includes critical checkpoints for plant growth and development. Instead of providing a conclusion of the present state of knowledge, I would like to pose some questions that might inspire the reader to design experiments specifically to solve them:

- Evidence is emerging that juvenile plant organs have a much less restricted solute movement than adult parts. What is the mechanism that controls the transition from a plasmodesmal diameter that allows the transport of free GFP (27 kDa) to a SEL of about 1 kDa? How is this transition regulated?
- Young tissues of the apical shoot meristem are not always well connected. Initiation of flowering is accompanied by a transient restriction of tracer movement into the apex. Does this transient change also change targeted transport, or can non-cell autonomous proteins override this restriction?

- Pre- and post-phloem transport of solutes is subject to modulations and physiological adjustments. Are these changes cell-autonomous adaptations or is there supercellular coordination involved?
- Apoplasmic loaders have minor veins that are not completely isolated. What is the role of plasmodesmata that connect this SECCC with the surrounding cells?
- The function of the transport phloem would be endangered if the lateral loss of assimilates amounted to high values. This is particularly relevant for trees. Loss can occur across the plasma membrane and via lateral plasmodesmata. There is evidence for an effective retrieval mechanism in the transport phloem, and sucrose transporters have been found all the way from source to sink phloem. Is the activity of these transporters sufficient to cope with the lateral apoplasmic and symplasmic loss? If so, what about the cost–benefit for the long-distance transport process? What percentage of the photoassimilates, loaded in source leaves, eventually arrives in sink organs?
- The lateral plasmodesmata in the transport phloem might well be closed by a pressure valve mechanism. Is this mechanism actively controlled, or is it automatically opened whenever the turgor difference between sieve tubes and phloem parenchyma drops?

## References


Ayre, B.G., Blair, J.E. and Turgeon, R. (2003a) 'Functional and phylogenetic analyses of a conserved regulatory program in the phloem of minor veins', *Plant Physiology* **133**, 1229–1239.

Ayre, B.G., Keller, F. and Turgeon, R. (2003b) 'Symplasmic continuity between companion cells and the translocation stream: long-distance transport is controlled by retention and retrieval mechanisms in the phloem', *Plant Physiology* **131**, 1518–1528.

Baluŝka, F., Cvrckova, F., Kendrick-Jones, J. and Volkmann, D. (2001) 'Sink plasmodesmata as gateways for phloem unloading. Myosin VIII and calreticulin as molecular determinants of sink strength?', *Plant Physiology* **126**, 39–46.

Beebe, D.U. and Russin, W.A. (1999) 'Plasmodesmata in the phloem-loading pathway, in *Plasmodesmata. Structure, Function, Role in Cell Communication* (eds A.J.E. van Bel and W.J.P. van Kesteren), Springer, Berlin, Germany, pp. 261–293.

Borisjuk, L., Rolletschek, H., Wobus, U. and Weber, H. (2003) 'Differentiation of legume cotyledons as related to metabolic gradients and assimilate transport into seeds', *Journal of Experimental Botany* **54**, 503–512.

Bush, D.R. (1993) 'Proton-coupled sugar and amino-acid transporters in plants', *Annual Review of Plant Physiology and Plant Molecular Biology* **44**, 513–542.

Complainville, A., Brocard, L., Roberts, I. *et al.* (2003) 'Nodule initiation involves the creation of a new symplasmic field in specific root cells of *Medicago* species', *Plant Cell* **15**, 2778–2791.

Ding, B., Parthasarathy, M.V., Niklas, K. and Turgeon, R. (1988) 'A morphometric analysis of the phloem-unloading pathway in developing tobacco leaves', *Planta* **176**, 307–318.

Drazeta, L., Lang, A., Hall, A.J., Volz, R.K. and Jameson, P.E. (2004) 'Causes and effects of changes in xylem functionality in apple fruit', *Annals of Botany* **93**, 275–282.

Ehlers, K. and Kollmann, R. (1996) 'Formation of branched plasmodesmata in regenerating *Solanum nigrum*-protoplasts', *Planta* **199**, 126–138.

Ehlers, K. and Kollmann, R. (2001) 'Primary and secondary plasmodesmata: structure, origin, and functioning', *Protoplasma* **216**, 1–30.

Erwee, M.G. and Goodwin, P.B. (1985) 'Symplastic domains in extrastelar tissues of *Egeria densa* Planch', *Planta* **163**, 9–19.

Evert, R.F. and Russin, W.A. (1993) 'Structurally, phloem unloading in the maize leaf cannot be symplasmic', *American Journal of Botany* **80**, 1310–1317.

Fisher, D.B. and Cash-Clark, C.E. (2000) 'Sieve tube unloading and post-phloem transport of fluorescent tracers and proteins injected into sieve tubes via severed aphid stylets', *Plant Physiology* **123**, 125–137.

Fisher, D.B., Wu, Y. and Ku, M.S.B. (1992) 'Turnover of soluble proteins in the wheat sieve tube', *Plant Physiology* **100**, 1433–1441.

Flora, L.L. and Madore, M.A. (1996) 'Significance of minor-vein anatomy to carbohydrate transport', *Planta* **198**, 171–178.

Gamalei, Y.V. (1991) 'Phloem loading and its development related to plant evolution from trees to herbs', *Trees* **5**, 50–64.

Giaquinta, R.T., Lin, W., Sadler, N.L. and Franceschi, V.R. (1983) 'Pathway of phloem unloading of sucrose in corn roots', *Plant Physiology* **72**, 362–367.

Gisel, A., Barella, S., Hempel, F.D. and Zambryski, P.C. (1999) 'Temporal and spatial regulation of symplasmic trafficking during development in *Arabidopsis thaliana* apices', *Development* **126**, 1879–1889.

Gisel, A., Hempel, F.D., Barella, S. and Zambryski, P.C. (2002) 'Leaf-to-shoot apex movement of symplasmic tracer is restricted coincident with flowering in *Arabidopsis*', *Proceedings of the National Academy of Sciences USA* **99**, 1713–1717.

Glockmann, C. and Kollmann, R. (1996) 'Structure and development of cell connections in the phloem of *Metasequoia glyptostroboides* needles. 1. Ultrastructural aspects of modified primary plasmodesmata in Strasburger cells', *Protoplasma* **193**, 191–203.

Goggin, F.L., Medville, R. and Turgeon, R. (2001) 'Phloem loading in the tulip tree. Mechanisms and evolutionary implications', *Plant Physiology* **124**, 891–899.

Golecki, B., Schulz, A. and Thompson, G.A. (1999) 'Translocation of structural P-proteins in the phloem', *The Plant Cell* **11**, 127–140.

Goodwin, P.B. and Cantrill, L.C. (1999) Use and limitations of fluorochromes for plasmodesmal research, in *Plasmodesmata. Structure, Function, Role in Cell Communication* (eds A.J.E. van Bel and W.J.P. van Kesteren) Springer, Berlin, Germany, pp. 67–84.

Gottwald, J.R., Krysan, P.J., Young, J.C., Evert, R.F. and Sussmann, M.R. (2000) 'Genetic evidence for the *in planta* role of phloem-specific plasma membrane sucrose transporters', *Proceedings of the National Academy of Sciences USA* **97**, 13979–13984.

Grabski, S., de Feijter, A.W. and Schindler, M. (1993) 'Endoplasmic reticulum forms a dynamic continuum for lipid diffusion between contiguous soybean root cells', *The Plant Cell* **5**, 25–38.

Haritatos, E., Ayre, B.G. and Turgeon, R. (2000a) 'Identification of phloem involved in assimilate loading in leaves by the activity of the galactinol synthase promoter', *Plant Physiology* **123**, 929–937.

Haritatos, E., Medville, R. and Turgeon, R. (2000b) 'Minor vein structure and sugar transport in *Arabidopsis thaliana*', *Planta* **211**, 105–111.

Haupt, S., Duncan, G.H., Holzber, S. and Oparka, K.J. (2001) 'Evidence for symplasmic phloem unloading in sink leaves of barley', *Plant Physiology* **125**, 209–218.

Haywood, V., Kragler, F. and Lucas, W.J. (2002) 'Plasmodesmata: pathways for protein and ribonucleoprotein signaling', *The Plant Cell* **Supplement 2002**, S303–S325.

Hoffmann-Thoma, G., van Bel, A. and Ehlers, K. (2001) 'Ultrastructure of minor-vein phloem and assimilate export in summer and winter leaves of the symplasmically loading evergreen *Ajuga reptans* L., *Aucuba japonica* Thumb., and *Hedera helix* L.', *Planta* **212**, 231–242.

Horner, H.T., Lersten, N.L. and Wirth, C.L. (1994) 'Quantitative survey of sieve tube distribution in foliar terminal veins of ten dicot species', *American Journal of Botany* **81**, 1267–1274.

Imlau, A., Truernit, E. and Sauer, N. (1999) 'Cell-to-cell and long-distance trafficking of the green fluorescent protein in the phloem and symplasmic unloading of the protein into sink tissue', *The Plant Cell* **11**, 309–322.

Itaya, A., Ma, F., Qi, Y. *et al.* (2002) 'Plasmodesma-mediated selective protein traffic between "symplasmically isolated" cells probed by a viral movement protein', *The Plant Cell* **14**, 1267–1274.

Jacobsen, K.R., Fisher, D.G., Maretzki, A. and Moore, P.H. (1992) 'Developmental changes in the anatomy of the sugarcane stem in relation to phloem unloading and sucrose storage', *Botanica Acta* **105**, 70–80.

Kempers, R., Ammerlaan, A. and van Bel, A. (1998) 'Symplasmic constriction and ultrastructural features of the sieve element/companion cell complex in the transport phloem of apoplasmically and symplasmically phloem-loading species', *Plant Physiology* **116**, 271–278.

Knop, C., Stadler, R., Sauer, N. and Lohaus, G. (2004) 'AmSUT1, a sucrose transporter in collection and transport phloem of the putative symplasmic phloem loader *Alonsoa meridionalis*', *Plant Physiology* **134**, 204–214.

Kollmann, R. and Glockmann, C. (1985) 'Studies on graft unions. 1. Plasmodesmata between cells of plants belonging to different unrelated taxa', *Protoplasma* **124**, 224–235.

Kollmann, R. and Glockmann, C. (1990) Sieve elements of graft unions, in *Sieve Elements* (eds H.D. Behnke and R.D. Sjolund), Springer, Berlin, Germany, pp. 219–237.

Kollmann, R. and Glockmann, C. (1999) Multimorphology and nomenclature of plasmodesmata in higher plants, in *Plasmodesmata. Structure, Function, Role in Cell Communication* (eds A.J.E. van Bel and W.J.P. van Kesteren), Springer, Berlin, Germany, pp. 149–172.

Kühn, C. (2003) 'A comparison of the sucrose transporter systems of different plant species', *Plant Biology* **5**, 215–232.

Kühn, C., Franceschi, V.R., Schulz, A., Lemoine, R. and Frommer, W.B. (1997) 'Macromolecular trafficking indicated by localization and turnover of sucrose transporters in enucleate sieve elements', *Science* **275**, 1298–1300.

Kühn, C., Hajirezaei, M.R., Fernie, A.R. *et al.* (2003) 'The sucrose transporter StSUT1 localizes to sieve elements in potato tuber phloem and influences tuber physiology and development', *Plant Physiology* **131**, 102–113.

Kühn, C., Quick, W.P., Schulz, A., Riesmeier, J.W., Sonnewald, U. and Frommer, W.B. (1996) 'Companion cell-specific inhibition of the potato sucrose transporter SUT1', *Plant, Cell and Environment* **19**, 1115–1123.

Lehmann, J. (1973) 'Zur Lokalisierung von Dehydrogenasen des Energiestoffwechsels im Phloem von *Cucurbita pepo* L.', *Planta* **111**, 187–198.

Ma, F. and Peterson, C.A. (2001) 'Frequencies of plasmodesmata in *Allium cepa* L. roots: implications for solute transport pathways', *Journal of Experimental Botany* **52**, 1051–1061.

Martens, H.J. and Schulz, A. (2003) 'Functionality of plasmodesmata between mesophyll cells and sieve element/companion cell complexes'. *International Conference on Phloem Transport*, September 1–5, 2003, Bayreuth, Germany, Proceedings, p. 170.

Martens, H.J. and Schulz, A. (2004) 'Intra- and intercellular trafficking via the ER-membrane studied by FRAP', *Abstracts of the Fifth International Plasmodesmata Conference*, Pacific Grove, California, USA, p. 130.

Martens, H.J., Hansen, M. and Schulz, A. (2004) 'Caged probes: a novel tool in studying symplasmic transport in plant tissues', *Protoplasma* **223**, 63–66.

Meshcheryakov, A., Steudle, E. and Komor, E. (1992) 'Gradients of turgor, osmotic pressure, and water potential in the cortex of the hypocotyl of growing *Ricinus* seedlings', *Plant Physiology* **98**, 840–852.

Münch, E. (1930) *Die Stoffbewegungen in der Pflanze*, Gustav Fischer, Jena, Germany.

Oparka, K.J. (2004) 'Getting the message across: how do plant cells exchange macromolecular complexes?', *Trends in Plant Science* **9**, 33–41.

Oparka, K.J. and Prior, D.A.M. (1992) 'Direct evidence for pressure-generated closure of plasmodesmata', *Plant Journal* **2**, 741–750.

Oparka, K.J. and Santa Cruz, S. (2000) 'The great escape: phloem transport and unloading of macromolecules', *Annual Review of Plant Physiology and Plant Molecular Biology* **51**, 323–347.

Oparka, K.J. and Turgeon, R. (1999) 'Sieve elements and companion cells – traffic control centers of the phloem', *The Plant Cell* **11**, 739–750.

Oparka, K.J., Duckett, C.M., Prior, D.A.M. and Fisher, D.B. (1994) 'Real-time imaging of phloem unloading in the root tip of *Arabidopsis*', *Plant Journal* **6**, 759–766.

Oparka, K.J., Prior, D.A.M. and Wright, K.M. (1995) 'Symplasmic communication between primary and developing lateral roots of *Arabidopsis thaliana*', *Journal of Experimental Research* **46**, 187–197.

Oparka, K.J., Roberts, A.G., Boevink, P. *et al.* (1999) 'Simple, but not branched, plasmodesmata allow the nonspecific trafficking of proteins in developing tobacco leaves', *Cell* **97**, 743–754.

Orlich, G., Hofbrückl, M. and Schulz, A. (1998) 'A symplasmic flow of sucrose contributes to phloem loading in *Ricinus* cotyledons', *Planta* **206**, 108–116.

Patrick, J.W. (1997) 'Phloem unloading: sieve element unloading and post-sieve element transport', *Annual Review of Plant Physiology and Plant Molecular Biology* **48**, 191–222.

Patrick, J.W. and Offler, C.E. (1996) 'Post-sieve element transport of photoassimilates in sink regions', *Journal of Experimental Botany* **53**, 1–14.

Patrick, J.W. and Offler, C.E. (2001) 'Compartmentation of transport and transfer events in developing seeds', *Journal of Experimental Botany* **52**, 551–564.

Patrick, J.W., Zhang, W., Tyerman, S.D., Offler, C.E. and Walker, N.A. (2001) 'Role of membrane transport in phloem translocation of assimilates and water', *Australian Journal of Plant Physiology* **28**, 695–707.

Petersen, M.L.C., Thompson, G.A. and Schulz, A. (2001) 'Long-distance and cell-to-cell transport of phloem proteins', *Fourth International Conference Plasmodesma 2001*, Cape Town, South Africa, August 20–24, 2001, Abstracts, p. 89.

Pradel, K.S., Ulrich, C.I., Santa Cruz, S. and Oparka, K.J. (1999) 'Symplasmic continuity in *Agrobacterium tumefaciens* induced tumours', *Journal of Experimental Botany* **50**, 183–192.

Qi, Y., Pélissier, T., Itaya, A., Hunt, E., Wassenegger, M. and Ding, B. (2004) 'Direct role of a viroid RNA motif in mediating directional RNA trafficking across a specific cellular boundary', *The Plant Cell* **16**, 1741–1752.

Ramsberger-Gleixner, M., Geiger, D., Hedrich, R. and Sauer, N. (2004) 'Differential expression of sucrose transporter and polyol transporter genes during maturation of common plantain companion cells', *Plant Physiology* **134**, 147–160.

Roberts, A.G., Santa Cruz, S., Roberts, I.M., Prior, D.A.M., Turgeon, R. and Oparka, K.J. (1997) 'Phloem unloading in sink leaves of *Nicotiana benthamiana*: comparison of a fluorescent solute with a fluorescent virus', *The Plant Cell* **9**, 1381–1396.

Roberts, I.M., Boevink, P., Roberts, A.G., Sauer, N., Reichel, C. and Oparka, K.J. (2001) 'Dynamic changes in the frequency and architecture of plasmodesmata during the sink–source transition in tobacco leaves', *Protoplasma* **218**, 31–44.

Roberts, I.M., Wang, D., Thomas, C.I. and Maule, A.J. (2003) 'Pea seed-borne mosaic virus seed transmission exploits novel symplasmic pathways to infect the pea embryo and is, in part, dependent upon chance', *Protoplasma* **222**, 31–43.

Sauter, J.J. and Kloth, S. (1986) 'Plasmodesmal frequency and radial translocation in ray cells of poplar (*Populus canadensis* Moench "robusta")', *Planta* **168**, 377–380.

Schulz, A. (1986) 'Wound phloem in transition to bundle phloem in primary roots of *Pisum sativum* L. II. The plasmatic contact between wound-sieve tubes and regular phloem', *Protoplasma* **130**, 27–40.

Schulz, A. (1990) Wound-sieve elements, in *The Sieve Element – Comparative Structure, Induction and Development* (eds H.-D. Behnke and R.D. Sjolund), Springer, Berlin, Germany, pp. 199–217.

Schulz, A. (1994) 'Phloem transport and differential unloading in pea seedlings after source and sink manipulations', *Planta* **192**, 239–248.

Schulz, A. (1995) 'Plasmodesmal widening accompanies the short-term increase in symplasmic phloem unloading of pea root tips under osmotic stress', *Protoplasma* **188**, 22–37.

Schulz, A. (1998) 'The phloem. Structure related to function', *Progress in Botany* **59**, 429–475.

Schulz, A. (1999) Physiological control of plasmodesmal gating, in *Plasmodesmata. Structure, Function, Role in Cell Communication* (eds A.J.E. van Bel and W.J.P. van Kesteren), Springer, Berlin, Germany, pp. 173–204.

Schulz, A., Kühn, C., Riesmeier, J.W. and Frommer, W.B. (1998) 'Ultrastructural effects in potato leaves due to antisense-inhibition of the sucrose transporter indicate an apoplasmic mode of phloem loading', *Planta* **206**, 533–543.

Sondergaard, T., Schulz, A. and Palmgren, M.G. (2004) 'Energization of transport processes in plants: roles of the plasma membrane $H^+$-ATPase', *Plant Physiology* **136**, 2475–2482.

Stadler, R., Brandner, J., Schulz, A., Gahrtz, M. and Sauer, N. (1995) 'Phloem loading by the PmSUC2 sucrose carrier from *Plantago major* occurs into companion cells', *The Plant Cell* **7**, 1545–1554.

Steinberg, G. and Kollmann, R. (1994) 'A quantitative analysis of the interspecific plasmodesmata in the non-division walls of the plant chimera *Laburnocytisus adamii* (Poit.) Schneid.', *Planta* **192**, 75–83.

Terry, B.R. and Robards, A.W. (1987) 'Hydrodynamic radius alone governs the mobility of molecules through plasmodesmata', *Planta* **171**, 145–157.

Terry, B.R., Matthews, E.K. and Haseloff, J. (1995) 'Molecular characterization of recombinant green fluorescent protein by fluorescence correlation microscopy', *Biochemical and Biophysical Research Communications* **217**, 21–27.

Thompson, G.A. (1999) P-protein trafficking through plasmodesmata, in *Plasmodesmata. Structure, Function, Role in Cell Communication* (eds A.J.E. van Bel and W.J.P. van Kesteren), Springer, Berlin, Germany, pp. 295–313.

Thompson, G.A. and Schulz, A. (1999) 'Macromolecular trafficking in the phloem', *Trends in Plant Science* **4**, 354–360.

Tomos, A.D. and Sharrock, R.A. (2001) 'Cell sampling and analysis (SiCSA): metabolites measured at single cell resolution', *Journal of Experimental Botany* **52**, 623–630.

Turgeon, R. (1996) 'Phloem loading and plasmodesmata', *Trends in Plant Science* **1**, 418–423.

Turgeon, R. and Medville, R. (1998) 'The absence of phloem loading in willow leaves', *Proceedings of the National Academy of Sciences USA* **95**, 12055–12060.

Turgeon, R. and Medville, R. (2004) 'Phloem loading: a reevaluation of the relationship between plasmodesmatal frequencies and loading strategies', *Plant Physiology* **136**, 3795–3803.

Turgeon, R., Medville, R. and Nixon, K.C. (2001) 'The evolution of minor vein phloem and phloem loading', *American Journal of Botany* **88**, 1331–1339.

van Bel, A.J.E. (1990) 'Xylem–phloem exchange via rays: the undervalued route of transport', *Journal of Experimental Botany* **41**, 631–644.

van Bel, A.J.E. (2003) 'Transport phloem: low profile, high impact', *Plant Physiology* **131**, 1509–1510.

van Bel, A.J.E., Ammerlaan, A.M.H. and Vandijk, A.A. (1994) 'A 3-step screening procedure to identify the mode of phloem loading in intact leaves – evidence for symplasmic and apoplasmic phloem loading associated with the type of companion cell', *Planta* **192**, 31–39.

van Bel, A.J.E. and Gamalei, Y.V. (1992) 'Ecophysiology of phloem loading in source leaves', *Plant Cell and Environment* **152**, 265–270.

van Bel, A.J.E. and Knoblauch, M. (2000) 'Sieve element and companion cell: the story of the comatose patient and the hyperactive nurse', *Australian Journal of Plant Physiology* **27**, 477–487.

van Bel, A.J.E. and van Rijen, H.V.M. (1994) 'Microelectrode-recorded development of the symplasmic autonomy of the sieve element/companion cell complex in the stem phloem of *Lupinus luteus* L.', *Planta* **192**, 165–175.

van Dongen J.T., Ammerlaan, A.M.H., Wouterloud, M., van Aelst, A.C. and Borstlap, A.C. (2003) 'Structure of the developing pea seed coat and the post-phloem transport pathway of nutrients', *Annals of Botany* **91**, 729–737.

Van Ieperen, W., Volkov, V.S. and Van Meeteren, U. (2003) 'Distribution of xylem hydraulic resistance in fruiting truss of tomato influenced by water stress', *Journal of Experimental Botany* **54**, 317–324.

Viola, R., Roberts, A.G., Haupt, S. *et al.* (2001) 'Tuberization in potato involves a switch from apoplasmic to symplasmic phloem unloading', *The Plant Cell* **13**, 385–398.

Volk, G.M., Turgeon, R. and Beebe, D.U. (1996) 'Secondary plasmodesmata formation in the minor-vein phloem of *Cucumis melo* L. and *Cucurbita pepo* L.', *Planta* **199**, 425–432.

Wang, N. and Fisher, D.B. (1994) 'The use of fluorescent tracers to characterize the post-phloem transport pathway in maternal tissues of developing grains', *Plant Physiology* **104**, 17–27.

Warmbrodt, R.D. (1985) 'Studies on the root of *Hordeum vulgare* L. – ultrastructure of the seminal root with special reference to the phloem', *American Journal of Botany* **72**, 414–432.

Wood, R.M., Offler, C.E. and Patrick, J.W. (1997) 'The cellular pathway of short-distance transfer of photosynthates and potassium in the elongating stem of *Phaseolus vulgaris* L. A structural assessment', *Annals of Botany* **79**, 89–100.

Wood, R.M., Patrick, J.W. and Offler, C.E. (1998) 'The cellular pathway of short-distance transfer of photosynthates and potassium in the elongating stem of *Phaseolus vulgaris* L. A physiological assessment', *Annals of Botany* **82**, 337–345.

Wright, K.M. and Oparka, K.J. (1997) 'Metabolic inhibitors induce symplasmic movement of solutes from the transport phloem of *Arabidopsis* roots', *Journal of Experimental Botany* **48**, 1807–1814.

Wright, K.M., Roberts, A.G., Martens, H.J., Sauer, N. and Oparka, K.J. (2003) 'Structural and functional vein maturation in developing tobacco leaves in relation to AtSUC2 promoter activity', *Plant Physiology* **131**, 1555–1565.

Yoo, B.-C., Kragler, F., Varkonyi-Gasic, E., Haywood, V., Archer-Evans, S., Lee, Y.M., Lough, T.J. and Lucas, W.J. (2004) 'A systemic small RNA signaling system in plants', *The Plant Cell* **16**, 1979–2000.

Zhang, L.-Y., Peng, Y.-B., Pelleschi-Travier, S. *et al.* (2004) 'Evidence for apoplasmic phloem unloading in developing apple fruit', *Plant Physiology* **135**, 574–586.

# 8 Plasmodesmata and the phloem: conduits for local and long-distance signaling

Robert L. Gilbertson, Maria R. Rojas and William J. Lucas

## 8.1 Introduction

The development of terrestrial plants necessitated the evolution of a system for effective long-distance transport systems to allow exchange of water, nutrients and other molecules among diverse and distantly located tissues and organs. This was accomplished through the evolution of the vascular system, comprised of two major components, the xylem and phloem (Figure 8.1). The mature xylem is comprised of a system of tubes or pipes composed of nonliving tracheary elements and/or vessels that transport water and mineral nutrients from the roots to above-ground parts of the plant. The phloem evolved to handle the transport of photoassimilates and an extensive array of other macromolecules, some of which are signals involved in regulatory and defense pathways that are activated throughout the body of the plant (systemic pathways; Figure 8.1) (Thompson and Schulz, 1999; Oparka and Santa Cruz, 2000; Zambryski and Crawford, 2000; Ruiz-Medrano *et al.*, 2001). It is important to note that, although the plant vascular system represents the functional equivalent of the animal circulatory system, a fundamental difference is that in the plant the system is noncirculatory in nature.

## 8.2 Phloem is well suited for long-distance trafficking of macromolecules

### *8.2.1 Sieve elements are the building blocks of a network of dynamic living tubes for macromolecular trafficking*

During the course of evolution, the phloem has developed into a highly specialized long-distance transport tissue, characterized by specialized cell types and plasmodesmata (PD) and a capacity for dynamic symplasmic interactions with the surrounding nonvascular tissues (Esau, 1969; Roberts and Oparka, 2003; van Bel, 2003). In flowering plants, the mature phloem is composed of two major cell types: sieve elements (SEs) and companion cells (CCs). Although strikingly different in structure and function, SEs and CCs are ontogenetically related, being derived from the same precursor, or mother cell. The SEs are a unique type of living cell, evolved to serve as specialized conduits for long-distance transport of nutrients

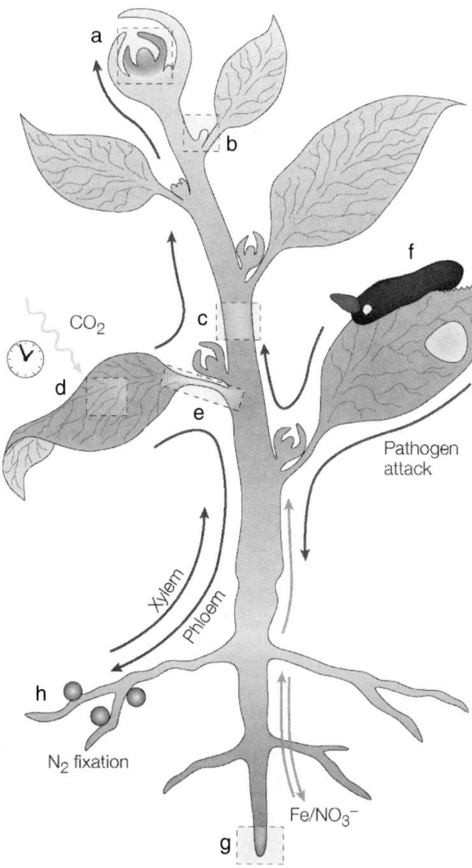


**Figure 8.1** Plants use a combination of local and long-distance signaling systems to integrate functions at the whole-plant level. Plasmodesmata (PD) establish specific symplasmic domains within the phloem that allow both for cell-to-cell and long-distance delivery of metabolites and information molecules. Input signals perceived by mature leaves are transmitted into the vascular system through the symplasm. Sets of specific signaling molecules are generated and enter the phloem in response to environmental conditions (d; light, circadian rhythms, temperature, mineral nutrient and water availability) or pathogen challenge (f; virus, bacteria, fungi and insects). These signals are exported through the long-distance translocation pathway of the petiole (e) and stem (c) and delivery to distant organs such as the shoot apical meristem (a), axillary meristems (b) or root tissues (g and h). This long-distance signaling system is thought to allow control over such processes as morphological traits on developing leaves, partitioning of nutrients (sugars and amino acids), adaptive responses to environmental inputs and resistance to pathogen challenge. In this way, the phloem acts as an essential conduit to inform distantly located tissues and organs as to the appropriateness of their ongoing developmental and physiological programs. Signals from the root system (mineral nutrient and water availability, and so on) can move to the vegetative regions of the plant through the xylem; the transpiration stream is also essential for the delivery of information from the root to the shoot. [Reproduced from Lucas and Lee (2004).]

and macromolecules. The development of SEs involves a process of targeted intracellular degradation or reduction (possibly a special type of programmed cell death, or apoptosis) in which most cellular components (e.g. nucleus, cytoskeleton, vacuoles and ribosomes) are degraded. Remaining cellular components include the endoplasmic reticulum (ER) which becomes smooth and aggregated along the SE periphery, mitochondria and two types of plastids: those with proteinaceous (P-type plastids) or starch (S-type plastids) inclusions (Esau, 1969; van Bel and Knoblauch, 2000; van Bel, 2003). These components may comprise a functional unit, the parietal layer, located at the periphery of SEs (Esau, 1969; Oparka and Turgeon, 1999). The mature SE is an elongated tube-like enucleate cell with specialized (thickened) cross-walls or sieve plates, in which the interconnecting PD become highly modified and enlarged into numerous plasma-membrane-lined pores, ranging in diameter from 200 nm to several micrometres (Esau and Cheadle, 1959; Esau, 1969). SEs are arranged, end-to-end, at the sieve plates, forming an extensive contiguous system of sieve tubes that function in the long-distance transport of both nutrients and macromolecules throughout the body of the plant. The large size of the sieve plate pores allow the passage of a wide array of macromolecules (Knoblauch and van Bel, 1998), including various obligate plant pathogens such as viruses, viroids and cell-wall-less bacteria (phytoplasmas), which use the phloem for long-distance spread within the plant.

### *8.2.2 Specialized companion cells maintain living enucleate SEs*

To help maintain their metabolic function, SEs are associated with specialized parenchyma cells referred to as CCs (Esau, 1969; van Bel and Knoblauch, 2000). The CCs are the opposite of the enucleate SEs; they have relatively large nuclei with nucleoli, numerous mitochondria, vacuoles and ribosomes. Thus, the protoplasm of CCs suggests high rates of metabolic activity, probably reflecting a role for these cells in providing essential materials for the upkeep and function of the SEs. Some of the metabolic activity in CCs is also associated with the synthesis and loading, transport and unloading of photoassimilates and other macromolecules.

### *8.2.3 SEs and CCs are interconnected by specialized PD*

A close functional relationship between SEs and CCs is indicated by the specialized, branched PD that interconnect these cell types. These PD are deltoid in shape, with the expanded, branched portion facing the CC side cell wall, and a simple pore facing the SE side; this has led to these PD being called pore-plasmodesma units (PPUs; van Bel and Kempers, 1997). The SE–CC PD also have important functional properties such as high or upregulated molecular size exclusion limits (SELs). These attributes are consistent with a role in the trafficking and regulation of macromolecules entering the SEs. The SE–CC complex represents a functional unit and, depending on the tissue type and developmental stage, it is often symplasmically isolated from adjacent nonvascular cells (i.e. few PD interconnecting CC with surrounding nonvascular cells). In these cases, CCs would be providing factors for SE function and maintenance, which would likely involve significant macromolecular trafficking through SE–CC PD.

In the following sections, we will address various aspects of macromolecular trafficking in the phloem, including the role of SE–CC PD. We will examine: (i) the types of macromolecules that are found in the phloem sap, (ii) their respective role (if any) in long-distance macromolecular trafficking and/or signaling and (iii) the interaction of these molecules with PD to engage this macromolecular pathway. We will review the current understanding of long-distance signaling in the phloem in terms of selected systemic signaling pathways. As will be evident from information presented in this chapter, technical challenges related to the size and structure of phloem cells continue to make it difficult to directly probe macromolecular trafficking in the phloem, necessitating the use of an integrated approach with a diversity of experimental approaches.

## 8.3 Macromolecular trafficking in the phloem

### *8.3.1 Source to sink: there is a strong directional movement in the phloem translocation stream*

A major function of the phloem is the long-distance transport of photoassimilates, and the role of PD in this phenomenon is addressed in Chapter 7. However, as photoassimilate transport impacts on the trafficking of all macromolecules in the phloem, a brief synopsis is provided. Photoassimilate transport involves a pressure-flow mechanism where sugars are loaded into SEs in net carbon exporting source tissues (e.g. mature photosynthesizing leaves and stems), and removed in net carbon importing sink tissues (e.g. shoot and root apices; Figure 8.1). A turgor pressure gradient established between source and sink tissues drives the movement of photoassimilate in a source-to-sink direction. Phloem loading and unloading can be apoplasmic (moving through the cell wall via the action of transporters in the plasma membrane) or symplasmic (involving PD), depending on the plant species (van Bel, 2003).

In terms of symplasmic photoassimilate transport, the phloem can be subdivided into three functional zones: the collection phloem (in source tissues), the transport phloem (in tissues in between source and sink tissues, such as stems) and release phloem (in sink tissues) (van Bel, 2003). The relative function of these components in photoassimilate transport influences the size and symplasmic status of CCs. For example, the relative CC : SE cell volume is high in collection phloem and low in transfer and release phloem; this presumably reflects the relative metabolic demands on CCs for local photoassimilate transfer and, possibly, macromolecular trafficking. It is important to recognize that the nature of the SE–CC complex is dynamic and undergoes changes, e.g. even within an individual leaf during the sink-to-source transition.

### *8.3.2 Do all long-distance signals in the phloem move in a source-to-sink direction?*

In addition to photoassimilate, various other endogenous (macro) molecules (e.g. hormone-like molecules, peptides/proteins and nucleic acids) traffic in the phloem, as well as a number of phytopathogens. Some of these endogenous factors are phloem-mobile, long-distance signals involved in systemic defense pathways [e.g. systemic acquired resistance (SAR) and post-transcriptional gene silencing] and development (e.g. flowering). However, it seems that, given the remarkable specialization of the phloem for photoassilimilate transfer, other systems relying on the phloem for long-distance signaling must also have source-to-sink directionality. This attribute would seem to be compatible for defense pathway signaling, where it is essential to protect new growth, and developmental signaling, where meristematic tissues are targeted (Figure 8.1). Long-distance transport of pathogens in the phloem, such as plant viruses, also occurs in a source-to-sink direction (Bennett, 1956; Gilbertson and Lucas, 1996; Roberts *et al.*, 1997; Nelson and van Bel, 1998; Sudarshana *et al.*, 1998). This allows viruses to establish important infection domains, such as vascular progenitor cells in shoot apical meristems (strong sinks) and vascular and nonvascular cells in unloading phloem of sink leaves, a tissue undergoing enhanced macromolecular trafficking driven by photoassimilate delivery (Gilbertson and Lucas, 1996; Roberts *et al.*, 1997; Sudarshana *et al.*, 1998). Alternatively, the phloem transport stream can be redirected, e.g. by the establishment of new sinks during nodulation of legume roots by rhizobia (Complainville *et al.*, 2003).

### *8.3.3 P-proteins and the wound response*

The presence of proteins in SEs has long been recognized. One of the first proteins to be described in phloem SEs was the phloem protein (P-protein), which occurs in SEs of some flowering plants. In SEs, P-protein occurs in fibrillar or tubular forms. P-protein may act as a clotting or sealing agent, preventing damaged SEs from releasing large quantities of phloem exudates (van Bel, 2003). The release of P-protein, and changes in P-protein properties, occurs in response to changes in turgor pressure mediated by wounding, presumably allowing for rapid blocking of damaged SEs. In some cases, the P-protein-mediated blockage can be reversed in a $Ca^{2+}$-mediated process, possibly revealing a more dynamic role for P-proteins in addition to simple blockage of the sieve plate pores (Knoblauch *et al.*, 2001). The prominent presence of P-proteins in the phloem may indicate additional, yet to be defined, roles for these proteins in macromolecular trafficking in the phloem.

### *8.3.4 A large and diverse population of proteins can be detected in SEs*

Analyses of phloem sap of various plants, including monocots (Nakamura *et al.*, 1993) and dicots (Fisher *et al.*, 1992; Schobert *et al.*, 1998), have revealed the presence of approximately 100–200 soluble proteins, many of which are in the 30–60 kDa range

(Hayashi *et al.*, 2000). More recently, mass spectrophotometry analyses of phloem proteins have supported the concept that over 100 proteins are moving within the phloem (Hoffman-Benning *et al.*, 2002). A number of lines of evidence suggests that these are *bone fide* phloem proteins, rather than contaminants coming from damaged nonvascular cells. These include: (i) detection of similar proteins in phloem sap collected from cut stems or severed aphid stylets, (ii) the profile of 'phloem proteins' is different from that of total plant proteins and (iii) heterografting experiments. Results of intergeneric and interspecific heterografting experiments with cucurbits clearly showed that phloem proteins, distinctive of the rootstock species, could be detected in sieve tube exudates (phloem sap) of grafted scions of a different species (Golecki *et al.*, 1998). Fisher *et al.* (1992) used radiolabeling studies to show that phloem proteins were undergoing continual turnover, which seemed to parallel photoassimilate transport in the phloem. This indicated that phloem protein movement occurs in a source-to-sink direction and that some of these proteins may be a source of amino acids in sink tissues (Kuhn *et al.*, 1997; Thompson and Schulz, 1999). The large number of proteins in the phloem translocation stream and their dynamic nature raises questions about their sites of synthesis, entry into the SEs and their function in the phloem.

### *8.3.5 Synthesis and trafficking of macromolecules across the SE–CC PD*

Given that CCs are well suited for protein synthesis, and have specialized PD connections with SEs, it would seem logical that CCs are a major source of these phloem proteins (Figure 8.2). Indeed, studies of P-proteins (Bostwick *et al.*, 1992), RPP13-1 (a phloem thioredoxin; Ishiwatari *et al.*, 1998) and CmPP36 (a cytochrome reductase; Xoconostle-Cázares *et al.*, 2000) have shown CC-specific patterns of expression. Presumably, these and other phloem proteins would gain access to SEs by trafficking across the specialized PD interconnecting SE–CCs. However, the size range of many of these proteins (~10–100 kDa) seemed to preclude passive diffusion across the SE–CC PD, at least based upon experiments showing that the SEL of mesophyll cell PD is less than 1 kDa. This enigma was solved, in part, when microinjection experiments revealed that the SEL of SE–CC PD in *Vicia faba* stems was approximately 20–30 kDa (Kempers and van Bel, 1997; van Bel, 2003). The high SEL of SE–CC PD would presumably allow for a greater level of macromolecular trafficking of proteins and possibly other macromolecules from CC into SE. Unfortunately, accessing the SE–CC complex for *in vivo* studies, such as microinjection, remains technically difficult, making it hard to directly measure the SEL of SE–CC PD in various tissues and plant species. Thus, there is a real need to develop new approaches to investigate macromolecular trafficking in SE–CC complexes.

### *8.3.6 Green fluorescent protein as a reporter for both trafficking in the phloem and SE–CC SEL*

Compelling evidence that proteins synthesized in CCs can be delivered into SEs, and that SE–CC PD have a higher SEL than PD in mesophyll tissue came from

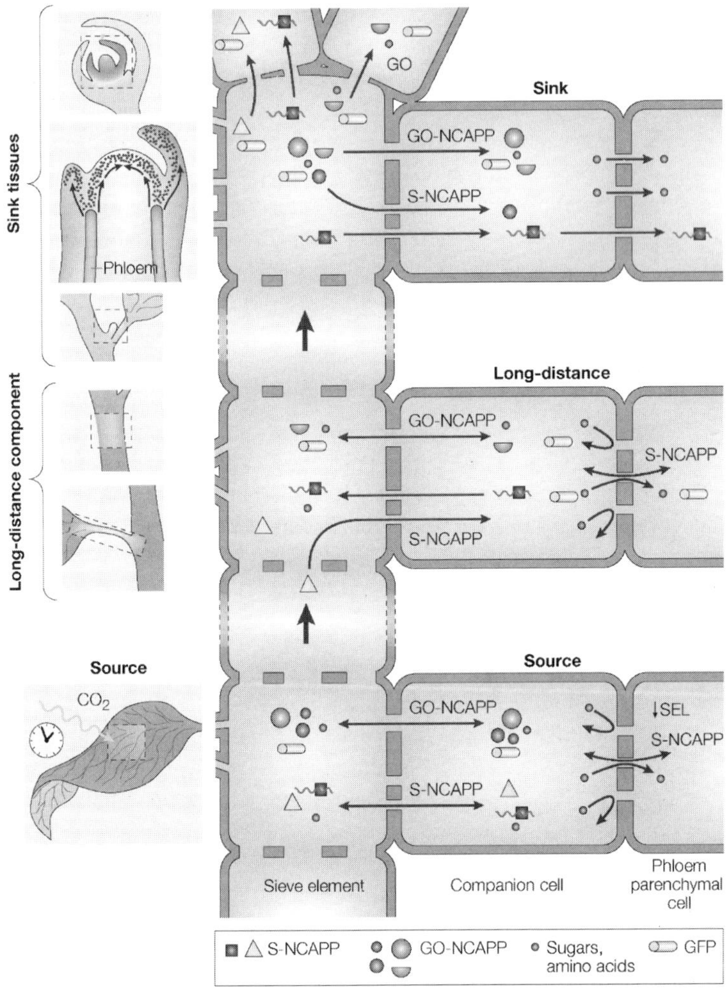


**Figure 8.2** Role of plasmodesmata and the phloem in the delivery of systemically acting non-cell autonomous (macro)molecules. Proteins and RNA are delivered through the phloem translocation pathway. The phloem of mature leaves functions as the site for nutrient loading (sugars and amino acids) for delivery to developing tissues; these two regions of the plant are referred to as source and sink tissues, respectively. Diffusion of nutrients from the phloem sieve-element–companion-cell (CC–SE) complex into the surrounding phloem parenchyma cells and beyond is regulated by modulation of the PD size exclusion limit (SEL). Within the source and long-distance tissues, the SEL of PD that connect the CC–SE appears to be increased from 800 Da to around 35 kDa. This presumably reflects the extensive exchange of non-cell-autonomous proteins (NCAPs) between the CC and SE. Such NCAP trafficking could occur by a selective NCAP pathway (S-NCAPP), or via a gate open (GO) NCAP pathway (GO-NCAPP) where protein diffusion is regulated by the modulation of the CC–SE PD. Along this same region of the plant, the exchange of NCAPs across the boundary from CCs to the phloem parenchyma probably occurs by a trafficking system that minimizes the release of metabolites, perhaps via the S-NCAPP. In sink tissues, the S-NCAPP is shown to mediate the delivery of NCAPs and RNA to specific tissues (see Sink tissues, middle image). Additionally, localized unloading of NCAPs through the GO-NCAPP could function to remove errant molecules for degradation. [Reproduced from Lucas and Lee (2004).]

studies of transgenic *Arabidopsis* and tobacco plants expressing the green fluorescent protein (GFP) reporter, under control of CC-specific promoters [e.g. the *Arabidopsis* sucrose transporter promoter (AtSUC2) or the *Commelina yellow mottle virus* (CoYMV) promoters]. In these plants, free GFP (~27 kDa) was synthesized in CCs, moved into SEs, and was translocated to sink tissues, where it was unloaded into surrounding nonvascular sink tissues (Imlau *et al.*, 1999; Figure 8.2). However, a 54-kDa dimeric form of GFP expressed in transgenic tobacco plants from the CoYMV promoter was confined to CCs, indicating that the SEL of SE–CC PD may have been too low to allow free diffusion of this protein (Itaya *et al.*, 2002). These studies, although performed in various plant species and with different CC-specific promoters, indicated that the SEL of SE–CC PD is between 27–54 kDa. Interestingly, an approximately 59 kDa fusion between the *Cucumber mosaic virus* (CMV) 3a movement protein (MP) and GFP, expressed in transgenic tobacco with the CoYMV promoter, effectively trafficked from CCs, indicating the capacity of the viral MP to interact with SE–CC PD to allow trafficking of a protein with a mass greater than the 'basal' PD SEL. The presence of many phloem proteins greater than 30 kDa in SEs would seem to suggest a broad capacity for such proteins to interact with CC–SE PD, or the presence of an endogenous macromolecular trafficking system that delivers a diversity of proteins into the phloem translocation stream (Figure 8.2). Proteins less than approximately 30 kDa may readily diffuse through SE–CC PD into SEs or, depending on their characteristics, need to be trafficked via the endogenous system. Clearly, phloem proteins must vary in terms of their capacity to interact with SE–CC PD. In this regard, it is important to keep in mind that reporter proteins most likely do not act like endogenous phloem proteins (Thompson and Schulz, 1999; Haywood *et al.*, 2002).

### *8.3.7 Phloem proteins can increase the SEL of mesophyll PD*

Because of the technical challenges associated with performing microinjection in phloem cells, one approach to assessing the capacity of phloem proteins to interact with PD has been to conduct microinjection experiments in mesophyll cells. A number of endogenous and viral proteins, known or suspected to traffic in the phloem, increased the SEL of mesophyll PD and mediated their own cell-to-cell movement or that of fluorescent probes, albeit to different extents and capacities (Balachandran *et al.*, 1997; Ishiwatari *et al.*, 1998; Rojas *et al.*, 2001; Lee *et al.*, 2003). The results of these studies support the hypothesis that phloem proteins can interact with mesophyll PD, and by deduction SE–CC PD, to mediate their transport into SEs. The finding that the MPs of the phloem-limited *Tomato yellow leaf curl virus* showed a limited capacity to interact with mesophyll PD may reveal a specialized capacity of some proteins to interact with SE–CC PD (Rojas *et al.*, 2001). It is also not clear whether the SEL of SE–CC PD is intrinsically higher than other PD and/or whether the SEL is upregulated during the course of extensive macromolecular trafficking from CC to SE.

### 8.3.8 *What is the function of the nonstructural phloem proteins?*

Insight into the role of phloem proteins has come from the cloning and characterization of the genes encoding these proteins, and inferences into function based on homologies with previously characterized proteins. This approach has revealed a wide range of proteins in the phloem sap that represent a diversity of functions (Hayashi *et al.*, 2000). In addition to P-proteins and sucrose transporters involved in phloem loading/unloading, other types of proteins have been detected. These include enzymes such as proteinases, proteinase inhibitors (PIs), protein kinases and ubiquitin that are likely involved in protein modification and degradation. Redox regulating enzymes such as thioredoxin h, glutaredoxin and glutathione reductase also have been detected (Hayashi *et al.*, 2000). The biological relevance of the redox regulating enzymes in the phloem was recently supported by the detection of a stress-associated functional antioxidant defense system in SEs of cucurbit plants (Walz *et al.*, 2002). This system may be involved in the maintenance and/or regulation of SEs, through the scavenging of reactive oxygen species (ROS), or it could be involved in ROS-mediated signaling in the phloem.

### 8.3.9 *Phloem proteins as molecular chaperones*

As mentioned above, some phloem proteins have the capacity to interact with PD to increase the SEL. Such proteins could themselves serve as long-distance signals, or they could mediate the transport of long-distance signaling molecules such as other proteins or nucleic acids (Figure 8.2). The identification of CmPP16, a protein from the phloem sap of pumpkin plants (*Cucurbita maxima*), that interacted with PD to mediate its cell-to-cell movement as well as that of single-stranded (ss) RNA, indicated that some phloem proteins may mediate movement of other macromolecules into SEs via interactions with SE–CC PD (Xoconostle-Cázares *et al.*, 1999). Furthermore, the CmPP16 has functional similarities with a viral MP, thus supporting the concept that some viral MP genes were derived from plant genes encoding components of an endogenous RNA trafficking system. In this scenario, genes were acquired by viruses, probably through RNA::RNA recombination, and altered to encode an MP that allowed the viral RNA genome to utilize the endogenous RNA trafficking network for cell-to-cell and long-distance transport (Lucas and Gilbertson, 1994).

Further evidence for the existence of a trafficking system for entry of non-cell autonomous proteins (NCAPs) into the phloem came from the identification of a protein from *Nicotiana tabacum* that mediates and regulates protein trafficking through PD, possibly via targeting to the ER (Lee *et al.*, 2003). This protein, named NON-CELL AUTONOMOUS PATHWAY PROTEIN1 (NtNCAPP1), was inferred to be involved in cell-to-cell movement of CmPP16 and *Tobacco mosaic virus* (TMV) MP through PD, based on results of microinjection experiments showing that an *N*-terminal NtNCAPP1 deletion mutant blocked their cell-to-cell movement. This dominant-negative mutant form of NtNCAPP1 did not interfere with movement of the CMV 3a MP or the KNOTTED 1 (KN1) transcription factor, indicating specificity in this pathway. Transgenic plants expressing this NtNCAPP1 mutant protein, or plants silenced for NtNCAPP1 expression, developed abnormal phenotypes, consistent with perturbation of plant development and an involvement in local and/or long-distance signaling. Similar abnormal phenotypes in plants transgenic for certain viral MPs also have been attributed to interference of normal macromolecular transport processes (Latham *et al.*, 1997; Hou *et al.*, 2000; Dunoyer *et al.*, 2004).

Finally, the finding that the phloem proteins may bind to and mediate trafficking of viroid RNAs in the phloem indicates that chaperone-type proteins may be involved in trafficking of nucleic acids in plants (Ding *et al.*, 2003). It will be important to further characterize these RNA binding proteins, as well as other chaperone-type proteins from phloem sap of other plants, and to investigate their roles in long-distance signaling.

## 8.4 Nucleic acids in the phloem and their role in long-distance signaling

### *8.4.1 Why do most plant viruses have RNA genomes?*

DNA and RNA each have the potential to serve as information macromolecules involved in long-distance signaling. Insight into the preferred form of nucleic acid involved in such processes may be gained from examining the nature of the nucleic acid genomes of plant-infecting viruses. In animals, invertebrates and prokaryotes, many families of viruses with single-stranded (ss) or double-stranded (ds) DNA genomes have evolved, whereas in plants, the preponderance of families (or groups) have ssRNA genomes (van Regenmortel *et al.*, 2000). Indeed, only a single family of plant viruses has an dsDNA genome (the *Caulimoviridae* family), whereas only two possess an ssDNA genome (the *Geminiviridae* and *Nanoviridae*). Furthermore, it has recently been established that limits on the genome size of bipartite geminiviruses, during systemic infection, are imposed at the level of cell-to-cell movement through PD – a process mediated by the viral MP (Gilbertson *et al.*, 2003). This led to the hypothesis that few types of plant-infecting DNA viruses have evolved because of constraints imposed by an endogenous trafficking system that utilizes RNA as a non-cell autonomous signal and ensures that DNA acts in a cell-autonomous manner.

### *8.4.2 Going with the flow: phloem-mediated delivery of viral reporters*

The movement of viral genomes through PD is the subject of Chapter 9. However, several points regarding long-distance movement of viruses and macromolecular trafficking in the phloem should be mentioned here. First, early studies using physiological and ultrastructural approaches indicated that long-distance movement of plant viruses occurs through the phloem and in the direction of sink tissues (e.g. Bennett, 1956; Figure 8.1). Recent studies performed with GFP-tagged viruses have confirmed the pathway and tropism associated with the establishment of new infections in developing tissues (Roberts *et al.*, 1997; Sudarshana *et al.*, 1998). Second, the long-distance spread of the viral genome is dependent on one or more proteins, usually an MP and/or the capsid protein (CP), but the precise nature of the viral form(s) moving in the phloem is less clear (Carrington *et al.*, 1996; Gilbertson and Lucas, 1996; Nelson and van Bel, 1998). In some cases, long-distance movement occurs in the absence of CP, indicating that a non-virion nucleoprotein complex can effectively move in the phloem (e.g. Sudarshana *et al.*, 1998). It is also possible that viruses may populate the phloem with more than one infectious form, e.g. one form for insect transmission (virions) and another for systemic infection (nonvirion nucleoprotein complex) (Gilbertson and Lucas, 1996; Rojas *et al.*, 2001).

Regardless of the actual nature of the infectious agent, viruses have evolved forms that are effectively transported via the phloem, and are able to transmit their infectious (often detrimental) nucleic acid signal to cell types in which new infections are initiated (e.g. phloem progenitor cells). These viral nucleic acid signals require 'chaperone' proteins for such long-distance movement, and these proteins may play a protective role and/or facilitate the trafficking of viral nucleic acids across PD boundaries and into susceptible cells. A better understanding of the nature of these complexes and the role of plant proteins involved in viral long-distance transport should provide additional insights into the endogenous RNA trafficking pathway in plants.

### *8.4.3 Endogenous mRNAs in the phloem: possible long-distance developmental signals*

Plant viruses demonstrate that RNA can be an effective long-distance information macromolecule and, furthermore, that the phloem is an ideal system for long-distance transport. It stands to reason that endogenous RNAs are also moving along this long-distance RNA transport system, and possibly transmitting a positive (developmental) nucleic acid signal to developing (sink) tissues/organs (Wu *et al.*, 2002; Figure 8.2). These endogenous RNAs, presumably synthesized in CCs, must cross SE–CC PD to gain access to the enucleate SEs. One of the first mRNAs detected in SEs was that of the potato *SUCROSE TRANSPORTER-1* (*SUT1*) gene. The localization of *SUT-1* mRNA to SEs and CCs, taken together with the absence of *SUT1* mRNA in SEs of transgenic plants having a CC-specific antisense suppression of *SUT1* mRNA, strongly suggested that these transcripts were synthesized in CCs and transported into SEs across SE–CC PD (Kühn *et al.*, 1997). However, it is also conceivable that some of the *SUT1* RNA in the phloem was derived from nucleate progenitor cells, during the development of mature SEs. The transition between nucleate SE progenitors and mature SEs has also been proposed as a critical point where viral genomes, transported long distance via the phloem, can initiate new infections as well as release infectious forms back into SEs (Gilbertson and Lucas, 1996).

Many hundreds of different RNA molecules have been detected in pumpkin phloem sap by RT-PCR analysis, and extensive control experiments established that these were *bona fide* phloem RNAs rather than contaminants associated with tissue wounding (Ruiz-Medrano *et al.*, 1999). Sequence analyses of randomly selected phloem RNA cDNA clones revealed similarities with the *CmPP16* chaperone gene and genes encoding a number of regulatory factors. One of the phloem mRNAs encoded an NAC-domain-type transcription factor; the gene was named *CmNACP-1* (Ruiz-Medrano *et al.*, 1999). Heterografting experiments with pumpkin and squash, which have differently sized *CmPP16* and *CmNACP-1* mRNAs, established that mRNAs generated in rootstocks can pass across the graft union, be transported through SEs and be detected in phloem sap from the scion tissues (Ruiz-Medrano *et al.*, 1999; Xoconostle-Cázares *et al.*, 1999). The detection of *CmNACP-1* mRNA in cucumber meristems, by *in situ* RT-PCR, raised the possibility that mRNAs moving long-distance in the phloem could be developmental signals that target meristematic tissues.

Direct evidence that mRNA, moving in the phloem, can be a long-distance developmental signal was provided by a study with the dominant tomato mutant phenotype, *Mouse ears* (*Me*). This phenotype is due to high-level expression of a gene fusion between a metabolic gene (*PHOSPHOFRUCTOKINASE, PFP*) and a developmental homeobox gene (*LeT6*, a tomato homolog of the KN-1 gene), which leads to extra orders of leaf compounding (Kim *et al.*, 2001). Wild-type tomato scions grafted on to *Me* rootstocks developed *Me*-like phenotypic characteristics, and *PFP-LeT6* mRNA (but not DNA) was detected in scion phloem and meristem tissues. These results are consistent with long-distance transport of functional *PFP-LeT6* mRNA in the phloem as well as delivery to appropriate target (meristematic) tissues. However, the possibility that the PFP-LeT6 protein serves as the long-distance signal was not excluded by these experiments. In any event, these results also raise the possibility that mRNA is a phloem mobile long-distance signaling molecule that can influence plant development.

### *8.4.4 Phloem translocation of small RNAs*

Systemic RNA silencing is the subject of Chapter 10, and so only a limited discussion of the long-distance signaling component of this process is relevant here. Post-transcriptional gene silencing (PTGS) in plants [a form of RNA interference (RNAi)] is a sequence-specific RNA surveillance mechanism whereby plants can defend themselves against rogue or alien RNAs, such as those of plant viruses and transposable elements (Mlotshwa *et al.*, 2002). PTGS-like systems also appear to moderate gene expression as part of normal plant development. PTGS is triggered by dsRNA, which is not typically found in plants, and results in the sequence-specific RNA degradation mediated by an RNAaseIII-like enzyme (Dicer in animals and Dicer-like in plants). Dicer cleaves dsRNAs into small dsRNAs of 21–25 nt, which are collectively referred to as small interfering (si) RNAs and are hallmarks of PTGS. These siRNAs serve to direct an RNA-induced silencing complex (RISC) to the target RNA, mediating its degradation and signal amplification.

In plants, PTGS is a non-cell-autonomous phenomenon in that it can be induced locally and then spread systemically. The first evidence of long-distance transport of a PTGS signal came from grafting experiments showing that a silencing signal, from transgenic rootstocks silenced for the transgene, moved across a graft union and induced silencing in a non-silenced scion transgenic for the same transgene. Silencing in the scion was specific to the sequence silenced in the rootstock, indicating specificity in the signal, such as that found in nucleic acids (Palauqui *et al.*, 1997; Jorgensen *et al.*, 1998; Mlotshwa *et al.*, 2002). A similar long-distance systemic spread of silencing was observed in transgenic plants when transgene DNA was introduced by agroinfiltration or particle bombardment (Palauqui and Balzergue, 1999). The movement of the silencing signal occurred in a source-to-sink direction consistent with photoassimilate transport, and this was elegantly demonstrated by the pattern of systemic silencing of the GFP reporter in transgenic plants (Voinnet *et al.*, 1998). Evidence that spread was through PD came from a failure to observe silencing in mature guard cells that had lost their PD connections (Mlotshwa *et al.*, 2002).

### *8.4.5 Are small RNAs the long-distance signal involved in PTGS?*

An obvious question was: what is the nature of the long-distance signal that mediates this systemic response? An attractive candidate was siRNA because these molecules: (i) are consistently associated with silencing, (ii) carry the necessary sequence specificity and (iii) are relatively small in size making them potentially amenable to cell-to-cell and long-distance transport. Furthermore, in animal systems, siRNAs induce RNAi (Elbashir *et al.*, 2001). However, there are examples where silencing appears to occur in the absence of siRNAs (Mlotshwa *et al.*, 2002).

Results of a recent study, involving biochemical and *in vivo* approaches, indicate that small ssRNAs (20–25 nt) may be long-distance PTGS signal in the phloem, and that trafficking of these RNAs may involve small RNA binding proteins (Yoo *et al.*, 2004). Populations of small RNAs were detected in phloem sap of four plant species; the nature of these populations varied depending on environmental conditions, the part of the plant examined and viral infection. Bioinformatics and other analyses revealed a variety of mobile small RNAs in pumpkin phloem sap, including various micro (mi)RNAs, and siRNAs associated with transposons and viral silencing and infection. Heterografting experiments confirmed that an siRNA associated with silencing was trafficking in the phloem and could induce silencing of the target gene in the scion.

A small (27 kDa) protein (CmPSRP1) was identified from phloem sap that preferentially binds to small ssRNAs, and is specifically expressed in the vasculature. Microinjection experiments, performed in mesophyll cells of *Nicotiana benthamiana* leaves, indicated that neither ss- or ds-fluorescently labeled forms of the 25-nt RNA (~8 and ~16 kDa, respectively) alone can move freely across PD. However, cell-to-cell movement of these RNAs was facilitated by CmPSRP1, but not by other proteins (TMV MP or the KN1 transcription factor) known to increase the PD SEL and mediate cell-to-cell trafficking of RNA. Thus, CmPSRP1 may specifically mediate cell-to-cell and long-distance transmission of the siRNA PTGS signal. Here, it should also be mentioned that it is possible that transport of siRNAs through SE–CC PD may be different from that occurring across mesophyll PD; e.g. the 20–30 kDa SEL of SE–CC PD may allow for free movement of siRNAs through PD. Collectively, these findings provide insight into the nature of the endogenous long-distance signaling system in the phloem that is associated with gene silencing, and should serve as the framework for future studies of this important pathway.

### *8.4.6 Do miRNAs serve as long-distance developmental signals?*

It is clear that the phloem carries a dynamic population of small RNAs, including mi/siRNAs. Whereas siRNAs are involved in targeting alien RNAs for degradation, the 20–22 nt miRNAs are complimentary to and target endogenous mRNAs. These miRNAs have been shown to control important developmental processes through regulation of gene expression by modulating transcript stability or translation (Pasquinelli and Ruvkun, 2002; Kidner and Martienssen, 2003). A large diversity of miRNAs have been detected in plants (and animals), and the apparent conservation of miRNA precursors suggests that these RNAs may be extremely important in gene regulation. Based on the detection of miRNAs in phloem sap (Yoo *et al.*, 2004), it appears that these molecules could also serve as long-distance signals involved in plant development (Figure 8.2).

## 8.5 Developmental regulation influences the number and properties of PD: impacts on long-distance trafficking in the phloem

It is becoming clear that the number and/or properties of PD (e.g. SEL) can vary among different tissues and even among cells in a given tissue (symplasmic domains), and that this is likely to be under developmental control (Gisel *et al.*, 1999; Oparka and Turgeon, 1999; Zambryski and Crawford, 2000). This dynamic aspect of PD function will greatly impact the nature and extent of local and long-distance signaling in specific tissues at different stages of plant growth (Figure 8.2). In the case of the phloem, this is exemplified by striking differences in PD number and function in collection, transfer and release phloem, especially in plants that use symplasmic loading (van Bel, 2003). There are far more PD connections between vascular and nonvascular cells in source (collection) and sink (release) phloem versus long-distance (transfer or stem) phloem, reflecting the higher levels of macromolecular trafficking associated with phloem loading/unloading of photoassimilates and the maintenance of the osmotic gradient.

Changes in PD-mediated macromolecular trafficking, over the course of plant development, have been revealed in studies of long-distance transport of viral reporters. For example, the capacity of a bipartite geminivirus GFP reporter to exit phloem tissues was found to be a function of the tissue type and stage of development (Sudarshana *et al.*, 1998). In general, the long-distance movement of GFP-tagged viruses confirmed that this process occurs in a source-to-sink pattern (i.e. passively) and that the greatest capacity to exit phloem and initiate new infections occurs within sink tissues (Roberts *et al.*, 1997; Sudarshana *et al.*, 1998). Thus, these plant viral reporters provide useful tools with which to probe the changes in PD-mediated phloem-associated macromolecular trafficking that appear to occur during plant development.

In contrast to viruses, which seem to simply move long distance in a source-to-sink direction with the normal flow of photoassimilate, other plant–microbe interactions can lead to changes in long-distance trafficking in the phloem. An example is the establishment of nodules in roots of *Medicago truncatula* by nitrogen-fixing rhizobia bacteria (Figure 8.1). This involves the establishment of nodule primordia, via cellular reprogramming, and their subsequent development into strong sinks (Complainville *et al.*, 2003). Vascular tissues associated with nodule primordia have increased numbers of PD and PD branching, indicating increased symplasmic trafficking of photoassimilates and probably other phloem constituents. Direct evidence for altered macromolecular trafficking in nodule primordia came from results with AtSUC2-GFP transgenic plants. Roots of these plants typically show symplasmic isolation of vascular and cortical tissues, as revealed by the confinement of GFP to the vasculature; however, the establishment of nodule primordia changed the nature of the symplasmic domains in roots, which allowed GFP to move from the phloem into cortical cells. These modifications are all consistent with nodule primordia becoming strong sinks, and represent an example of a plant–microbe interaction where endogenous signaling can be altered to modify the delivery of molecules to specific sink tissues (Figure 8.1).

The importance of altered (macro)molecular trafficking, in respect to nodule development, was also indicated by the finding that more nodules developed on roots of plants transgenic for a viral MP, suggesting the MP-mediated upregulation of PD SEL favored nodulation Complainville *et al.*, 2003. It is possible that local signaling, with contributions from the bacterial symbiont, is involved in the initiation of nodule primordia, including an increase in PD SEL. As development of the primordia proceeds, this may cause further redirection of other long-distance signals associated with photoassimilate translocation and meristem development. It will be important to better define the nature of the plant and symbiont signals and the pathways that are involved in this process. It is possible that similar signals and pathways are involved in the induction of SAR-responses upon colonization of roots by growth-promoting rhizobacteria (van Loon *et al.*, 1998; Iavicoli *et al.*, 2003).

## 8.6 Signaling and systemic defense responses

Plants defend themselves against insect and pathogen attack in various ways. An important component of plant protection against both insects and pathogens are inducible local and/or systemic responses. Systemic responses are those that develop in tissues distant from the point of insect or pathogen attack (Figure 8.1). Activation of these systemic responses involves the transmission of a long-distance signal(s) that appear to move in the phloem. Three such systems have been studied in detail: (i) the systemic wound response, activated in response to insect feeding; (ii) SAR, activated by plant pathogen invasion; and (iii) induced systemic resistance, activated by colonization of roots by rhizobacteria bacteria. The latter two systems will be mentioned only briefly in this chapter. Emphasis will be placed on long-distance signaling and the wound response, particularly in terms of the nature of the long-distance signal and the role of PD in signal transduction.

### *8.6.1 Signaling in the wound response: the challenges of identifying the long-distance signal*

Many types of plants respond to insect herbivory by the upregulation of various defense-related genes throughout the body of the plant. For example, in members of the Solanaceae family, defense-associated proteins, such as protease inhibitors (PIs) that can interfere with the capacity of insects to digest their plant-derived meal, are expressed in response to insect feeding (Pearce *et al.*, 1991). The expression of defense proteins occurs in wounded and nonwounded tissues, including those distant from the point of damage, indicating a systemic response. This notion was further supported by grafting experiments in which defense proteins accumulated in unwounded scions that were grafted on to wounded rootstocks (Pearce, 1991). Investigation into the nature of the long-distance signaling molecule led to the discovery of systemin, one of the first peptide signals characterized in plants. Systemin is a proline-rich 18-amino-acid polypeptide that was isolated from leaves of wounded tomato plants, and it was named based on having a role in the systemic wound response (Ryan and Pearce, 2003). Indeed, a number of lines of experimental evidence suggested that systemin was the actual long-distance signal and that it was moving in the phloem. First, application of nanomolar quantities of systemin to cut tomato stems led to the expression of PIs in leaves of young tomato plants within minutes, consistent with the timeframe associated with the wound response. Second, $^{14}$C-labeled systemin was detected in the phloem, and *p*-chloromercuribenzene sulfonic acid – an inhibitor of phloem loading – prevented the movement of radiolabeled systemin (Narváez-Vásquez *et al.*, 1995). These findings raised numerous other questions regarding systemin as the long-distance signal of the systemic wound response, including how and where it is synthesized, how it enters the phloem and how it is involved in the expression of defense proteins in unwounded tissues.

### *8.6.2 The systemin receptor: unexpected homolog of the brassinosteroid hormone receptor*

Systemin is produced from prosystemin, a 200-amino-acid precursor protein, expressed in the vasculature, and specifically in the phloem parenchyma (Narváez-Vásquez *et al.*, 2004). The finding that blocking prosystemin production in transgenic tomatoes via antisense RNA led to an inhibition in the systemic wound response signal and increased susceptibility to insect herbivory was further evidence of a key role for systemin in this response. A model was proposed where, upon wounding, systemin is released from prosystemin in the vasculature and moves into SEs. Once in SEs, systemin would move with the translocation stream, in a source–sink manner, to unwounded tissues where it would initiate expression of PIs and defense proteins (Figure 8.3; Ryan, 2000).

The recent identification of a systemin receptor, a 160-kDa plasma membrane protein, SR160 (Scheer and Ryan, 2002), revealed that signaling in the systemin-associated systemic wound response may be more complex. The SR160 protein is a leucine-rich repeat receptor-like kinase protein (LRR-RLK) that, rather unexpectedly, turned out to be the tomato ortholog of the *Arabidopsis* brassinosteroid receptor, BRI1 (Montoya *et al.*, 2002). This finding revealed that the tomato BRI1 (SR160) recognized two types of ligands and, perhaps more importantly, a possible linkage between signaling mediated by steroid and peptide ligands (Wang and He, 2004).

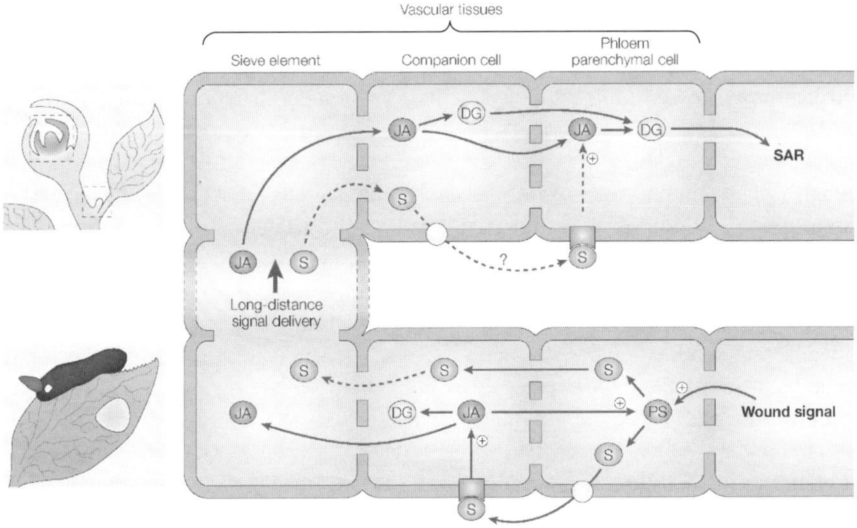


**Figure 8.3** Systemic defense is mediated by the long-distance movement of information (macro)-molecules. Wound-induced signaling, following attack by chewing insects, is used to illustrate the principle of phloem-mediated systemic defense against pathogens. Prosystemin (PS) is made in the phloem parenchyma cells and is cleaved into an 18-amino-acid peptide hormone, systemin (S) in response to incoming wound signals. Systemin is then released into the apoplasm where it binds to a leucine-rich repeat kinase on the plasma membrane. The S-activated form of this protein kinase upregulates the jasmonate (JA) signaling pathway, thereby leading to transcription of a spectrum of defense genes (DGs). In the current model, JA (210 Da) diffuses into the sieve tube system through the CC–SE PD and is then translocated in the phloem to developing sink tissues. Phloem delivery of JA along this pathway then induces the same DG cascade leading to systemic protection against the invading pathogenic agent. [Reproduced from Lucas and Lee (2004).]

The interaction of systemin with the BRI1/SR160 receptor leads to activation of a range of rapid signaling events, including activation of the octadecanoid pathway. This latter pathway results in the generation of the signal molecule, jasmonic acid (JA), via 12-oxo-phytodienoic acid (OPDA) from linolenic acid (LeA) released from cell membranes. Interestingly, JA had been previously shown to induce the expression of PIs associated with the systemic wound response. These findings suggested a key role for JA in the systemic wound response pathway, perhaps even as the long-distance signal.

### *8.6.3 Is JA the long-distance signal involved in the systemic wound response?*

The question of the role of systemin and JA in the systemic wound response signaling was elegantly addressed through the use of grafting experiments with wild-type and wound-response mutant tomato lines (Li *et al.*, 2002; Lee and Howe, 2003; Stratmann, 2003). Mutant lines deficient in JA perception (*jai-1*$^2$ plants) failed to express PI genes when grafted on to wounded wild-type or prosystemin transgenic rootstocks. However, wounded *jai-1*$^2$ plants delivered the wound signal when used as rootstocks with wild-type scions. Mutant plants deficient in JA biosynthesis (*spr2* plants) expressed PI genes when grafted on to wild-type or prosystemin rootstocks (consistent with competence for JA perception); however, *spr2* plants could not deliver the wound signal to wild-type scions when used as rootstocks (and wounded), consistent with failure to generate the long-distance signal. Finally, mutant plants deficient in systemin-specific signaling (*spr1* plants; cannot perceive systemin) underwent the systemic wound response (expressed PIs) when grafted on to wounded wild-type rootstocks, indicating that systemin was not necessary for long-distance signaling. The local wound response in *spr1* plants was not greatly reduced suggesting that systemin plays a role in the production and/or movement of the long-distance signal. Together, these results indicate that JA, or a related molecule, is the long-distance signal, and that systemin amplifies the signal at the wound site (Figure 8.3). JA serving as the long-distance signal is consistent with many types of plants having a systemic wound response system in the absence of systemin-like peptides (found only in solanaceous plants).

### *8.6.4 A JA biosynthesis pathway in the phloem is poised for a rapid wounding response*

Assuming that JA (or related pentacyclic oxylipins) is the long-distance signal involved in the systemic wound response, then how does it gain access to the phloem translocation pathway. JA is generated from LeA, as part of a lipid-based signaling cascade, and a major source of LeA is chloroplast membranes. A number of plastid-localized enzymes are involved in this pathway, including lipoxygenase (LOX), allene oxide synthase (AOS) and allene oxide cyclase (AOC). There is some evidence for a JA biosynthesis pathway in the phloem, including localization of AOC to phloem and the specific accumulation of OPDA and JA in vein and phloem tissue compared with laminar tissues (Hause *et al.*, 2000; Stenzel *et al.*, 2003; Hause *et al.*, 2004). More recently, the LOX, AOS and AOC enzymes were localized to the SE, with AOC further localized to P-type plastids in SEs and CCs (Hause *et al.*, 2004).

*In situ* hybridization analyses further revealed that AOC mRNA accumulated in the CC, whereas that of AOS localized in CC and SE. This suggested that AOC (~26 kDa) is translated in CCs and moves, via PD, into SEs; whereas for AOS, the mRNA would seem to traffic through PD and AOS may be synthesized on the parietal ER in SEs. The AOC protein is within the size range of proteins that may be able to freely move through the SE–CC PD, whereas the approximately-60-kDa AOS may be greater than the CC–SE PD SEL, necessitating mRNA movement or selective trafficking of AOS protein. The presence of a JA biosynthesis pathway in the phloem would allow for the rapid release of JA into the phloem translocation stream, possibly mediated by the breakdown of plastids, where it would be delivered to unwounded tissues in a source-to-sink direction (Figures. 8.1 and 8.3).

### *8.6.5 Prosystemin in phloem parenchyma cells: compartmentalization of different signals?*

What then is the role played by systemin? Based upon genetic studies, systemin is an upstream signal that initiates and possibly amplifies the JA signal. The localization of the systemin precusor, prosystemin, to vascular phloem parenchyma cells of minor veins and leaf midribs (Narváez-Vásquez and Ryan, 2004) may reveal a vascular compartmentalization of molecules involved in the systemic wound response, thereby guarding against the release of the JA signal unless wounding occurs. In this scenario, wounding would trigger a local signal, comprised of systemin released from prosystemin in vascular phloem parenchyma cells and possibly moving apoplasmically to CCs, and JA generated via the systemin-mediated interaction with the BRI1/SR160 receptor and/or the octadecanoid pathway upregulated by the availability of LeA from damaged membranes. This local signaling would result in the generation and amplification of the JA long-distance signal, including JA production in SEs, and the subsequent JA-mediated long-distance signaling (Figure 8.3). However, many questions remain to be answered including the precise role of systemin in long-distance signaling, the role of PD in macromolecular trafficking of systemin, why systemin has been detected in the phloem and the significance of the nuclear localization of prosystemin in nuclei of phloem parenchyma cells.

### *8.6.6 Systemic acquired resistance: similar challenges in the identification of the long-distance signal*

Interesting parallels can be drawn between the elucidation of the long-distance signals involved in the systemic wound response and SAR. Increased levels of salicylic acid (SA) were long associated with local and systemic SAR, leading to the hypothesis that SA was the long-distance signaling agent. Evidence for this came from experiments showing that transgenic plants defective in SA failed to develop SAR, and that SAR could be induced in plants by treatment with SA analogs (Dong, 2001). However, as with the systemic wound response signal, grafting experiments revealed that SA was not the long-distance signal but rather a local signal involved with the generation of the long-distance signal (Vernooij *et al.*, 1994). The precise nature of the SAR long-distance signal remains to be established, but recent evidence suggests it may be a lipid-derived signal. First, an *Arabidopsis* T-DNA insertion mutant, defective in development of induced resistance, carried a mutation in a gene encoding a potential apoplasmic lipid-transfer protein essential for SAR (Maldonado *et al.*, 2002). Second, an SA-binding protein, involved in local resistance and SAR to TMV, was shown to be a lipase. This led to the hypothesis that, upon binding with SA, this lipase is activated and releases a lipid-derived mobile SAR signal. Thus, a model is emerging in which SA acts locally to mediate the generation of a lipid-derived mobile signal that may then move through the phloem, possibly together with SA, to mediate SAR in distant tissues.

### *8.6.7 A need to better understand trafficking of lipid-based signals*

The systemic wound response and SAR pathways have been extensively studied; yet, the precise identification of the long-distance signals involved in these pathways has been challenging. Through a combination of genetic, biochemical and molecular approaches, a key role for hormone-like signal molecules, such as JA, is now emerging. Further evidence is needed to establish that JA is the long-distance signal, such as radiolabeling experiments to follow movement in the plant. More insight into the nature of the JA biosynthesis pathway in the phloem, such as the localization and further identification of the components of this pathway, is needed in order to understand how the JA signal enters the translocation stream and exits into target cells. Given the relatively small size of JA, it will be of particular interest to elucidate the role of PD in movement across various boundaries, especially the SE–CC. In addition, a role for other signals in the systemic wound response or SAR, such as changes in electrical potential, hydrogen peroxide ($H_2O_2$) and nitric oxide (NO) cannot be discounted.

### *8.6.8 Cross-talk in systemic signaling systems*

It also will be important to understand the cross-talk that goes on between the JA-mediated systemic wound signaling and other JA-associated pathways, including those involved in plant development (Figure 8.1). The finding that a common receptor can recognize systemin and brassinosteroid signals indicates that these pathways may have evolved from a similar pathway to mediate distinct responses (Wang and He, 2004). Alternatively, there may be a benefit of linking the wound response and developmental signaling in an overall plant response to wounding. It has also been suggested that there is a division of labor in terms of the broad type of invader targeted by defense response pathways, with SA-mediated pathways affecting so-called biotrophic or obligate pathogens (e.g. viruses and fungi such as powdery and downy mildews), JA/ethylene-mediated pathways target necrotrophic pathogens and herbivorous insects and PTGS targeting viruses. Obviously, determining signals and responses that are specific to each of these pathways could be complicated by cross-talk between and among pathways (Pieterse and van Loon, 2004).

## 8.7 Prospectives

In higher plants, the phloem has evolved into an efficient system for long-distance transport of a diversity of molecules, including a variety of signaling agents involved in key processes such as defense and development (Figure 8.1). Specialized PD interconnect the SE–CC and mediate macromolecular trafficking. In addition to transporting endogenous macromolecules, viruses and other microbes have evolved to use the phloem for long-distance transport throughout the host. However, in all of these cases, the precise nature of the long-distance signal and the components of trafficking system(s) used to enter and exit the phloem translocation pathway are not fully understood. The challenges associated with dissecting these pathways and identifying the signal(s) that are moving long distance in the phloem were discussed in this chapter and exemplified for the systemic wound response.

It is clear that the phloem is a very active highway for the translocation of many macromolecules. A wide range of proteins occur in the phloem and these appear to be involved in a diversity of processes. In terms of long-distance signaling, most of these proteins do not appear to be signals *per se*, but are either downstream components of enzymatic pathways involved in these responses or chaperones mediating movement or amplification of signals. Because phloem proteins are synthesized in CCs, they must traffic across SE–CC PD. Some phloem proteins can mediate their own cell-to-cell transport, whereas others may rely on specialized chaperone proteins for such passage. More needs to be known about the specificity of protein trafficking across SE–CC PD, in terms of the type of proteins that move and the pathway(s) that allow for movement. More information is also needed on the properties of SE–CC PD and the extent to which results obtained from mesophyll PD apply to these specialized PD. In order to obtain this information, new innovative approaches will need to be developed in order to overcome the technical challenges associated with accessing these SE–CC PD for *in vivo* studies.

Evidence is emerging that small RNAs (si/miRNAs) are important long-distance signals, both in defense (siRNA in gene silencing) and development (miRNAs); the role of mRNAs is less clear. These small RNAs appear to be too large to pass freely across PD, and may rely on RNA-binding proteins for local and/or long-distance movement. RNA-binding proteins may play a particularly important role in long-distance movement of RNA signals (Federoff, 2002), and some viral MPs may well represent variants of what were originally endogenous non-cell-autonomous RNA-binding proteins. Evidence continues to accumulate that the phloem is an endogenous 'information superhighway' and that RNA is a pivotal information macromolecule. The identification of hundreds of RNAs in phloem sap indicates that there is much work ahead to sort out the complexity of potential RNA signaling pathways, including the role of small RNAs versus miRNAs. Understanding of these pathways could be greatly facilitated by genomic/proteomic-based analyses of phloem sap.

A number of themes seem to be emerging in this field of study. One is a new appreciation of the role of lipid-based signals, such as JA and other related molecules, in long-distance signaling through the phloem. A phloem-based JA biosynthesis pathway may provide an important mechanism for rapid release of JA into SEs, although a better understanding of this and other potential metabolic pathways localized to SEs is needed. The role of PD in JA signaling in the phloem is unclear and needs to be determined as movement of this small molecule could be via an apoplasmic and/or symplasmic route. Second, there is likely to be considerable cross-talk and overlap in the systemic signals and pathways used in defense and development, as exemplified by the common receptor for systemin and brassinosteroid and commonalities in the siRNA and miRNA pathways (Pieterse and van Loon, 2004; Wang and He, 2004). This will complicate the dissection of such pathways but, on the other hand, findings from one pathway may be applicable to other pathways. Third, it is becoming clear that local signal amplification and relay is critical for the generation and maintenance of long-distance signaling in the phloem, and PD could play an important role in these processes. In some cases, the pathway involved in local amplification of the signal may be different from that involved in the generation of the long-distance signal. Continued progress in understanding the complexity of long-distance signaling systems in the phloem and the role of PD will require an integrated approach involving a diversity of tools such as: (i) grafting and heterografting studies ('old' methods that are still very useful for studying phloem transport), (ii) various types of transgenic plants (including those engineered to express macromolecules specifically in the phloem via phloem-specific promoters), (iii) *in vivo* approaches (e.g. microinjection and PTGS-based methods), (iv) genetic approaches (e.g. identification and characterization of signaling mutants and PTGS-targeting of candidate genes), (v) viral reporter systems and (vi) genomic, proteomic and metabolomic analyses of phloem contents. Ultimately, the information gained from these studies of macromolecular trafficking in the phloem will lead to new approaches to modifying plants to improve yield and nutrition and to better handle biotic and abiotic stresses.

## References

Balachandran, S., Xiang, Y., Schobert, C., Thompson, G.A. and Lucas, W.J. (1997) 'Phloem sap proteins from *Cucurbita maxima* and *Ricinus communis* have the capacity to traffic cell to cell through plasmodesmata', *Proceedings of the National Academy of Sciences USA* **94**, 14150–14155.

Bennett, C.W. (1956) 'Biological relations of plant viruses', *Annual Review of Plant Physiology* **7**, 143–170.

Bostwick, D.E., Dannenhoffer, J.M., Skaggs, M.I., Lister, R.M., Larkins, B.A. and Thompson, G.A. (1992) 'Pumpkin phloem lectin genes are specifically expressed in companion cells', *The Plant Cell* **4**, 1539–1548.

Carrington, J.C., Kasschau, K.D., Mahajan, S.K. and Schaad, M.C. (1996) 'Cell-to-cell and long-distance transport of viruses in plants', *The Plant Cell* **8**, 1669–1681.

Complainville, A., Brocard, L., Roberts, I. *et al.* (2003) 'Nodule initiation involves the creation of a new symplasmic field in specific root cells of *Medicago* species', *The Plant Cell* **15**, 2778–2791.

Ding, B., Itaya, A. and Qi, Y. (2003) 'Symplasmic protein and RNA traffic: regulatory points and regulatory factors', *Current Opinion in Plant Biology* **6**, 596–602.

Dong, X. (2001) 'Genetic dissection of systemic acquired resistance', *Current Opinion in Plant Biology* **4**, 309–314.

Dunoyer, P., Lecellier, C.-H., Parizotto, E.A., Himer, C. and Voinnet, O. (2004) 'Probing the microRNA and small interfering RNA pathways with virus-encoded suppressors of RNA silencing', *The Plant Cell* **16**, 1235–1255.

Elbashir, S.M., Lendeckel, W. and Tuschl, T. (2001) 'RNA interference is mediated by 21- and 22-nucleotide RNAs', *Genes and Development* **15**, 188–200.

Esau, K. (1969) The Phloem, in *Handbuch der Pflanzenanatomie, Vol. 5. no. 2* (eds W. Zimmerman, P. Ozenda, and H.D. Wulff), Gebruder Borntraeger, Berlin, pp. 363–372.

Esau, K. and Cheadle, V.I. (1959) 'Size of pores and their contents in sieve elements of dicotyledons', *Proceedings of the National Academy of Sciences USA* **45**, 156–162.

Fedoroff, N.V. (2002) 'RNA-binding proteins in plants: the tip of an iceberg?', *Current Opinion in Plant Biology* **5**, 452–459.

Fisher, D.B., Wu, Y. and Ku, M.S.B. (1992) 'Turnover of soluble proteins in the wheat sieve tube', *Plant Physiology* **100**, 1433–1441.

Gilbertson, R.L. and Lucas, W.J. (1996) 'How do viruses traffic on the "vascular highway"?', *Trends in Plant Science* **1**, 260–268.

Gilbertson, R.L., Sudarshana, M.R., Jiang, H., Rojas, M.R. and Lucas, W.J. (2003) 'Limitations on geminivirus genome size imposed by plasmodesmata and viral-encoded movement protein: insights into DNA trafficking', *The Plant Cell* **15**, 1–15.

Gisel, A., Barella, S., Hempel, F.D. and Zambryski, P.C. (1999) 'Temporal and spatial regulation of symplastic trafficking during development, in *Arabidopsis thaliana* apices', *Development* **126**, 1879–1889.

Golecki, B., Schulz, A., Carstens-Behrens, U. and Kollman, R. (1998) 'Evidence for graft transmission of structural phloem proteins or their precursors in heterografts of Cucurbitaceae', *Planta* **206**, 630–640.

Hause, B., Hause, G., Kutter, C., Miersch, O. and Wasternack, C. (2004) 'Enzymes of jasmonate biosynthesis occur in tomato sieve elements', *Plant Cell Physiology* **44**, 643–648.

Hause, B., Stenzel, I., Miersch, O. *et al.* (2000) 'Tissue specific oxylipin signature, of tomato flowers: allene oxide cyclase is highly expressed in distinct flower organs and vascular bundles', *Plant Journal* **24**, 113–126.

Hayashi, H., Fukuda, A., Suzui, N. and Fujimaki, S. (2000) 'Proteins in the sieve element-companion cell complexes: their detection, localization and possible function', *Australian Journal of Plant Physiology* **27**, 489–496.

Haywood, V., Kragler, F. and Lucas, W.J. (2002) 'Plasmodesmata: pathways for protein and ribonucleoprotein signaling', *The Plant Cell* **14**, S303–S325.

Hoffman-Benning, S., Gage, D.A., McIntosh, L., Kende, H. and Zeevaart, J.A.D. (2002) 'Comparison of peptides in the phloem sap of flowering and non-flowering *Perilla* and lupine plants using microbore HPLC followed by matrix-assisted laser desorption/ionization time-of-flight mass spectrometry', *Planta* **216**, 140–147.

Hou, Y.-M., Sanders, R., Ursin, V.M. and Gilbertson, R.L. (2000) 'Transgenic tomato plants expressing geminivirus movement proteins: abnormal phenotypes and delayed infection by *Tomato mottle virus* in transgenic tomatoes expressing the *Bean dwarf mosaic virus* BV1 and BC1 proteins', *Molecular Plant–Microbe Interactions* **13**, 297–308.

Iavicoli, A., Boutet, E., Buchala, A. and Metraux, J.-P. (2003) 'Induced systemic resistance in *Arabidopsis thaliana* in response to root inoculation with *Pseudomonas fluorescens* CHA0', *Molecular Plant–Microbe Interactions* **16**, 851–858.

Imlau, A., Truernit, E. and Sauer, N. (1999) 'Cell-to-cell and long-distance trafficking of the green fluorescent protein in the phloem and symplastic unloading of the protein into sink tissues', *The Plant Cell* **11**, 309–322.

Ishiwatari, Y., Fujiwara, T., McFarland, K.C. *et al.* (1998) 'Rice phloem thioredoxin h has the capacity to mediate its own cell-to-cell transport through plasmodesmata', *Planta* **205**, 12–22.

Itaya, A., Ma, F., Qi, Y. *et al.* (2002) 'Plasmodesmata-mediated selective protein traffic between "symplastically isolated" cells probed by a viral movement protein', *The Plant Cell* **14**, 2071–2083.

Jorgensen, R.A., Atkinson, R.G., Forster, R.L.S. and Lucas, W.J. (1998) 'An RNa-based information superhighway in plants', *Science* **279**, 1486–1487.

Kempers, R. and van Bel, A.J.E. (1997) 'Symplasmic connections between sieve element and companion cell in the stem phloem of *Vicia faba* L. have a size exclusion limit of at least 10 kDa', *Planta* **201**, 195–201.

Kidner, C.A. and Martienssen, R.A. (2003) 'Macro effects of microRNAs in plants', *Trends in Genetics* **19**, 13–16.

Kim, M., Canio, W., Kessler, S. and Sinha, N. (2001) 'Developmental changes due to long-distance movement of a homeobox fusion transcript in tomato', *Science* **293**, 287–289.

Knoblauch, M. and van Bel, A.J.E. (1998) 'Sieve tubes in action', *The Plant Cell* **10**, 35–50.

Knoblauch, M., Peters, W.S., Ehlers, K. and van Bel, A.J.E. (2001) 'Reversible calcium-regulated stopcocks in legume sieve elements', *The Plant Cell* **13**, 1221–1230.

Kühn, C., Franceschi, V.R., Schulz, A., Lemoine, R. and Frommer, W.B. (1997) 'Macromolecular trafficking indicated by localization and turnover of sucrose transporters in enucleate sieve elements', *Science* **275**, 1298–1300.

Latham, J.R., Saunders, K., Pinner, M. and Stanley, J. (1997) 'Induction of plant cell division by beet curly top virus gene C4', *Plant Journal* **11**, 1273–1283.

Lee, G.I. and Howe, G.A. (2003) 'The tomato mutant *spr1* is defective in systemin perception and the production of a systemic wound signal for defense gene expression', *Plant Journal* **33**, 567–576.

Lee, J.Y., Yoo, B.-C., Rojas, M.R., Gomez-Ospina, N., Staehelin, A. and Lucas, W.J. (2003) 'Selective trafficking of non-cell autonomous proteins mediated by NCAPP1', *Science* **299**, 392–396.

Li, L., Li, C., Lee, G.I. and Howe, G.A. (2002) 'Distinct roles for jasmonate synthesis in the systemic wound response of tomato', *Proceedings of the National Academy of Sciences USA* **99**, 6416–6421.

Lucas, W.J. and Gilbertson, R.L. (1994) 'Plasmodesmata in relation to viral movement within leaf tissues', *Annual Review of Phytopathology* **32**, 387–411.

Lucas, W.J. and Lee, J.-Y. (2004) 'Cell-to-cell communication in plants: plasmodesmata as a supracellular control network', *Nature Reviews Molecular Cell Biology* **5**, 712–726.

Maldonado, A.M., Doerner, P., Dixon, R.A., Lamb, C.J. and Cameron, R.K. (2002) 'A putative lipid transfer protein involved in systemic acquired resistance signaling in *Arabidopsis*', *Nature* **419**, 399–403.

Mlotshwa, S., Voinnet, O., Mette, M.F. *et al.* (2002) 'RNA silencing and the mobile silencing signal', *The Plant Cell* **14**, S289–S301.

Montoya, T., Nomura, T., Farrar, K., Kaneta, T., Yokota, T. and Bishop, G. (2002) 'Cloning of the tomato *curl3* gene highlights the putative dual role of the leucine-rich repeat receptor kinase tBRI1/SR160 in plant steroid hormone and peptide hormone signaling', *The Plant Cell* **14**, 3163–3176.

Nakamura, S.I., Hayaski, H., Mori, S. and Chino, M. (1993) 'Protein phosphorylation in sieve tubes of rice plants', *Plant Cell Physiology* **34**, 927–933.

Narváez-Vásquez, J. and Ryan, C.A. (2004) 'The cellular localization of prosystemin: a functional role for phloem parenchyma in systemic wound signaling', *Planta* **218**, 360–369.

Narváez-Vásquez, J., Pearce, G., Orozco-Cardenas, M.L., Franceschi, V.R. and Ryan, C.A. (1995) 'Autoradiographic and biochemical evidence for the systemic translocation of systemin in tomato plants', *Planta* **195**, 593–600.

Nelson, R.S. and van Bel, A.J.E. (1998) 'The mystery of virus trafficking into, through and out of vascular tissue', *Progress in Botany* **59**, 476–533.

Oparka, K.J. and Santa Cruz, S. (2000) 'The great escape: phloem transport and unloading of macromolecules', *Annual Review of Plant Physiology and Plant Molecular Biology* **51**, 323–347.

Oparka, K.J. and Turgeon, R. (1999) 'Sieve elements and companion cells – traffic control centers of the phloem', *The Plant Cell* **11**, 739–750.

Palauqui, J.C. and Balzergue, S. (1999) 'Activation of systemic acquired silencing by localized introduction of DNA', *Current Biology* **9**, 59–66.

Palauqui, J.C., Elmayan, T., Pollien, J.-M. and Vaucheret, H. (1997) 'Systemic acquired silencing: transgene-specific post-transcriptional silencing is transmitted by grafting from silenced stocks to non-silenced scions', *EMBO Journal* **16**, 4738–4745.

Pasquinelli, A.E. and Ruvkun, G. (2002) 'Control of developmental timing by microRNAs and their targets', *Annual Review of Cell and Developmental Biology* **18**, 495–513.

Pearce, G., Strydom, D., Johnson, S. and Ryan, C.A. (1991) 'A polypeptide from tomato leaves induces wound-inducible inhibitor proteins', *Science* **253**, 895–898.

Pieterse, C.M.J. and van Loon, L.C. (2004) 'NPR1: the spider in the web of induced resistance signaling pathways', *Current Opinion in Plant Biology* **7**, 456–464.

Roberts, A.G. and Oparka, K.J. (2003) 'Plasmodesmata and the control of symplasmic transport', *Plant Cell and Environment* **26**, 103–124.

Roberts, A.G., Santa Cruz, S., Roberts, I.M., Prior, D.A.M., Turgeon, R. and Oparka, K.J. (1997) 'Phloem loading in sink leaves of *Nicotiana benthamiana*: comparison of a fluorescent solute with a fluorescent virus', *The Plant Cell* **9**, 1381–1396.

Rojas, M.R., Jiang, H., Salati, R. *et al.* (2001) 'Functional analysis of proteins involved in movement of the monopartite begomovirus, *Tomato yellow leaf curl virus*', *Virology* **291**, 110–125.

Ryan, C.A. (2000) 'The systemin signaling pathway: differential activation of plant defensive genes', *Biochimica et Biophysica Acta* **1477**, 112–121.

Ryan, C.A. and Pearce, G. (2003) 'Systemins: a functionally defined family of peptide signals that regulate defensive genes in Solanaceae', *Proceedings of the National Academy of Sciences USA* **100**, 14577–14580.

Ruiz-Medrano, R., Xoconostle-Cázares, B. and Lucas, W.J. (1999) 'Phloem long-distance transport of *CmNACP* mRNA: implications for supracellular regulation in plants', *Development* **126**, 4405–4419.

Ruiz-Medrano, R., Xoconostle-Cázares, B. and Lucas, W.J. (2001) 'The phloem as a conduit for inter-organ communication', *Current Opinion in Plant Biology* **4**, 202–209.

Scheer, J.M. and Ryan C.A. (2002) 'The systemin receptor SR160 from *Lycopersicon peruvianum* is a member of the LRR receptor kinase family', *Proceedings of the National Academy of Sciences USA* **99**, 9585–9590.

Schobert, C., Baker, L., Szederkenyi, J. *et al.* (1998) 'Identification of immunologically related proteins in sieve-tube exudates collected from monocotyledonous plants', *Planta* **206**, 245–252.

Stenzel, I., Hause, B., Maucher, H. *et al.* (2003) 'Allene oxide cyclase dependence of the wound response and vascular bundle specific generation of jasmonates in tomato-amplification in wound-signaling', *Plant Journal* **33**, 577–589.

Stratmann, J.W. (2003) 'Long distance run in the wound response – jasmonic acid is pulling ahead', *Trends in Plant Science* **8**, 247–250.

Sudarshana, M.R., Wang, H.L., Lucas, W.J. and Gilbertson, R.L. (1998) 'Dynamics of bean dwarf mosaic geminivirus cell-to-cell and long-distance movement in *Phaseolus vulgaris* revealed, using the green fluorescent protein', *Molecular Plant–Microbe Interactions* **11**, 277–291.

Thompson, G.A. and Schulz, A. (1999) 'Macromolecular trafficking in the phloem', *Trends in Plant Science* **4**, 354–360.

van Bel, A.J.E. (2003) 'The phloem, a miracle of ingenuity', *Plant Cell and Environment* **26**, 125–149.

van Bel, A.J.E. and Kempers, R. (1997) 'The pore/plasmodesm unit; key element in the interplay between sieve element and companion cell', *Progress in Botany* **58**, 278–291.

van Bel, A.J.E. and Knoblauch, M. (2000) 'Sieve element and companion cell: the story of the comatose patient and the hyperactive nurse', *Australian Journal of Plant Physiology* **27**, 477–487.

van Loon, L.C., Bakker, P.A.H.M. and Pieterse, C.M.J. (1998) 'Systemic resistance induced by rhizosphere bacteria', *Annual Review of Phytopathology* **36**, 453–483.

van Regenmortel, M.H.V., Fauquet, C.M., Bishop, D.H.L. *et al.* (2000) 'Virus taxonomy: classification and nomenclature of viruses', *Seventh Report of the International Committee on Taxonomy of Viruses*, Academic Press, San Diego, CA.

Vernooij, B., Friedrich, L., Morse, A. *et al.* (1994) 'Salicylic acid is not the translocated signal responsible for inducing systemic acquired resistance but is required in signal transduction', *The Plant Cell* **6**, 1583–1592.

Voinnet, O., Vain, P., Angell, S. and Baulcombe, D.C. (1998) 'Systemic spread of sequence-specific transgene RNA degradation in plants is initiated by localized introduction of ectopic promoterless DNA', *Cell* **95**, 177–187.

Walz, C., Juenger, M., Schad, M. and Kehr, J. (2002) 'Evidence for the presence of activity of a complete antioxidant defense system in mature sieve tubes', *Plant Journal* **31**, 189–197.

Wang, Z.-Y. and He, J.-X. (2004) 'Brassinosteroid signal transduction – choices of signals and receptors', *Trends in Plant Science* **9**, 91–96.

Wu, X.L., Weigel, D. and Wigge, P.A. (2002) 'Signaling in plants by intercellular RNA and protein movement', *Genes and Development* **16**, 151–156.

Xoconostle-Cázares, B., Ruiz-Medrano, R. and Lucas, W.J. (2000) 'Proteolytic processing of CmPP36, a protein from the cytochrome b(5)-reductase family, is required for entry into the phloem translocation pathway', *Plant Journal* **24**, 735–747.

Xoconostle-Cázares, B., Xiang, Y., Ruiz-Medrano, R. *et al.* (1999) 'Plant paralog to viral movement protein potentiates transport of mRNA into the phloem', *Science* **283**, 94–98.

Yoo, B.-Y., Kragler, F., Varkonyi-Gasic, E. *et al.* (2004) 'A systemic small RNA signaling system in plants', *The Plant Cell* **16**, 1–22.

Zambryski, P. and Crawford, K. (2000) 'Plasmodesmata: gatekeepers for cell-to-cell transport of developmental signals in plants', *Annual Review of Cell and Developmental Biology* **16**, 393–421.

# 9 Movement of viruses to and through plasmodesmata

Richard S. Nelson

## 9.1 Introduction

The study of virus movement in plants has evolved from a spectator sport to a 'hands-on' analysis of the viral and host participants in the process. Early studies of virus movement involved the extraction of suspect plant tissue and inoculation of the extract to an indicator host, to determine if virus was present in (i.e. had moved into) the sampled tissue. There was no ability to identify the specific viral or host participants that controlled movement, or the mechanism(s) by which they worked. Today, virus spread can be followed in near real-time, through the use of a chimeric virus expressing a fluorescent label, such as green fluorescent protein (GFP), thereby marking its position in the plant. Our ability to locate viruses in plant cells, coupled with our ability to alter the expression of specific virus- and host-encoded genes, has led to our understanding that plasmodesmata (PD) play a critical role in virus movement. In addition, the expression of individual virus or host proteins as fusions with GFP has allowed the analysis of their individual functions during virus spread. Since the year 2000, research in this area has progressed rapidly beyond tracking virus movement, toward determining the activities of individual viral components and characterizing how these viral components interact with each other and the host factors that control movement.

For those unfamiliar with plant viruses and their movement through the host, a brief history of this research is given here. Virus movement through plant tissue is viewed as a two-step process: local spread from the initially infected cell (i.e. cell-to-cell movement) followed by vascular-mediated spread to distant plant tissues (i.e. vascular-mediated movement). This chapter will only deal with local movement through PD. The reader interested in a more detailed understanding of findings on virus cell-to-cell and vascular movement up to the year 2000 should consult numerous books and reviews published around that year which address the subject (e.g. Derrick and Nelson, 1999; Lazarowitz and Beachy, 1999; Reichel *et al.*, 1999; McLean and Zambryski, 2000; Tzfira *et al.*, 2000; Aaziz *et al.*, 2001; Hull, 2002; Oparka and Roberts, 2001). These reviews have relevant information that may not be referenced in more current reviews of the subject due to space limitations. Throughout the remainder of this chapter, only more current reviews and current primary literature will be cited for each topic.

As noted above, virus movement was first tracked indirectly through biological assays. Later, virus movement was determined through immunocytochemical studies with dead tissue and, most recently, through fluorescence-based studies with live tissue using reporter genes expressed from the virus genome. Incorporation of reporter genes into virus genomes was made possible during the molecular revolution of the 1980s when several viruses were cloned, and their genomes shown to be infectious when inoculated from plasmids or transcripts thereof. The development of commercially available confocal scanning laser microscope systems that could visualize fluorescent reporter genes in near real time was essential for the unambiguous determination of the subcellular location of virally expressed proteins. During the mid 1990s, the molecular identity of host genes involved in virus movement became a more easily achievable goal due to the ability to screen mutant lines and clone the genes necessary for virus movement through a combination of classical and molecular genetic approaches. The technology breakthroughs described above allowed the identification of viral proteins that altered the size exclusion limits (SELs) of, and moved through, PD. These proteins also allowed the movement of nonviral macromolecules between cells (see Chapter 5). These viral proteins became known as movement proteins (MPs) and are defined by these traits. The co-localization and/or association of these proteins with host cytoskeletal proteins led to models of intra- and inter-cellular virus movement based on proposed interactions between these viral and host proteins. The presence of actin within PD and the effect of a microfilament (MF) antagonist, cytochalasin D, on the SELs of the PD further supported theories that cytoskeletal elements were required for virus movement. The host membrane system, generally, was not included in these early models describing virus movement, although it had long been known that the host plasma membrane and the endoplasmic reticulum (ER) also spanned the PD (see Chapter 1).

## 9.2 Virus cell-to-cell movement

As for most activities involving viruses, both viral and host factors are necessary for successful virus cell-to-cell movement through PD. These factors are necessary either for virus movement or to defeat host defenses that might prevent virus movement. Cell-to-cell movement does not consist of simply getting through PD. The process starts with the translation of the viral RNA (vRNA)/DNA, replication of the viral genome and the formation of an intracellular transport form of viral and host factors that then targets the PD. Once at the PD, other viral and host factors become necessary for transport of infectious virus through the PD, and the establishment of new replication and transport complexes in the next cell. Mutations in host or viral genes necessary for any step in this process would, at the whole plant level, yield the same phenotype (i.e. no systemic or local symptoms). Virologists, with more and more success, have created mutants that unlink virus movement from virus replication (replication being an indirect and confounding process for those interested in virus movement). In many instances, they have identified viral proteins required for movement, but which do not support their own movement or the movement of their encoding RNA between cells. These proteins are not classical MPs since they do not support their own movement and/or do not alter the SEL of PD.

While such proteins were suspected to be present in the 1990s, recent work has provided critical experimental evidence to support this position. The requirement of multiple virus-encoded proteins for virus movement throughout all virus genera is one of the major additions to the literature in the last 4 years.

In the following sections, I will review findings on the cell-to-cell movement of viruses representing different genera where a large number of results have been published. The movement process of viruses not within these genera will be mentioned if they differ substantially from these, or if they supply novel information pertinent to models delineated for the heavily studied genera. This is not an exhaustive review, so readers are encouraged to read recent general reviews (e.g. Haywood *et al.*, 2002; Roberts and Oparka, 2003) in addition to the reviews listed in each section for more information on other viruses.

### *9.2.1 Potexviruses and Hordeiviruses*

These genera represent a large group of positive-stranded RNA viruses that require three partially overlapping open reading frames (ORFs) for their cell-to-cell movement. These overlapping ORFs have been referred to as the 'triple gene block' or TGB (reviewed in Morozov and Solovyev, 2003). The TGB-encoded proteins are referred to as TGBp1, TGBp2 and TGBp3 in order of the appearance of their encoding ORFs on the viral genome. Many studies have been published over the last four years describing the cell-to-cell movement requirements of TGB-containing viruses. Models describing their cell-to-cell movement are complicated, but work with these viruses is beginning to serve as a basis for comparison with all other virus genera.

TGBp1 contains NTPase and helicase sequence motifs similar to those found in the large ORFs associated with virus replication from genera not containing TGBs (e.g. the 125–130-kDa proteins encoded by viruses in the genus, *Tobamovirus*; Koonin and Dolja, 1993). TGBp1s from members of the *Hordeivirus* and related genera contain an extended N-terminal domain compared with those from members of the *Potexvirus* genus. These extended domains are rich in arginine/lysine clusters and are conserved between the *Hordeivirus* and related genera members (Morozov and Solovyev, 2003). TGBp2s are conserved in sequence and size among all TGB-containing virus genera. They contain two hydrophilic segments predicted to allow the protein to take a U-shaped transmembrane orientation with the N- and C-termini extending into the cytoplasm (Solovyev *et al.*, 1996). TGBp3s, like TGBp1s, differ substantially in sequence and size between the potexviruses and hordeiviruses. TGBp3s from potex-like viruses contain a single predicted transmembrane segment while those from hordei-like viruses contain two predicted transmembrane segments (Solovyev *et al.*, 1996; Krishnamurthy *et al.*, 2003).

The functions of the TGB proteins in cell-to-cell movement have been studied using biochemical and cell-biological methods. TGBp1s bind ssRNA nonspecifically and cooperatively (reviewed in Morozov and Solovyev, 2003). TGBp1s from hordei-like viruses contain an additional strong noncooperative RNA-binding activity associated with their extended N-terminal domain (Donald *et al.*, 1997; Kalinina *et al.*, 2001). TGBp1s from both potex- and hordei-like viruses display helicase activity *in vitro* (Kalinina *et al.*, 2002). Although the functional relationship of helicases from different superfamilies is not clear, it is interesting that the cellular eIF4-A helicase – a member of the superfamily II helicases – can unwind RNA substrates in both directions similarly to the superfamily I TGBp1s (Du *et al.*, 2002; Kalinina *et al.*, 2002).

The RNA-binding activity of TGBp1 is considered important for the movement of hordei- and potex-like viruses. Support for this theory comes from the finding that the potexviral TGBp1 is co-translocated with the virus genomic RNA and the viral coat protein (CP) during cell-to-cell movement (Morozov *et al.*, 1997; Lough *et al.*, 1998, 2000, 2001). Potex-like viruses, unlike hordei-like viruses, often require the presence of their CP in order to move cell-to-cell within particular hosts (see review by Callaway *et al.*, 2001). TGBp1 of the potexvirus, *Potato virus X*, induces an energy-dependent conformational change in the virion *in vitro* (Atabekov *et al.*, 2000). It was theorized that, as for other helicases that remodel ribonucleoprotein (RNP) complexes in cells, TGBp1 may disrupt existing protein–RNA interactions within an RNP to allow new protein–RNA interactions to occur, perhaps necessary for changing from viral replication to movement competent RNPs (reviewed in Morozov and Solovyev, 2003). The proposed energy requirement for virus movement both to and through the PD could be satisfied through the activity of the NTPase activity of the TGBp1 helicases (Morozov and Solovyev, 2003).

Through fusion of GFP with TGBp1, the location of TGBp1 in cells was visualized in the absence of TGBp2 and TGBp3. The subcellular location of TGBp1 differs between potex- and hordei-like viruses. Potex-like TGBp1s move cell to cell (Yang *et al.*, 2000 and references therein) unlike the hordei-like TGBp1s which do not move to the cell periphery (Erhardt *et al.*, 1999; Lawrence and Jackson, 2001). It is interesting that the smaller TGBp1 from the potexvirus allows an activity (i.e. cell-to-cell movement) that the larger TGBp1s from the hordei-like viruses do not allow. One explanation for this observation is that the additional sequences on the hordei-like TGBp1s prevent their targeting to the cell periphery without the interaction of additional viral proteins. It is interesting that the additional N-terminal sequence on the hordei-like TGBp1s is required to support cell-to-cell movement of hordeiviruses (Donald *et al.*, 1997 and references therein; Solovyev *et al.*, 1999). This is the same function attributed to the CP of potex-like viruses (Lough *et al.*, 2001 and references therein). The CP of *Potato virus X*, when expressed in the presence of the TGBp1 from that virus, however, prevents intercellular movement of TGBp1 (Yang *et al.*, 2000). It is therefore possible that the N-terminal extension of TGBp1 from hordei-like viruses functions similarly to the CP of potex-like viruses in being necessary for cell-to-cell movement while preventing the targeting of TGBp1 to the cell periphery in the absence of other viral factors.

The functional analysis of the TGBp2s and TGBp3s has also progressed significantly over the last 4 years. GFP-tagged TGBp2s were localized to the cortical ER network (Solovyev *et al.*, 2000; Cowan *et al.*, 2002; Mitra *et al.*, 2003; Zamyatnin *et al.*, 2004). When highly expressed, a sub-population of the TGBp2s associates with mobile vesicles resembling Golgi stacks (Solovyev *et al.*, 2000; reviewed in Morozov and Solovyev, 2003). For *Potato virus X*, the association of TGBp2 with a membrane is required, but not sufficient, for virus movement (Mitra *et al.*, 2003). This targeting to ER and Golgi-like stacks is controlled by the C-terminal hydrophobic segment of TGBp2 (Solovyev *et al.*, 2000; Zamyatnin *et al.*, 2002). Recently, it was determined that both potex- and hordei-like TGBp2s can facilitate the movement of free GFP between adjacent epidermal cells (Tamai and Meshi, 2001a). It was proposed that the ability of the TGBp2 to allow GFP spread resides not in a direct effect on GFP or its encoding RNA by TGBp2, but on an interaction between TGBp2 and a host protein, TIP (Fridborg *et al.*, 2003). These researchers determined that TIP not only associated with TGBp2 but with β-1,3-glucanase, an enzyme previously implicated in altering plasmodesmal SEL through control of callose degradation (Iglesias and Meins, 2000). Indeed, limitation of spread of PVX during a hypersensitive response is accompanied by heavy callose deposition near PD (Allison and Shalla, 1974). Thus, it was proposed that TGBp2, through TIP, prevents callose build-up at the neck region of PD by allowing greater β-1,3-glucanase activity, thereby maintaining 'open' PD for virus movement.

GFP-labeled TGBp3s were associated with membrane bodies near the cell wall (Solovyev *et al.*, 2000; Cowan *et al.*, 2002). TGBp3 synthesis results in the formation of new membrane bodies connected to the cortical ER network (Zamyatnin *et al.*, 2002; Gorshkova *et al.*, 2003). The TGBp3 is thus able to increase the level of a specific domain within the ER located near the cell wall of the host cells (reviewed in Morozov and Solovyev, 2003). During expression of a GFP-TGBp3 from a hordeivirus in transgenic *N. benthamiana* plants, the fusion protein formed punctate bodies directly opposite each other in adjacent cells (Gorshkova *et al.*, 2003). These bodies colocalized with callose, a marker associated with PD, thus showing that the structural link for these bodies may reside near PD (Gorshkova *et al.*, 2003). Association of the potexvirus TGBp3 with membranes was shown to require the intact, predicted transmembrane domain (Krishnamurthy *et al.*, 2003). Targeting of the hordei-like virus TGBp3 to the cell wall area may involve an oligomerization sequence in the central hydrophilic region and the C-terminal transmembrane domain (Cowan *et al.*, 2002; Gorshkova *et al.*, 2003; reviewed in Morozov and Solovyev, 2003). Mutations in the hordeivirus TGBp3 central hydrophilic region or the C-terminal predicted transmembrane region yielded a reticulate network of small bodies on cortical ER tubules. Thus, mutations in either of the signal regions of the hordeivirus TGBp3 allowed protein accumulation at what appear to be ER sites, but these protein bodies cannot further traffic to the cell wall (Morozov and Solovyev, 2003).

The expression of TGB proteins individually in cells cannot provide information about interactions between these proteins during virus movement, as noted above for the function of the TGBp1 in the absence or presence of the viral CP. The unique nature of the TGB proteins (small and not involved in virus replication) simplifies the expression and analysis of results when these proteins are transiently co-expressed in cells. Expression of TGBp2 in the presence of TGBp3 resulted in the colocalization of this protein with TGBp3 in the peripheral bodies formed during TGBp3 expression alone (Solovyev *et al.*, 2000; Zamyatnin *et al.*, 2002). Although it is possible that a direct interaction between TGBps 2 and 3, as has been shown with the *Potato mop top virus* (Cowan *et al.*, 2002), causes this co-targeting, it is also possible that the targeting requires only that the TGBp2 be associated with particular membranes that are targeted to the cell periphery with the appearance of TGBp3 (Morozov and Solovyev, 2003). Specifically, it may be that the Golgi-like mobile vesicles formed by TGBp2 (discussed earlier) are retargeted by TGBp3 and that any protein that positions itself in these bodies therefore can be retargeted. It has been shown that the *Poa semilatent virus* TGBp3 can target unrelated membrane-bound proteins involved in the movement of viruses that do not express TGBs (Zamyatnin *et al.*, 2002).

TGBp2 and TGBp3 together likely target hordei-like TGBp1 to PD. Evidence for this comes from studies showing that TGBp1 required TGBp2 and TGBp3 from the homologous virus to traffic to PD (Erhardt *et al.*, 2000; Lough *et al.*, 2000; Lawrence and Jackson, 2001; Zamyatnin *et al.*, 2004). The transport of the hordei-like TGBp1 may require energy as mutations in the NTPase/helicase domains of the TGBp1 abolished its targeting to PD (Erhardt *et al.*, 2000; Lawrence and Jackson, 2001; Zamyatnin *et al.*, 2004).

In summary, viruses with TGB proteins appear to utilize a sophisticated method to transport vRNA from the site of synthesis to the PD for transport between cells. A model describing the cell-to-cell movement of TGB-containing viruses is provided in Figure 9.1a. For the potex- and hordei-like viruses, vRNA is bound by TGBp1 (for potex-like viruses the CP is also involved) to form a movementcompetent RNP complex. Then, either through interaction with TGBps 2 and 3 or alone, these RNPs traffic to the periphery of the cell. The interaction of intracellular movement forms of these viruses with the cytoskeleton has not been studied extensively (see note at end of chapter). In the region of the PD, TGBp3 may aid in positioning the transport RNP and TGBp2 at the PD. TGBp2 may bind the vRNA at this time (Cowan *et al.*, 2002) as well as regulate callose deposition (Fridborg *et al.*, 2003). At the cell wall near the PD the *N*-terminal extension of TGBp1 of the hordei-like viruses, or the CP of the potex-like viruses, may then function to allow transport of the vRNA between cells. This could be achieved through a conformational change in the vRNA : viral protein complex allowing new or altered binding by TGBp1 or the CP. Other viral factors such as the large replicase-associated ORF may affect movement, but none have yet been implicated. Host-factors other than the ER and TIP proteins have not been studied with these viruses (see note at end of chapter).

### *9.2.2 Tobamoviruses*

Cell-to-cell movement of *Tobacco mosaic virus* (TMV), a member of the genus *Tobamovirus*, was the first virus to yield its secrets about cell-to-cell movement at the molecular level. *Tobamoviruses*, like the TGB-containing viruses, are composed

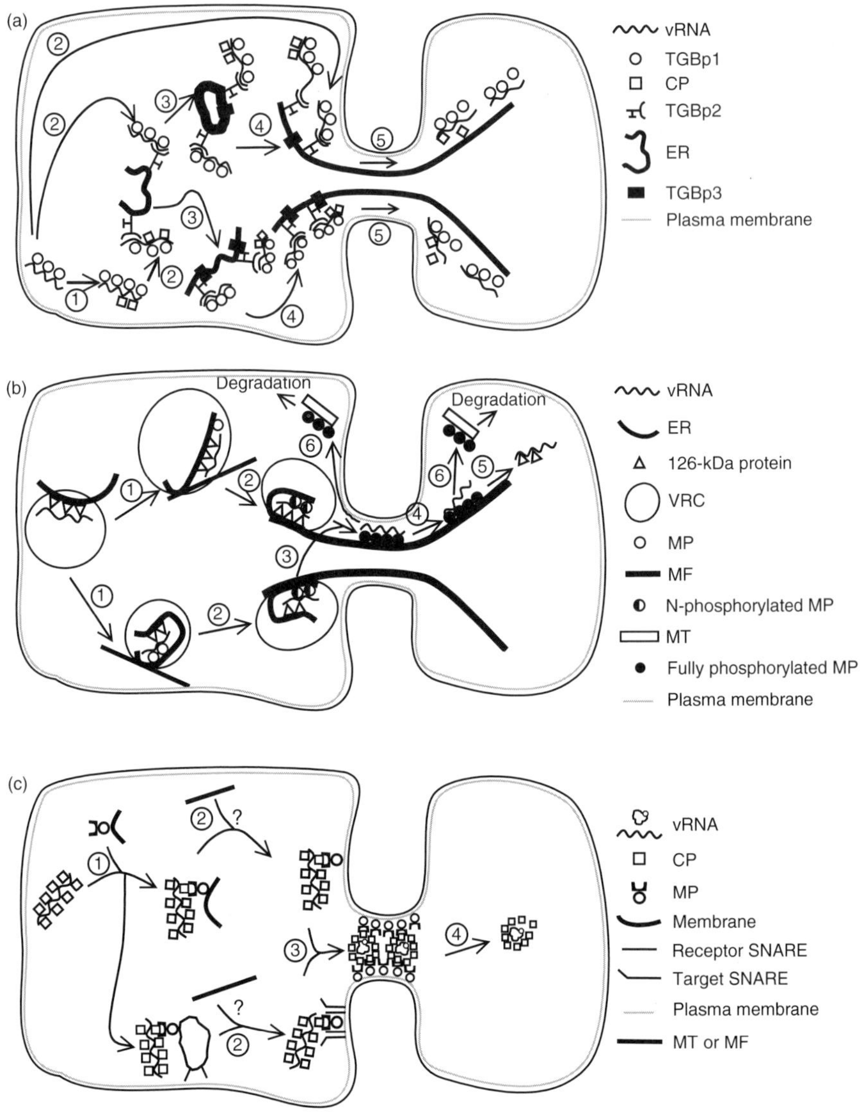
(a)
vRNA
TGBp1
CP
TGBp2
ER
TGBp3
Plasma membrane
(b)
Degradation
Degradation
vRNA
ER
126-kDa protein
VRC
MP
MF
N-phosphorylated MP
MT
Fully phosphorylated MP
Plasma membrane
(c)
?
?
vRNA
CP
MP
Membrane
Receptor SNARE
Target SNARE
Plasma membrane
MT or MF

of positive-stranded RNA (reviewed in Hull, 2002). However, unlike the latter group, the *Tobamoviruses* do not contain a TGB, and unlike potex-like viruses they do not need their CP for cell-to-cell spread in the host. In landmark experiments, the cell-to-cell movement of TMV was shown to require a virally encoded 30-kDa protein translated from a subgenomic RNA (reviewed in Heinlein, 2002). This protein was later referred to as an MP and shown to have the following characteristics: an ability to (i) interact with PD to increase their SEL, (ii) bind ssRNA and (iii) transport itself between cells (reviewed in Haywood *et al.*, 2002). MPs are defined by these characteristics and it is often noted in the literature that TMV has a single MP. While this is true if one understands the definition of an MP, it can be misinterpreted by those not directly involved in studying virus movement. For example, it should be remembered that for the hordei- and potex-like viruses, other proteins are required for their cell-to-cell movement that do not meet these criteria (e.g. TGBp2 and TGBp3 of the *Hordeiviruses*). Often these proteins are referred to as being required for virus cell-to-cell movement, but they are not necessarily MPs *sensu strictu*.

---

**Figure 9.1** Models for cell-to-cell movement of plant viruses using the triple-gene block (TGB), tobamovirus or tubule-forming strategies. (a) Schematic showing a model for cell-to-cell movement of viruses encoding TGBs. Progeny viral RNA (vRNA) bound to the TGBp1 protein, in the absence or presence of the viral coat protein (CP; step 1), binds TGBp2 protein attached to the endoplasmic reticulum (ER) to form a movement–competent ribonucleoprotein (RNP) complex (step 2). This RNP complex then directly or indirectly, through a membrane targeting complex, interacts with the TGBp3 protein to be positioned near the PD within the cell (steps 3 and 4). The schematic shows an association with the ER, however, the plasma membrane may be involved. The TGBp3 protein may aid in positioning the RNP complex for transport through the PD. At, or near, the PD, the TGBp2 protein may downregulate callose deposition. Near, or within, the PD, the *N*-terminal extension of the TGBp1 of hordeiviruses or the CP of potexviruses may function to allow the transport of the vRNA between cells through a conformational change in the viral-RNA–viral-protein complex. (b) Schematic showing a model for cell-to-cell movement of tobamoviruses. After translation, the 126-kDa protein binds both the vRNA and the ER, forming a cytoplasmic body within the cell, the viral replication complex (VRC). The movement protein (MP) is translated and associates with the ER and possibly the vRNA within the VRC (step 1). The VRC associates with the ER either through the 126-kDa protein or the MP. An association of the VRC with microfilaments (MFs) occurs at this time, either directly through viral proteins or indirectly through the ER associated with the viral proteins. Only the indirect association is shown at this step due to space limitations. The VRC then is transported to the region of the PD through its interaction with MFs (step 2). The association of the vRNA with the 126-kDa protein ends (step 3). Phosphorylation of the MP occurs either within the cytoplasm, the cell wall or both, allowing transport through PD and subsequent translation of the vRNA in the new cell (steps 2, 3 and 4). Phosphorylation of the *C*-terminus of the MP directs the MP into a degradation pathway mediated through its attachment to microtubules (MTs; step 6). (c) Schematic showing a model for cell-to-cell movement of viruses utilizing tubules. vRNA associated with CP after translation then associates with the MP, itself associated with a membrane of unknown origin (step 1). The complex then moves, either as a vesicle directed to the PD through targeting proteins such as those from the SNARE family, or through other unknown targeting signals to the cell periphery. The requirement for the cytoskeleton in transporting an MP–vRNA complex is unclear since the nepovirus, *Grapevine fanleaf virus*, requires these elements for proper delivery of its MP to the cell wall, while *Cowpea mosaic virus* and *Cauliflower mosaic virus* do not require them (step 2). At, or near, the PD, the vesicular or nonvesicular membranes fuse with the plasma membrane and the attached MP directs the CP-associated vRNA through the PD (step 3). The vRNA is then released in the next cell to initiate new virus replication and movement (step 4).

Researchers should be aware of this technical issue and not assume that a particular virus described as having a single MP means that that protein is the only one necessary for the cell-to-cell movement of the virus. The finding that TMV encodes a protein in addition to its MP that is necessary for its cell-to-cell transport (Hirashima and Watanabe, 2001, 2003), but does not appear to move itself cell-to-cell (Jianzhong Liu, Yiming Bao and Richard Nelson, unpublished results) is an example of the importance of understanding the definitions of MPs and movement-required proteins.

The history of studies of TMV movement differed from those of TGB-containing viruses in that the search for host factors which interacted with viral components occurred early. TMV MP was found to colocalize with microtubules (MTs) or MFs (reviewed in Heinlein, 2002). Early models describing TMV viral movement suggested that the viral MP, bound to genomic vRNA, associated directly with MTs for transport to the region of the PD. In some models, it was suggested that an interaction of the MP–vRNA complex with MFs occurred in the region of the PD. Reports showing that disruption of MT arrays affected virus movement, and that mutations in an MP motif with similarity to an MT sequence mediating lateral contacts between MT protofilaments also decreased TMV cell-to-cell movement, supported this model (reviewed in Heinlein, 2002). More recent studies, to be discussed later, have led to significant modifications of this model, diminishing the role of MTs in movement to and through PD, but still allowing for their importance in other viral and host–cell activities.

Recent studies on TMV MP have focused on determining the effect of phosphorylation on the activity of the protein, and on identifying the host factors with which it interacts. TMV MP becomes phosphorylated *in vitro* and in protoplasts by protein kinases (reviewed in Tzfira *et al.*, 2000; also see Karger *et al.*, 2003). The TMV MP also becomes phosphorylated in transgenic plants expressing the protein, and in non-transformed plants infected with the virus (Waigmann *et al.*, 2000). In that study, phosphorylation of the MP proceeded through the activity of a serine/threonine-specific protein kinase associated with the cell wall of the host cell. Substitutions of amino acids within the *C*-terminus of the protein with amino acids that mimic the negative charge exhibited after phosphorylation of the native amino acids prevented the MP from altering the plasmodesmal SEL, but not its stability *in vivo* nor its ability to bind RNA and a pectin methyltransferase *in vitro*. The same MP mutations, when expressed from within an infectious virus clone, did not allow local virus movement in *Nicotiana tabacum*. It was theorized that the function of MP was negatively regulated by phosphorylation in *N. tabacum*: unphosphorylated MP is able to modify PD SEL for virus transport while phosphorylated MP is unable to accomplish this function. MPs from *Potato leafroll virus* (PLRV), *Tomato bushy stunt virus* (TBSV) and *Cucumber mosaic virus* (CMV) can also be phosphorylated (Sokolova *et al.*, 1997; Desvoyes *et al.*, 2002; Matsushita *et al.*, 2002b), providing support for the idea that MP activity across virus genera is regulated by phosphorylation.

Interestingly, the same TMV MP mutations that blocked local virus movement in *N. tabacum* did allow movement in *N. benthamiana*. This result was unusual since both *N. tabacum* and *N. benthamiana* phosphorylated the MP normally (Waigmann *et al.*, 2000). It was theorized that the plasmodesmal transport machineries were different for *N. tabacum* and *N. benthamiana*, and that this may explain why the latter species is infected by such a wide range of plant viruses (Waigmann *et al.*, 2000). They suggested that *N. benthamiana*, unlike *N. tabacum*, allows movement of both phosphorylated and unphosphorylated MP between cells. These authors did not state where in the transport process the systems differed. Lee and Lucas (2001) hypothesized that phosphorylation of the MP prevented its association with a host chaperone molecule necessary for transport of the MP–vRNA complex to the PD in *N. tabacum*. They suggested that an equivalent chaperone in *N. benthamiana* might bind phosphorylated MP normally and allow virus movement.

A second interpretation is that phosphorylation of the MP alters its binding to vRNA, thereby making it more accessible to interactions with host factors that destroy vRNA. Although mutant TMV MPs that mimicked the phosphorylated MP bound RNA normally *in vitro* (Waigmann *et al.*, 2000), RNA binding may be altered *in vivo*. It was shown that a wild-type MP–vRNA complex cannot be translated *in vitro* or establish infection in protoplasts, but when introduced into plants, it could initiate an infection (Karpova *et al.*, 1997). Thus, a change in the phosphorylation state of the MP at the cell wall may weaken the binding between the MP and vRNA and allow vRNA translation in a newly infected cell (Karpova *et al.*, 1997; reviewed in Lee and Lucas, 2001). It is possible that the alteration in translatability exhibited in the new cell may make the vRNA susceptible to host defense systems as well. It was recently shown that *N. benthamiana* is a natural mutant for an enzyme in the RNA silencing pathway, RNA-dependent RNA polymerase 1 (RdRP1, and now referred to as RDR1), known to provide resistance against TMV infection in the *Nicotiana* species (Xie *et al.*, 2001; Yang *et al.*, 2004). The absence of the RDR1 in *N. benthamiana* would allow the vRNA to be translated and replicated to high levels in the newly infected cell, whereas in *N. tabacum*, the vRNA, although also translatable, would be subject to degradation through the RNA silencing pathway. Susceptibility to a host defense system could also explain the chaperone model advanced by Lee and Lucas (2001) if the *N. tabacum* chaperone bound less tightly to the vRNA–MP complex after phosphorylation compared with the *N. benthamiana* chaperone, thereby allowing RDR1 access to the vRNA in *N. tabacum*. RNA silencing occurs in protoplasts (i.e. representing the first infected cell in a plant), but is not sufficient to fully block virus accumulation in these cells (Szittya *et al.*, 2003). Therefore, the surveillance system could be functioning in the initially infected cell but more efficiently in the recipient cell.

The phosphorylated amino acids required within the TMV MP for movement are not fully understood. Deletion of the *C*-terminal amino acid residues that are phosphorylation targets and which regulate TMV movement in tobacco (Waigmann *et al.*, 2000) are not necessary for TMV movement (Boyko *et al.*, 2000, and references therein). Loss of a phosphorylation site at $Thr^{104}$ in the TMV MP ($Thr \rightarrow Ala$) did not significantly alter TMV cell-to-cell movement (Karger *et al.*, 2003). Further complicating the situation is the finding that a specific amino acid residue in the

*N*-terminus of the MP (position 37) susceptible to phosphorylation is required to stabilize the MP and the MP's association with the viral replication complex (VRC; Kawakami *et al.*, 1999, 2003). The VRC has recently been shown to move to PD in TMV-infected cells (Kawakami *et al.*, 2004). It was determined, however, that modifications elsewhere within the MP, not involving phosphorylation, could function in place of serine 37 to allow virus movement (Kawakami *et al.*, 2003). Thus, phosphorylation of specific amino acid residues may be only one of several ways to allow the MP to function. To further complicate the situation, it may be that sequential phosphorylation of the MP due to the activities of multiple protein kinases at specific sites is necessary for regulation of MP activity (Lee and Lucas, 2001). Further work is necessary to define the function of each phosphorylation site relative to virus movement and host defense. Indeed, in regard to host defense modification, it was recently shown that the MP from *Tomato mosaic virus* (ToMV), a virus closely related to TMV, bound to a transcriptional coactivator, MBF1 (Matsushita *et al.*, 2002a). Orthologs of this protein have been shown to be upregulated during pathogen challenge (Goday *et al.*, 2001). Thus, it is possible that the MP could modulate expression of host genes to suppress defense pathways (Matsushita *et al.*, 2002a). The MP of TBSV also interacts with a host factor, a homeodomain protein, which is expressed at higher levels during virus infection (Desvoyes *et al.*, 2002). Homeodomain transcription factors such as KNOTTED 1 traffic between cells (Lee *et al.*, 2000, and references therein). Future experiments will be necessary to determine whether such host factors are necessary to support virus movement, inhibit host defenses or both, thereby allowing enhanced virus cell-to-cell movement.

The TMV MP has been shown to associate with the ER (Gillespie *et al.*, 2002, and references therein) and behaves as an intrinsic membrane protein (Brill *et al.*, 2000). In addition, expression of the MP during infection is greatest at the infection front where virally induced cytoplasmic bodies, containing MP, are paired opposite each other with PD in between (Szécsi *et al.*, 1999). The gating of PD is restricted to this area of the infection (Oparka *et al.*, 1997). In older areas of the infection, the MP is no longer associated with the virally-induced cytoplasmic bodies which themselves are no longer paired at the cell wall (Szécsi *et al.*, 1999). Thus, the MP is associated with structures at the infection front at or near the time of cell-to-cell transport.

Although the requirement of the MP for TMV cell-to-cell movement has extensive historical support, the identification of an additional TMV-encoded protein for virus movement was only recently demonstrated. The first indication that the TMV MP was insufficient for TMV cell-to-cell movement came from a study where an MP : GFP fusion in the presence of an infectious virus expressing an inactive MP moved cell-to-cell, but did not move in the absence of this virus (Tamai and Meshi, 2001b). The identity of a second viral protein necessary for this movement was identified through mutagenesis of the TMV replicase coding region (Hirashima and Watanabe, 2001). These authors determined that a mutation within the 126/183 5′co-terminal ORF yielded a virus that could replicate in protoplasts, but could not move cell-to-cell. This work managed to separate the requirements for replication and movement, respectively, for proteins believed to be involved in both activities. It is a lesson for those who study virus movement never to underestimate the multifunctional nature of viral proteins.

The involvement of the 126- and/or 183-kDa protein in TMV cell-to-cell movement brings the movement mechanism for this virus group into greater alignment with viruses expressing TGB proteins. The 126-kDa ORF contains a helicase-like domain in its carboxy terminus that is within the same superfamily of helicases that contains TGBp1 (Koonin and Dolja, 1993). Through mutagenesis studies, Hirashima and Watanabe (2003) showed that the helicase domain found within the N′co-terminal 126- and 183-kDa proteins was essential for movement, possibly through an interaction with a central nonconserved region within the N' co-terminal proteins (Hirashima and Watanabe, 2003). The 126-kDa protein has been shown to form a hetero-dimer with the 183-kDa protein and associate with the ER, likely through its association with membrane proteins (Watanabe *et al.*, 1999; Yamanaka *et al.*, 2000; dos Reis Figueira *et al.*, 2002; Hagiwara *et al.*, 2003; Tsujimoto *et al.*, 2003). The ability to oligomerize and associate with membranes may reflect a similarity in function between this protein and the TGB proteins which also show this activity. The 126-kDa protein and vRNA colocalize in virally induced cytoplasmic bodies (Mas and Beachy, 1999), and the protein binds vRNA at the 3′tRNA-like structure (Osman and Buck, 2003). Lastly, the 126-kDa protein has been shown to be a suppressor of gene silencing (Kubota *et al.*, 2003; Ding *et al.*, 2004). Thus, the 126-kDa protein shows further similarity to TGBp1 and TGBp2.

An analog of the 126-kDa protein, the 1a protein from *Brome mosaic virus* (BMV), also forms oligomers (O'Reilly *et al.*, 1997, and references therein) and targets ER membranes in plant and yeast cells (Restrepo-Hartwig and Ahlquist, 1996, 1999; den Boon *et al.*, 2001). In yeast, the 1a protein forms membrane-enveloped spherules that change shape and size according to the ratio of 1a and 2a proteins present (Schwartz *et al.*, 2004). The 2a protein is the analog of the carboxy terminus of the 183-kDa protein from TMV (Schwartz *et al.*, 2002). It will be interesting to determine if the 126-kDa analogs from different virus genera encoding single MPs (e.g. cucomoviruses, bromoviruses and tobraviruses) are also required for cell-to-cell movement of their viral genomes.

Considering that TMV now has at least two proteins necessary for cell-to-cell movement, and perhaps greater similarity with TGB-containing viruses to achieve movement, it is important to review what is known about the association of the TMV proteins with the cytoskeleton, an area heavily researched with TMV but less so with TGB-containing viruses. Both the MP and 126-kDa protein, respectively, associate or co-align with actin or MFs when expressed alone (McLean *et al.*, 1995; Liu, J.Z., Blancaflor, E.B. and Nelson, R.S., unpublished results) and the MP associates with MTs (reviewed in Heinlein, 2002). Results from multiple studies suggested that the association of the MP with MTs was important for TMV cell-to-cell movement (reviewed in Heinlein, 2002; Heinlein and Epel, 2004). However, another model now is being pursued due to recent findings. Gillespie *et al.* (2002) determined that disruption of the MT network, either through pharmacological or gene silencing technology, had no effect on the cell-to-cell spread of TMV. Second, it was shown that the TMV VRCs move within cells to the PD and that an MF antagonist, latrunculin B, prevents this movement (Kawakami *et al.*, 2004). Third, during virus replication, the size of the virus replication complexes correlate with the function of the 126/183-kDa proteins, suggesting that one or both of these proteins control the formation of these bodies (Liu, J.Z., Bao, Y.M and Nelson, R.S., unpublished results). Lastly, the VRCs, like the 126-kDa protein bodies, move along MFs during virus infection (Liu, J.Z., Blancaflor, E.B. and Nelson, R.S., unpublished results). Thus, the VRCs moving along MFs to the PD appear to be a potential vehicle by which TMV moves cell to cell.

The importance of the 126-kDa protein and the MP at each step of this movement process to and within the PD requires further analysis. Several possibilities exist, such as the 126-kDa protein functioning similarly to TGBp1 by preparing the vRNA for targeting to peripheral membrane bodies where the MP may then function. Alternatively, the MP functions to transport the VRCs to the cell periphery where the 126-kDa protein positions the bodies for PD transport (as postulated for rab-like proteins that transport macromolecular complexes or vesicle bodies; reviewed in Oparka, 2004). No interaction has been reported between the 126-kDa protein and the MP. Whether the 126-kDa protein, through an indirect interaction with the MF through the ER or directly with the MF, and/or the MP, directs VRC movement to the cell periphery is not known. Recently, it was shown that a host non-cell autonomous pathway protein from *N. tabacum*, NtNCAPP1, interacts with a host protein, CmPP16, found in phloem sap (Lee *et al.*, 2003). CmPP16 has some sequence and functional similarities with the TMV MP (Xoconostle-Cázares *et al.*, 1999). Expression of a dysfunctional NtNCAPP1 inhibited the movement of both CmPP16 and MP. An NtNCAPP1 : GFP fusion was located at the outer surface of the cortical ER membrane often near, but not in, PD. It was noted that TGBp1s share features with CmPP16, and thus with TMV MP (Morozov and Solovyev, 2003). The location of the NtNCAPP1 is similar to that of the TGBp2 and 3 complex. TGBp2 and 3, like the 126-kDa protein, associate with cell membranes and are necessary for cell-to-cell movement. It is intriguing to consider that the 126-kDa protein serves a similar role as these proteins in virus movement.

The involvement of MTs in virus infection had been postulated to be in the degradation of MP (Más and Beachy, 1999, and references within). This theory is supported by the determination that a mutant MP, $MP^{R3}$, which allowed enhanced cell-to-cell movement of the virus, had little affinity for MTs and greater resistance to proteasome-mediated degradation (Gillespie *et al.*, 2002). Further support for this theory was supplied by the finding that a host protein, MPB2C, which binds to MP and colocalizes with the MP at MTs, interferes with the movement function of the MP, but not with $MP^{R3}$ (Kragler *et al.*, 2003). For TMV, it appears that MFs and MTs have two different functions.

In summary, a model for TMV movement is emerging with some similarity to the one postulated for TGB-containing viruses (Figure 9.1b). The 126-kDa protein may function similarly to TGB2 and/or 3 to support movement of a VRC containing the MP to the cell wall area. It is, however, possible that the MP itself directs the movement of the VRC to the cell wall area where the 126-kDa protein then positions the complex close to the PD. Regardless of the TMV protein responsible for transport of the VRC to the wall, the VRCs utilize MFs to move within the cell. What occurs at the PD is a mystery, but it is possible that the vRNA–MP complex is altered through an NTPase activity supplied by the 126-kDa protein so that the MP can transport the vRNA–MP complex through the PD. This type of activity for the 126-kDa protein would mimic the proposed activity of TGBp1 at the cell wall for potex- and hordei-like viruses (Morozov and Solovyev, 2003). Phosphorylation of the MP likely regulates movement both within the cell and the PD. The targeting of TMV MP to MTs appears to result in targeting protein for degradation, but it may affect movement indirectly if the degradation system cannot function normally.

### *9.2.3 Comoviruses, Nepoviruses and Tospoviruses*

The *Comovirus*, *Nepovirus* and *Tospovirus* represent genera whose members move cell-to-cell by the passage of encapsidated viruses (virions) or RNP complexes through modified PD, often referred to as tubules, containing virus-encoded MP (Hull, 2002; Pouwels *et al.*, 2002a). This method of cell-to-cell movement is also displayed by viruses belonging to the genera *Bromovirus* and *Caulimovirus*, the latter containing DNA viruses (reviewed in Hull, 2002). The unique nature of their transport form between cells sets these groups apart from the tobamoviruses and TGB-containing viruses, but other aspects of their cell-to-cell movement may be mechanistically similar to viruses from these groups. Recent studies with tubule-forming viruses have centered on the association of virus proteins with host factors.

*Cowpea mosaic virus* (CPMV) forms extensive membranous vesicles that are closely associated with its replication (reviewed in Pouwels *et al.*, 2002a). Through specific labeling of the ER during virus infection, it was shown that the ER membrane undergoes dramatic morphological changes during infection (Carette *et al.*, 2000). The proliferated ER is surrounded and traversed by cytopathic structures containing replicative proteins and vRNA (Carette *et al.*, 2002a). The proliferation of the ER during virus infection can be mimicked by expression of the 32-kDa and 60-kDa proteins. The 32-kDa protein induces the proliferation of ER and the formation of small bodies in the cytoplasm near the nucleus, while the 60-kDa protein induces only the small bodies. The 60-kDa protein can form small membranous vesicles in the cytoplasm of insect cells and is postulated to be a helicase due to sequence similarities with other known helicases and its ability to bind ATP (reviewed in Pouwels *et al.*, 2002a). Even more interestingly, the *C*-terminal domain of this protein interacts in yeast with an *N*-ethyl-maleimide-sensitive-factor attachment protein receptor (SNARE)-like protein localized in the ER of yeast (Carette *et al.*, 2002b). SNARE proteins, both receptor and target family members, present on vesicle and plasma membranes interact specifically to permit the vesicle to fuse with the plasma membrane and allow discharge of the cargo (reviewed in Oparka, 2004). The 32- or 60-kDa protein are not classical MPs for CPMV, and their association with membranes and membrane-associated proteins and putative helicase activity suggests some functional similarity with TGB and 126-kDa proteins. It will be worthwhile to search for mutations in the 32- and 60-kDa proteins that allow virus accumulation, but not cell-to-cell movement.

The 48-kDa protein is considered to be the MP of CPMV as it is found within the tubules of the host during virus infection and is the only viral protein necessary for tubule formation (reviewed in Pouwels *et al.*, 2002a). An MP : GFP fusion was localized to fluorescent spots in the cell wall and was able to spread from cell-to-cell; however, virus expressing the MP : GFP fusion could not move cell-to-cell (Pouwels *et al.*, 2002a). Through mutational analysis of the MP, it was shown that the *N*-terminal and central region of the protein is involved in tubule formation while the *C*-terminus is necessary to interact with virus particles (reviewed in Pouwels *et al.*, 2002a; Bertens *et al.*, 2003). This requirement for an intact *C*-terminus to interact with virus particles likely explains why GFP fused to the *C*-terminus of the MP prevented virus particles from being present in tubules and blocked the cell-to-cell movement of the virus (Bertens *et al.*, 2003). Within the domain necessary for tubule-formation in the MP, small regions have been identified that are necessary for targeting the MP to the plasma membrane (Bertens *et al.*, 2000). Thus, the MP of CPMV has the capacity to interact with host membranes and to bind the virion through an apparent interaction with its CP.

Disruption of cytoskeletal elements did not interfere with the formation of tubules by the CPMV MP, although treatment with brefeldin A (BFA) did inhibit tubule formation (Pouwels *et al.*, 2002b). BFA inhibits the formation of Golgi from the ER (Ritzenthaler *et al.*, 2002, and references therein) and it was postulated that a decrease in the flow of Golgi-derived vesicles containing the MP may have been responsible for the loss of tubules (reviewed in Pouwels *et al.*, 2002a).

*Grapevine fanleaf virus* (GFLV), a nepovirus and another member of this group of viruses, was recently studied for its intracellular movement characteristics. The MP from this virus, fused with GFP, behaved as an intrinsic membrane protein and directly or indirectly associated with KNOLLE, a membrane protein expressed during cytokinesis and also a member of the SNARE family of proteins (Laporte *et al.*, 2003). As was determined in experiments with CPMV MP, treatment of tissue with BFA prevented tubule formation by the GFLV MP. Thus, both CPMV and GFLV require a functional endomembrane system for tubule formation, however, the intracellular location of the MPs after treatment was not the same. This may reflect the membranes with which these proteins initially associate during infection (Laporte *et al.*, 2003). The proper delivery of GFLV GFP : MP to the cell wall for tubule formation was shown to require MTs, although it was suggested that MFs could deliver the GFP : MP if MTs were depleted (Laporte *et al.*, 2003). Sites of tubule formation were always associated with the presence of calreticulin, a protein important for membrane adhesion to the cell wall through integrin-like proteins that contain RGD-binding motifs. Interestingly, the MP from *Turnip crinkle virus* interacts with an RGD-containing protein in yeast (Lin and Heaton, 2001).

For Tospoviruses, the NSm protein has some of the properties associated with MPs – including localization close to PD, association with the nucleocapsid and a tubule-forming capacity (Kormelink *et al.*, 1994; Storms *et al.*, 1995). NSm, more recently was shown to interact with plant DnaJ-like proteins (Soellick *et al.*, 2000; von Bargen *et al.*, 2001). DnaJ proteins function to import protein into organelles and regulate HSP70 chaperone activity (reviewed in Oparka, 2004). HSP70s from PD-rich wall fractions of *Curcubita* phloem exudates interact with PD and modify their SEL (Aoki *et al.*, 2002). These cell-wall-derived HSP70s contain a *C*-terminal sequence necessary for their cell-to-cell trafficking, a sequence not present in members of the cytosolic HSP70 family. The ATPase activity encoded in the *N*-terminal region of these proteins was not necessary for cell-to-cell trafficking. *Beet yellows virus* (BYV), a closterovirus, encodes an HSP70 homolog (HSP70h) that is associated with PD and is required for the formation of movement-competent viral RNP complexes (Medina *et al.*, 1999; Alzhanova *et al.*, 2001). Interestingly, HSP70h does not contain the *C*-terminal motif present in the cell-wall-associated HSP70s that traffic cell-to-cell (Aoki *et al.*, 2002). In addition, an intact putative catalytic site of the ATPase domain within HSP70h was required for cell-to-cell movement of the virus, while the ATPase domain within the HSP70 was not necessary for its cell-to-cell movement (Peremyslov *et al.*, 1999; Aoki *et al.*, 2002). For HSP70h, it is possible that host factors or other viral protein domains, including those present in HSP70h, provide the cell-to-cell movement function found in the *C*-terminus of the cell-wall-associated host-encoded HSP70s.

In summary, for this group of tubule-forming viruses, a model for their cell-to-cell transport that is rich in host and virus protein interactions can be presented (Figure 9.1c). During virus replication, the host membrane network is enriched in viral proteins (e.g. the 32- and 60-kDa proteins of CPMV). During this period, the MP of these viruses is produced that can interact with either the encapsidated virus, through its CP, or with an RNP complex. In addition, for at least the comoviruses and nepoviruses, these proteins may interact directly or indirectly with host proteins (e.g. members of the SNARE protein family) that target to the cell periphery. In either case, they become localized to membranes and associate with the virion. The presence of the 32- or 60-kDa protein within these same membrane-associated complexes has not been shown. MTs, or MFs after MT depletion, may target the MP–virion complex to the correct cell wall area, but they are not necessary for tubule formation. At the cell wall, vesicles may fuse with the plasma membrane near PD. The MP then lines the interior of the plasma membrane within the PD and the attached virion passes through. For tospoviruses, the interaction of the MP with DnaJ proteins may regulate host HSP70 activity thereby allowing virus cell-to-cell movement through the PD. The importance of HSP70s for movement of comoviruses and nepoviruses is not known. The association of the comovirus 32-kDa protein, 60-kDa protein and MP with membranes, and the putative helicase activity of the 60-kDa protein, suggests that there are similarities in function between these proteins and the MPs and movement-associated proteins encoded by the TGB-containing viruses and the tobamoviruses.

### 9.2.4 Other viruses

This section will detail recent findings on the cell-to-cell movement of viruses from other genera. The selected findings are presented to provide a deeper conceptual understanding of virus movement in general. The findings may have relevance only to a small group of viruses, but later may be shown to have significance to many or all viruses as we gather more information in this field.

Gene I of *Cauliflower mosaic virus* (CaMV) is the MP of this virus, and the *N*-terminal third of this protein was shown to interact with a host protein related to mammalian rab acceptors (Huang *et al.*, 2001). Rab proteins have GTPase activity and are important for determining vesicle transport specificity (reviewed in Oparka, 2004). Interactions between the MP and the rab-like protein in yeast correlate with mutations in the MP that destroy its ability to form tubules in protoplasts (Huang *et al.*, 2001, and references therein). The rab-like protein was not observed in tubules (Huang *et al.*, 2001), suggesting that its function was to transport or localize virus RNPs close to the PD. These proteins interact with members of the SNARE protein family (reviewed in Oparka, 2004). By fluorescence resonance energy transfer (FRET), it was shown that the rab-like protein and the MP were in close proximity. Similar to other tubule-forming viruses, it appears that interactions between the CaMV MP and host factors that associate with specific membrane complexes are important for the cell-to-cell movement of this DNA virus.

Viruses within the family Potyvirideae require their CP for cell-to-cell movement. The CP of *Potato virus A* (PVA), a potyvirus with a monopartite genome, is phosphorylated by a plant protein kinase (Ivanov *et al.*, 2001). More recently, the host enzyme was identified as protein kinase CK2 (Ivanov *et al.*, 2003). CK2-mediated phosphorylation of PVA CP inhibited CP binding to RNA *in vitro*. The site within the CP phosphorylated by CK2 was conserved among potyviruses. Thus, the CP of potyviruses, and the MP of tobamoviruses, are phosphorylated by protein kinase CK2. Indeed, the potyvirus CP has functions associated with both the MP and CP of the tobamovirus (Matsushita *et al.*, 2003). In addition, phosphorylation of the PVX CP, a protein required for cell-to-cell movement of this virus, apparently affects its RNA-binding capability (Atabekov *et al.*, 2001). These results indicate the importance of phosphorylation for virus movement throughout unrelated virus genera.

A paradigm highlighted in most chapters in this book is that movement of host macromolecules between cells is regulated to maintain a viable organism. Part of this regulated movement likely includes controlling the size of particles that pass between cells. Viruses have to 'live' in this world, and thus it is likely that the size and structure of their movement forms must conform to those allowed by the host. Gilbertson *et al.* (2003) recently showed that the size of the *Bean dwarf mosaic virus* (BDMV) genome was regulated during cell-to-cell movement but not during virus replication. BDMV is a DNA virus, and thus has evolved to utilize the RNA cell-to-cell transport system under specific constraints. It is likely that similar findings will appear for both RNA and DNA viruses.

For some plant viruses, it has been extremely difficult to identify viral proteins involved in cell-to-cell movement. The genus *Phytoreovirus* is one example of a group of viruses difficult to analyse. Phytoreoviruses are composed of 12 segments of dsRNA (Boccardo and Milne, 1984). The inability to produce infectious clones of these viruses makes it necessary to study their movement either by expression of individual viral ORFs in uninfected plants (similar to studies reported for TGB-containing viruses), or by studying the cell-to-cell movement of unrelated movement-defective viruses that express putative phytoreovirus MPs. The effect of putative viral MPs on the cell-to-cell movement of unrelated movement-defective viruses has been successfully determined for small RNA viruses (e.g. Fedorkin *et al.*, 2000). Li *et al.* (2004) recently took this approach to study the cell-to-cell movement of proteins encoded by all 12 segments of *Rice dwarf virus* (RDV). Only protein encoded by segment 6 (S6) complemented the movement of a PVX movement-defective virus. By immunocytochemistry, the S6 protein was identified in PD of rice infected with RDV. This is an effective procedure to identify proteins necessary for cell-to-cell movement of viruses for which infectious clones are not available.

### *9.2.5 Summary*

Results from studies conducted in the last 4 years have made simple models for virus cell-to-cell movement obsolete. The new models depicting virus cell-to-cell movement include multiple viral proteins and host proteins that interact with membranes, direct viruses where to go in the cell and regulate movement in increasingly complex manners. Mechanistic studies of virus cell-to-cell movement require analyses of infected tissue rather than expression of individual proteins in isolation from the infection process. These studies will require fluorescent reporters for multiple proteins, both virus and host-derived, that are easily observed and quantified. The reporters will need to be small, so that interference with protein function is minimal. The future is bright for those who identify and utilize such reporters to study virus movement.

Another trend emerging over the last 4 years is the similar function of proteins from viruses representing widely differing genera for cell-to-cell movement. Thus, the requirement of proteins with helicase domains for movement of TGB-containing viruses and tobamoviruses may indicate that a protein containing such a domain from a tubule-forming virus has a similar function. It is illuminating that so many viruses now have been shown to require more than one viral protein for vRNA cell-to-cell movement. Often these additional proteins have been associated with host membranes for transport to, and likely through, the PD. Thus, plant virologists will need to become increasingly familiar with the physiology of the host membrane system. Lastly, host proteins that aid virus movement across virus genera are being identified. The determination that protein kinase CK2 phosphorylates the MP from a tobamovirus and the CP from a potyvirus indicates that host factors will often have similar activities on viruses from different genera.

Although the identification of similar requirements for virus cell-to-cell movement among genera has been emphasized in this review, it is certain that each genus has unique requirements for their cell-to-cell movement. These differences should be studied carefully since they may reflect alternative methods used by the host to transport host macromolecules. A combination of both predicative and exploratory research will likely allow for the most rapid advances in this area.

## Note

TGBp2 and TGBp3 have recently been shown to localize with motile granules in the cell that use the ER-actin network for intracellular movement (*N. tabacum*) and TGBp2 and a Rab5 ortholog colocalize in intracellular vesicles (*N. benthamiana*; Haupt *et al.*, 2005).

## Acknowledgements

The author thanks Kiran Mysore and Elison Blancaflor for reviewing the text and Darla Snelson for final figures. Financial support for the preparation of this review was from the Samuel Roberts Noble Foundation.

## References

Aaziz, R., Dinant, S. and Epel, B.L. (2001) 'Plasmodesmata and plant cytoskeleton', *Trends in Plant Science* **6**, 326–330.

Allison, A.V. and Shalla, T.A. (1974) 'The ultrastructure of local lesions induced by *Potato virus X*: a sequence of cytological events in the course of infection', *Phytopathology* **64**, 784–793.

Alzhanova, D.V., Napuli, A.J., Creamer, R. and Dolja, V.V. (2001) 'Cell-to-cell movement and assembly of a plant closterovirus: roles for the capsid proteins and Hsp70 homolog', *EMBO Journal* **20**, 6997–7007.

Aoki, K., Kragler, F., Xoconostle-Cázares, B. and Lucas, W.J. (2002) 'A subclass of plant heat shock cognate 70 chaperones carries a motif that facilitates trafficking through plasmodesmata', *Proceedings of the National Academy of Sciences USA* **99**, 16342–16347.

Atabekov, J.G., Rodionova, N.P., Karpova, O.V., Kozlovsky, S.V. and Poljakov, V.Y. (2000) 'The movement protein-triggered *in situ* conversion of potato virus X virion RNA from a nontranslatable into a translatable form', *Virology* **271**, 259–263.

Atabekov, J.G., Rodionova, N.P., Karpova, O.V., Kozlovsky, S.V., Novikov, V.K. and Arkhipenko, M.V. (2001) 'Translational activation of encapsidated potato virus X RNA by coat protein phosphorylation', *Virology* **286**, 466–474.

Bertens, P., Heijne, W., Van der Wel, N., Wellink, J. and Van Kammen, A. (2003) 'Studies on the C-terminus of the *Cowpea mosaic virus* movement protein', *Archives of Virology* **148**, 265–279.

Bertens, P., Wellink, J., Goldbach, R. and Van Kammen, A. (2000) 'Mutational analysis of the cowpea mosaic virus protein', *Virology* **267**, 199–208.

Boccardo, G. and Milne, R.G. (1984) 'Plant reovirus group' in *Description of Plant Viruses, No. 294* (eds, A.F. Murant and B.D. Harrison), Commonwealth Mycological Institute and Association of Applied Biologists, Old Woking, England.

Boyko, V., Ferralli, J. and Heinlein, M. (2000) 'Cell-to-cell movement of TMV RNA is temperature-dependent and corresponds to the association of movement protein with microtubules', *Plant Journal* **22**, 315–325.

Brill, L.M., Nunn, R.S., Kahn, T.W., Yeager, M. and Beachy, R.N. (2000) 'Recombinant tobacco mosaic virus movement protein is an RNA-binding, $\alpha$-helical membrane protein', *Proceedings of the National Academy of Sciences USA* **97**, 7112–7117.

Callaway, A., Giesman-Cookmeyer, D., Gillock, E.T., Sit, T.L. and Lommel, S.A. (2001) 'The multifunctional capsid proteins of plant RNA viruses', *Annual Review of Phytopathology* **39**, 419–460.

Carette, J.E., Gühl, K., Wellink, J. and Van Kammen, A. (2002a) 'Coalescence of the sites of cowpea mosaic virus RNA replication into a cytopathic structure', *Journal of Virology* **76**, 6235–6243.

Carette, J.E., Stuiver, M., Van Lent, J., Wellink, J. and Van Kammen, A. (2000) 'Cowpea mosaic virus infection induces a massive proliferation of endoplasmic reticulum but not Golgi membranes and is dependent on *de novo* membrane synthesis', *Journal of Virology* **74**, 6556–6563.

Carette, J.E., Verver, J., Martens, J., van Kampen, T., Wellink, J. and Van Kammen, A. (2002b) 'Characterization of plant proteins that interact with cowpea mosaic virus "60K" protein in the yeast two-hybrid system', *Journal of General Virology* **83**, 885–893.

Cowan, G.H., Lioliopoulou, F., Ziegler, A. and Torrance, L. (2002) 'Subcellular localisation, protein interactions, and RNA binding of *Potato mop-top virus* triple gene block proteins', *Virology* **298**, 106–115.

den Boon, J., Chen, J. and Ahlquist, P. (2001) 'Identification of sequences in brome mosaic virus replicase protein 1a that mediated association with endoplasmic reticulum membranes', *Journal of Virology* **75**, 12370–12381.

Derrick, P.M. and Nelson, R.S. (1999) 'Plasmodesmata and long-distance virus movement' in *Plasmodesmata. Structure, Function, Role in Cell Communication* (eds A.J.E. van Bel and W.J.P. van Kesteren) Springer-Verlag, Berlin, pp. 315–339.

Desvoyes, B., Faure-Rabasse, S., Chen, M.-H., Park, J.-W. and Scholthof, H.B. (2002) 'A novel plant homeodomain protein interacts in a functionally relevant manner with a virus movement protein', *Plant Physiology* **129**, 1521–1532.

Ding, X.S., Liu, J., Cheng, N.-H. *et al.* (2004) '*The Tobacco mosaic virus* 126-kDa protein associated with virus replication and movement suppresses RNA silencing', *Molecular Plant–Microbe Interactions* **17**, 583–592.

Donald, R.G.K., Lawrence, D.M. and Jackson, A.O. (1997) 'The barley stripe mosaic virus 58-kilodalton $\beta$b protein is a multifunctional RNA binding protein', *Journal of Virology* **71**, 1538–1546.

dos Reis Figueira, A., Golem, S., Goregaoker, S.P. and Culver, J.N. (2002) 'A nuclear localization signal and a membrane association domain contribute to the cellular localization of the tobacco mosaic virus 126-kDa replicase protein', *Virology* **301**, 81–89.

Du, M.X., Johnson, R.B., Sun, X.L., Staschke, K.A., Colacino, J. and Wang, O.M. (2002) 'Comparative characterization of two DEAD-box RNA helicases in superfamily II: human translation-initiation factor 4A and hepatitis C virus non-structural protein 3 (NS3) helicase', *Biochemistry Journal* **363**, 147–155.

Erhardt, M., Morant, M., Ritzenthaler, C. *et al.* (2000) 'P42 movement protein of beet necrotic yellow vein virus is targeted by the movement proteins P13 and P15 to punctate bodies associated with plasmodesmata', *Molecular Plant–Microbe Interactions* **13**, 520–528.

Erhardt, M., Stussi-Garaud, C., Guilley, H., Richards, K.E., Honard, G. and Bouzoubaa, S. (1999) 'The first triple gene block protein of peanut clump virus localizes to the plasmodesmata during virus infection', *Virology* **264**, 220–229.

Fedorkin, O.N., Merits, A., Lucchesi, J. *et al.* (2000) 'Complementation of the movement-deficient mutations in potato virus X: potyvirus coat protein mediates cell-to-cell trafficking of C-terminal truncation but not deletion mutation of potexvirus coat protein', *Virology* **270**, 31–42.

Fridborg, I., Grainger, J., Page, A., Coleman, M., Findlay, K. and Angell, S. (2003) 'TIP, a novel host factor linking callose degradation with the cell-to-cell movement of *Potato virus X*', *Molecular Plant–Microbe Interactions* **16**, 132–140.

Gilbertson, R.L., Sudarshana, M., Jiang, H., Rojas, M.R. and Lucas, W.J. (2003) 'Limitations on geminivirus genome size imposed by plasmodesmata and virus-encoded movement protein: insights into DNA trafficking', *The Plant Cell* **15**, 2578–2591.

Gillespie, T., Boevink, P., Haupt, S. *et al.* (2002) 'Functional analysis of a DNA-shuffled movement protein reveals that microtubules are dispensable for the cell-to-cell movement of *Tobacco mosaic virus*', *The Plant Cell* **14**, 1207–1222.

Goday, A.V., Zanetti, M.E., Segundo, B.S. and Casalongué, C.A. (2001) 'Identification of a putative *Solanum tuberosum* transcriptional coactivator up-regulated in potato tubers by *Fusarium solani* f. sp. *eumartii* infection and wounding', *Physiologia Plantarum* **112**, 217–222.

Gorshkova, E.N., Erokhina, T.N., Stroganova, T.A. *et al.* (2003) 'Immunodetection and fluorescent microscopy of transgenically expressed hordeivirus TGBp3 movement protein reveals its association with endoplasmic reticulum elements in close proximity to plasmodesmata', *Journal of General Virology* **84**, 985–994.

Hagiwara, Y., Komoda, K., Yamanaka, T. *et al.* (2003) 'Subcellular localization of host and viral proteins associated with tobamovirus RNA replication', *EMBO Journal* **22**, 344–353.

Haupt, S., Cowan, G.H., Ziegler, A., Roberts, A.G., Oparka, K. and Torrance, L. (2005) 'Two plant-viral movement proteins traffic in the endocytic recycling pathway', *The Plant Cell* **17**, 164–181.

Haywood, V., Kragler, F. and Lucas, W.J. (2002) 'Plasmodesmata: pathways for protein and ribonucleoprotein signaling', *The Plant Cell* **14**, S303–S325.

Heinlein, M. (2002) 'Plasmodesmata: dynamic regulation and role in macromolecular cell-to-cell signaling', *Current Opinion in Plant Biology* **5**, 543–552.

Heinlein, M. and Epel, B.L. (2004) 'Macromolecular transport and signaling through plasmodesmata', *International Review of Cytology* **235**, 93–164.

Hirashima, K. and Watanabe, Y. (2001) 'Tobamovirus replicase coding region is involved in cell-to-cell movement', *Journal of Virology* **75**, 8831–8836.

Hirashima, K. and Watanabe, Y. (2003) 'RNA helicase domain of tobamovirus replicase executes cell-to-cell movement possibly through collaboration with its nonconserved region', *Journal of Virology* **77**, 12357–12362.

Huang, Z., Andrianov, V.M., Han, Y. and Howell, S.H. (2001) 'Identification of *Arabidopsis* proteins that interact with the cauliflower mosaic virus (CaMV) movement protein', *Plant Molecular Biology* **47**, 663–675.

Hull, R. (2002) *Matthews' Plant Virology*, Academic Press, San Diego, California, London.

Iglesias, V.A. and Meins, F., Jr. (2000) 'Movement of plant viruses is delayed in a β-1,3-glucanase-deficient mutant showing a reduced plasmodesmatal size exclusion limit and enhanced callose deposition', *The Plant Journal* **21**, 157–166.

Ivanov, K.I., Puustinen, P., Gabrenaite, R. *et al.* (2003) 'Phosphorylation of the potyvirus capsid protein by protein kinase CK2 and its relevance for virus infection', *The Plant Cell* **15**, 2124–2139.

Ivanov, K.I., Puustinen, P., Merits, A., Saarma, M. and Mäkinen, K. (2001) 'Phosphorylation down-regulates the RNA binding function of the coat protein of potato virus A', *The Journal of Biological Chemistry* **276**, 13530–13540.

Kalinina, N.O., Rakitina, D.A., Solovyev, A.G., Schieman, J. and Morozov, S.Y. (2002) 'RNA helicase activity of the plant virus movement proteins encoded by the first gene of the triple gene block', *Virology* **296**, 321–329.

Kalinina, N.O., Rakitina, D.A., Yelina, N.E. *et al.* (2001) 'RNA-binding properties of the 63 kDa protein encoded by the triple gene block of poa semilatent hordeivirus', *Journal of General Virology* **82**, 2569–2578.

Karger, E.M., Frolova, O. Yu., Fedorova, N.V. *et al.* (2003) 'Dysfunctionality of a tobacco mosaic virus movement protein mutant mimicking threonine 104 phosphorylation', *Journal of General Virology* **84**, 727–732.

Karpova, O.V., Ivanov, K.I., Rodionova, N.P., Dorokhov, Y.L. and Atabekov, J.G. (1997) 'Nontranslatability and dissimilar behavior in plants and protoplasts of viral RNA and movement protein complexes formed *in vitro*', *Virology* **230**, 11–21.

Kawakami, S., Hori, K., Hosokawa, D., Okada, Y. and Watanabe, Y. (2003) 'Defective tobamovirus movement protein lacking wild-type phosphorylation sites can be complemented by substitutions found in revertants', *Journal of General Virology* **77**, 1452–1461.

Kawakami, S., Padgett, H.S., Hosokawa, D., Okada, Y., Beachy, R.N. and Watanabe, Y. (1999) 'Phosphorylation and/or presence of serine 37 in the movement protein of tomato mosaic tobamovirus is essential for intracellular localization and stability *in vivo*', *Journal of Virology* **73**, 6831–6840.

Kawakami, S., Watanabe, Y. and Beachy, R.N. (2004) 'Tobacco mosaic virus infection spreads cell to cell as intact replication complexes', *Proceedings of the National Academy of Sciences USA* **101**, 6291–6296.

Koonin, E.V. and Dolja, V.V. (1993) 'Evolution and taxonomy of positive-strand RNA viruses – implications of comparative analysis of amino acid sequences', *Critical Reviews in Biochemistry and Molecular Biology* **28**, 375–430.

Kormelink, R., Storms, M., van Lent, J., Peters, D. and Goldbach, R. (1994) 'Expression and subcellular location of the $NS_M$ protein of tomato spotted wilt virus (TSWV), a putative viral movement protein', *Virology* **200**, 56–65.

Kragler, F., Curin, M., Trutnyeva, K., Gansch, A. and Waigmann, E. (2003) 'MPB2C, a microtubule-associated plant protein binds to and interferes with cell-to-cell transport of tobacco mosaic virus movement protein', *Plant Physiology* **132**, 1870–1883.

Krishnamurthy, K., Heppler, M., Mitra, R. *et al.* (2003) 'The *Potato virus X* TGBp3 protien associates with the ER network for virus cell-to-cell movement', *Virology* **309**, 135–151.

Kubota, K., Tsuda, S., Tamai, A. and Meshi T. (2003) '*Tomato mosaic virus* replication protein suppresses virus-targeted post-transcriptional gene silencing', *Journal of Virology* **77**, 11016–11026.

Laporte, C., Vetter, G., Loudes, A.-M. *et al.* (2003) 'Involvement of the secretory pathway and the cytoskeleton in intracellular targeting and tubule assembly of *Grapevine fanleaf virus* movement protein in tobacco BY-2 cells', *The Plant Cell* **15**, 2058–2075.

Lawrence, D.M. and Jackson, A.O. (2001) 'Interactions of the TGB1 protein during cell-to-cell movement of *Barley stripe mosaic virus*', *Journal of Virology* **75**, 8712–8723.

Lazarowitz, S.G. and Beachy, R.N. (1999) 'Viral movement proteins as probes for intracellular and intercellular trafficking in plants', *The Plant Cell* **11**, 535–548.

Lee, J.-L., Yoo, B.-C. and Lucas, W.J. (2000) 'Parallels between nuclear-pore and plasmodesmal trafficking of information molecules', *Planta* **210**, 177–187.

Lee, J.-Y. and Lucas, W.J. (2001) 'Phosphorylation of viral movement proteins – regulation of cell-to-cell trafficking', *Trends in Microbiology* **9**, 5–8.

Lee, J.-Y., Yoo, B.-C., Rojas, M.R., Gomez-Ospina, N., Staehelin, L.A. and Lucas, W.J. (2003) 'Selective trafficking of non-cell-autonomous proteins mediated by NtNCAPP1', *Science* **299**, 392–396.

Li, Y., Bao, Y.M., Wei, C.H. *et al.* (2004) 'Rice dwarf phytoreovirus segment S6-encoded nonstructural protein has a cell-to-cell movement function', *Journal of Virology* **78**, 5382–5389.

Lin, B. and Heaton, L.A. (2001) 'An *Arabidopsis thaliana* protein interacts with a movement protein of *Turnip crinkle virus* in yeast cells and *in vitro*', *Journal of General Virology* **82**, 1245–1251.

Lough, T.J., Emerson, S.J., Lucas, W.J. and Forster, R.L. (2001) 'Trans-complementation of long-distance movement of white clover mosaic virus triple gene block (TGB) mutants: phloem-associated movement of TGBp1', *Virology* **288**, 18–28.

Lough, T.J., Netzler, N.E., Emerson, S.J. *et al.* (2000) 'Cell-to-cell movement of potexviruses: evidence for a ribonucleic protein complex involving the coat protein and first triple gene block protein', *Molecular Plant–Microbe Interactions* **13**, 962–974.

Lough, T.J., Shash, K., Xoconostle-Cázares, B. *et al.* (1998) 'Molecular dissection of the mechanism by which potexvirus triple gene block proteins mediate cell-to-cell transport of infectious RNA', *Molecular Plant–Microbe Interactions* **11**, 801–814.

Más, P. and Beachy, R.N. (1999) 'Replication of tobacco mosaic virus on endoplasmic reticulum and role of the cytoskeleton and virus movement protein in intracellular distribution of viral RNA', *Journal of Cell Biology* **147**, 945–958.

Matsushita, Y., Miyakawa, O., Deguchi, M., Nishiguchi, M. and Nyunoya, H. (2002a) 'Cloning of a tobacco cDNA coding for a putative transcriptional coactivator MBF1 that interacts with the tomato mosaic virus movement protein', *Journal of Experimental Botany* **53**, 1531–1532.

Matsushita, Y., Ohshima, M., Yoshioka, K., Nishiguchi, M. and Nyunoya, H. (2003) 'The catalytic subunit of protein kinase CK2 phosphorylates *in vitro* the movement protein of *Tomato mosaic virus*', *Journal of General Virology* **84**, 497–505.

Matsushita, Y., Yoshioka, K., Shigyo, T., Takahashi, H. and Nyunoya, H. (2002b) 'Phosphorylation of the movement protein of cucumber mosaic virus in transgenic tobacco plants', *Virus Genes* **24**, 231–234.

McLean, B.G. and Zambryski, P.C. (2000) 'Interactions between viral movement proteins and cytoskeleton', in *Actin: A Dynamic Framework for Multiple Plant Cell Functions* (eds C. J. Staiger, F. Baluška, D. Volkmann and P.W. Barlow), Kluwer Academic Publishers, The Netherlands, pp. 517–540.

McLean, B.G., Zupan, J. and Zambryski, P.C. (1995) 'Tobacco mosaic virus movement protein associates with the cytoskeleton in tobacco cells', *The Plant Cell* **7**, 2101–2114.

Medina, V., Peremyslov, V.V., Hagiwara, Y. and Dolja, V.V. (1999) 'Subcellular localization of the HSP70-homolog encoded by *beet yellows closterovirus*', *Virology* **260**, 173–181.

Mitra, R., Krishnamurthy, K., Blancaflor, E., Payton, M., Nelson, R.S. and Verchot-Lubicz, J. (2003) 'The *Potato virus X* TGBp2 protein association with the endoplasmic reticulum plays a role in but is not sufficient for viral cell-to-cell movement', *Virology* **312**, 35–48.

Morozov, S.Y. and Solovyev, A.G. (2003) 'Triple gene block: modular design of a multifunctional machine for plant virus movement', *Journal of General Virology* **84**, 1351–1366.

Morozov, S.Y., Fedorkin, O.N., Jüttner, G., Schiemann, J., Baulcombe, D.C. and Atabekov, J.G. (1997) 'Complementation of a potato virus X mutant mediated by bombardment of plant tissues with cloned viral movement protein genes', *Journal of General Virology* **78**, 2077–2083.

Oparka, K.J. (2004) 'Getting the message across: how do plants cells exchange macromolecular complexes?', *Trends in Plant Science* **9**, 33–40.

Oparka, K.J. and Roberts, A.G. (2001) 'Plasmodesmata. A not so open-and-shut case', *Plant Physiology* **125**, 123–126.

Oparka, K.J., Prior, D.A.M., Santa Cruz, S., Padgett, H.S. and Beachy, R.N. (1997) 'Gating of epidermal plasmodesmata is restricted to the leading edge of expanding infection sites of tobacco mosaic virus (TMV)', *The Plant Journal* **12**, 781–789.

O'Reilly, E.K., Paul, J.D. and Kao, C.C. (1997) 'Analysis of the interaction of viral RNA replication proteins by using the yeast two-hybrid assay', *Journal of Virology* **71**, 7526–7532.

Osman, T.A. and Buck, K.W. (2003) 'Identification of a region of the tobacco mosaic virus 126- and 183 kilodalton replication proteins which binds specifically to the viral 3′-terminal tRNA-like structure', *Journal of Virology* **77**, 8669–8675.

Peremyslov, V.V., Hagiwara, Y. and Dolja, V.V. (1999) 'HSP70 homolog functions in cell-to-cell movement of a plant virus', *Proceedings of the National Academy of Sciences USA* **96**, 14771–14776.

Pouwels, J., Carette, J.E., Van Lent, J. and Wellink, J. (2002a) '*Cowpea mosaic virus:* effects on host cell processes', *Molecular Plant Pathology* **3**, 411–418.

Pouwels, J., Van Der Krogt, G.N.M., Van Lent, J., Bisseling, T. and Wellink, J. (2002b) 'The cytoskeleton and the secretory pathway are not involved in targeting the cowpea mosaic virus movement protein to the cell periphery', *Virology* **297**, 48–56.

Reichel, C., Maś, P. and Beachy, R.N. (1999) 'The role of the ER and cytoskeleton in plant viral trafficking', *Trends in Plant Sciences* **4**, 458–463.

Restrepo-Hartwig, M. and Ahlquist, P. (1999) 'Brome mosaic virus RNA replication proteins 1a and 2a colocalize and 1a independently localizes on the yeast endoplasmic reticulum', *Journal of Virology* **73**, 10303–10309.

Restrepo-Hartwig, M.A. and Ahlquist, P. (1996) 'Brome mosaic virus helicase- and polymerase-like proteins colocalize on the endoplasmic reticulum at sites of viral RNA synthesis', *Journal of Virology* **70**, 8908–8916.

Ritzenthaler, C., Nebenführ, A., Movafeghi, A. *et al.* (2002) 'Reevaluation of the effects of brefeldin A on plant cells using tobacco bright yellow 2 cells expressing Golgi-targeted green fluorescent protein and COPI antisera', *The Plant Cell* **14**, 237–261.

Roberts, A.G. and Oparka, K.J. (2003) 'Plasmodesmata and the control of symplastic transport', *Plant, Cell and Environment* **26**, 103–124.

Schwartz, M., Chen, J., Janda, M., Sullivan, M., den Boon, J. and Ahlquist, P. (2002) 'A positive-strand RNA virus replication complex parallels form and function of retrovirus capsids', *Molecular Cell* **9**, 505–514.

Schwartz, M., Chen, J., Lee, W.-M., Janda, M. and Ahlquist, P. (2004) 'Alternate, virus-induced membrane rearrangements support positive-strand RNA virus genome replication', *Proceedings of the National Academy of Sciences USA* **101**, 11263–11268.

Soellick, T.-R., Uhrig, J.F., Bucher, G.L., Kellman, J.-W. and Schreier, P.H. (2000) 'The movement protein $NS^m$ of tomato spotted wilt tospovirus (TSWV): RNA binding, interaction with the TSWV N protein, and identification of interacting plant proteins', *Proceedings of the National Academy of Sciences USA* **97**, 2373–2378.

Sokolova, M., Prüfer, D., Tacke, E. and Rohde, W. (1997) 'The potato leafroll virus 17K movement protein is phosphorylated by a membrane-associated protein kinase from potato with biochemical features of protein kinase C', *FEBS Letters* **400**, 201–205.

Solovyev, A.G., Savenkov, E.I., Agranovsky, A.A. and Morozov, S.Y. (1996) 'Comparisons of the genomic cis-elements and coding regions in RNA beta components of the hordeiviruses barley stripe mosaic virus, lychnis ringspot virus, and poa semilatent virus', *Virology* **219**, 9–18.

Solovyev, A.G., Savenkov, E.I., Grdzelishvili, V.Z. *et al.* (1999) 'Movement of hordeivirus hybrids with exchanges in the triple gene block', *Virology* **253**, 278–287.

Solovyev, A.G., Stroganova, T.A., Zamyatnin Jr., A.A., Fedorkin, O.N., Schieman, J. and Morozov, S.Y. (2000) 'Subcellular sorting of small membrane-associated triple gene block proteins: TGBp3-assisted targeting of TGBp2', *Virology* **269**, 113–127.

Storms, M.M.H., Kormelink, R., Peters, D., Van Lent, J.W.M. and Goldbach, R.W. (1995) 'The non-structural $NS^m$ protein of tomato spotted wilt virus induces tubular structures in plant and insect cells', *Virology* **214**, 485–493.

Szécsi, J., Ding, X.S., Lim, C.O. *et al.* (1999) 'Development of tobacco mosaic virus infection sites in *Nicotiana benthamiana', Molecular Plant–Microbe Interactions* **12**, 143–152.

Szittya, G., Silhavy, D., Molnar, A. *et al.* (2003) 'Low temperature inhibits RNA silencing-mediated defence by the control of siRNA generation', *EMBO Journal* **22**, 633–640.

Tamai, A. and Meshi, T. (2001a) 'Cell-to-cell movement of potato virus X: the role of p12 and p8 encoded by the second and third open reading frames of the triple gene block', *Molecular Plant–Microbe Interactions* **14**, 1158–1167.

Tamai, A. and Meshi, T. (2001b) 'Tobamoviral movement protein transiently expressed in a single epidermal cell functions beyond multiple plasmodesmata and spreads multicellularly in an infection-coupled manner', *Molecular Plant–Microbe Interactions* **14**, 126–134.

Tsujimoto, Y., Numaga, T., Ohshima, K. *et al.* (2003) '*Arabidopsis TOBAMOVIRUS MULTIPLICATION (TOM) 2* locus encodes a transmembrane protein that interacts with TOM1', *EMBO Journal* **22**, 335–343.

Tzfira, T., Rhee, Y., Chen, M.-H. and Citovsky, V. (2000) 'Nucleic acid transport in plant-microbe interactions: the molecules that walk through the walls', *Annual Review of Microbiology* **54**, 187–219.

von Bargen, S., Salchert, K., Paape, M., Piechulla, B. and Kellmann, J.-W. (2001) 'Interactions between the tomato spotted wilt virus movement protein and plant proteins showing homologies to myosin, kinesin and DnaJ-like chaperones', *Plant Physiology and Biochemistry* **39**, 1083–1093.

Waigmann, E., Chen, M.-H., Bachmaier, R., Ghoshroy, S. and Citovsky, V. (2000) 'Regulation of plasmodesmal transport by phosphorylation of tobacco mosaic virus cell-to-cell movement protein', *EMBO Journal* **19**, 4875–4884.

Watanabe, T., Honda, A., Iwata, A., Ueda, S., Hibi, T. and Ishihama, A. (1999) 'Isolation from tobacco mosaic virus-infected tobacco of a solubilized template-specific RNA-dependent RNA polymerase containing a 126K/183K protein heterodimer', *Journal of Virology* **73**, 2633–2640.

Xie, Z., Fan, B., Chen, C. and Chen, Z. (2001) 'An important role of an inducible RNA-dependent RNA polymerase in plant antiviral defense', *Proceedings of the National Academy of Sciences USA* **98**, 6516–6521.

Xoconostle-Cázares, B., Xiang, Y., Ruiz-Medrano, R. *et al.* (1999) 'Plant paralog to viral movement protein that potentiates transport of mRNA into the phloem', *Science* **283**, 94–98.

Yamanaka, T., Ohta, T., Takahashi, M. *et al.* (2000) 'TOM1, an *Arabidopsis* required for efficient multiplication of a tobamovirus, encodes a putative transmembrane protein', *Proceedings of the National Academy of Sciences USA* **97**, 10107–10112.

Yang, S.-J., Carter, S.A., Cole, A.B., Cheng, N.-H. and Nelson, R.S. (2004) 'A natural variant of a host RNA-dependent RNA polymerase is associated with increased susceptibility to viruses by *Nicotiana benthamiana*', *Proceedings of the National Academy of Sciences USA* **101**, 6297–6302.

Yang, Y., Ding, B., Baulcombe, D.C. and Verchot, J. (2000) 'Cell-to-cell movement of the 25K protein of *Potato virus X* is regulated by three other viral proteins', *Molecular Plant–Microbe Interactions* **13**, 599–605.

Zamyatnin, A.A., Jr., Solovyev, A.G., Sablina, A.A. *et al.* (2002) 'Dual-colour imaging of membrane protein targeting directed by poa semilatent virus movement protein TGBp3 in plant and mammalian cells', *Journal of General Virology* **83**, 651–662.

Zamyatnin, A.A., Jr., Solovyev, A.G., Savenkov, E.I. *et al.* (2004) 'Transient coexpression of individual genes encoded by the triple gene block of *Potato mop-top virus* reveals requirements for TGBp1 trafficking', *Molecular Plant —Microbe Interactions* **8**, 921–930.

# 10 Systemic RNA silencing

Manfred Heinlein

## 10.1 Introduction

RNA silencing is a newly discovered mechanism of genetic regulation in eukaryotes that acts through RNA-mediated sequence-specific interactions (Voinnet, 2002). In recent years, this mechanism has become an area of increasing interest, both because of its fascinating biology and also because of its power as an experimental tool (Hammond *et al.*, 2001b; Dykxhoorn *et al.*, 2003; Timmons *et al.*, 2003; Waterhouse and Helliwell, 2003; Burch-Smith *et al.*, 2004; Carpenter and Sabatini, 2004; Gupta *et al.*, 2004; Hannon and Conklin, 2004; Jones *et al.*, 2004; Kessler *et al.*, 2004; Love, 2004; Mittal, 2004; Morris *et al.*, 2004; Paddison *et al.*, 2004; Reynolds *et al.*, 2004). Originally described as co-suppression and post-transcriptional gene silencing (PTGS) in plants (Depicker and van Montagu, 1997; Marathe *et al.*, 2000; Meins, 2000), quelling in fungi (Pickford *et al.*, 2002), and RNA interference (RNAi) in animals (Fire *et al.*, 1998; Bosher and Labouesse, 2000; Plasterk and Ketting, 2000; Sharp, 2001), the mechanism of RNA silencing is now known to be related to a wide range of phenomena that commonly involve small RNA species and play fundamental roles in defense responses against viruses, in the control of transposon mobility, and in development (Carrington and Ambros, 2003; Finnegan and Matzke, 2003; Lai, 2003). RNA silencing is generally induced by double-stranded (ds)-RNA, which can originate from various sources such as transgenes, viral replication intermediates, or experimentally introduced dsRNA sequences.

RNA silencing degrades dsRNA and can suppress the expression of homologous genes. Short RNA molecules that result from the processing of dsRNA act as specificity determinants of the silencing system by becoming incorporated into different silencing effector complexes so as to guide these complexes to homologous nucleic acids for suppression. A remarkable feature of RNA silencing in plants, as well as in *Caenorhabditis elegans* (Timmons and Fire, 1998) and planaria (Newmark *et al.*, 2003), is its ability to act beyond the cells in which it is initiated (Voinnet and Baulcombe, 1997). RNA thus acts at both the single cell level (local silencing) and the whole plant level (systemic silencing). Local silencing suppresses homologous genes in the cells in which dsRNAs accumulate, and generates silencing signals that are mobile and carry sequence-specific information able to confer suppression of homologous RNA molecules in adjacent cells. During degradation of homologous RNA, new signal molecules are produced that allow RNA silencing to continue its spread from cell to cell (short distance) and systemically (systemic silencing)

(Mlotshwa *et al.*, 2002). This ability of the silencing signal to spread and replicate is reminiscent of infectious viruses. Indeed, like viruses, RNA silencing appears to move cell-to-cell via plasmodesmata. This chapter reviews current knowledge of systemic RNA silencing and the nature of the silencing signal.

## 10.2 The RNA silencing pathway

The biochemical process of RNA silencing (Figure 10.1) is mediated by proteins that are conserved between species (Hannon, 2002). dsRNA molecules that are experimentally introduced (Waterhouse and Helliwell, 2003), or endogenously generated from transgenes or during viral replication, are processed by proteins called Dicers or Dicer-like (DCL) enzymes. These proteins are homologs of *Escherichia coli* RNaseIII and are found in all organisms (Zamore *et al.*, 2000; Bernstein *et al.*, 2001; Ketting *et al.*, 2001; Knight and Bass, 2001). Cleavage of dsRNA by Dicer activity results in the formation of functionally different short RNAs called micro-RNAs (miRNAs) and short interfering RNAs (siRNAs) (Denli and Hannon, 2003). miRNAs are 21–22-nt single-stranded RNA molecules that are processed from endogenous hairpin RNA precursor transcripts derived from miRNA genes (Pasquinelli and Ruvkun, 2002) that are present in high numbers in humans, animals and plants (Bartel, 2004; Griffiths-Jones, 2004). siRNAs are chemically similar to miRNAs, although they typically range in size between 21 and 26 nt and are characterized by 2-nt 3′overhangs (Hamilton *et al.*, 2002; Llave *et al.*, 2002a; Cerutti, 2003; Tang *et al.*, 2003; Voinnet, 2002). They are associated with both post-transcriptional forms of RNA silencing and transcriptional silencing involving DNA methylation and chromatin modification (Finnegan and Matzke, 2003). siRNAs were first found in plants that exhibited transgene-mediated RNA silencing (Hamilton and Baulcombe, 1999). Later, it was demonstrated that siRNAs can derive from various sources. Endogenous sources include retroelements and transposons, other highly repetitive sequences, pseudogenes, intergenic regions and expressed genes (Hamilton *et al.*, 2002; Llave *et al.*, 2002a; Mette *et al.*, 2002). Exogenous siRNAs can arise from transgenes and from viruses (Hamilton and Baulcombe, 1999; Mette *et al.*, 2000). siRNAs are generally processed from precursors containing extensive dsRNA structure, such as transcripts containing inverted repeats or intermediates formed during RNA virus replication (Hannon, 2002). The intrinsic relationship of siRNAs to the RNA silencing mechanism has been established without doubt since it has been shown that synthetic siRNA duplexes induce RNA silencing in various organisms, both *in vivo* and *in vitro* (Parrish *et al.*, 2000; Caplen *et al.*, 2001; Elbashir *et al.*, 2001a,b; Nykanen *et al.*, 2001; Klahre *et al.*, 2002). The presence of siRNAs thus serves as a reliable molecular marker for induced silencing in response to endogenous and exogenous triggers.

In many animals, both miRNAs and siRNAs are formed by the activity of the same Dicer enzyme (Grishok *et al.*, 2001; Hutvagner *et al.*, 2001; Ketting *et al.*, 2001; Knight and Bass, 2001; Provost *et al.*, 2002; Zhang *et al.*, 2002; Myers *et al.*, 2003).

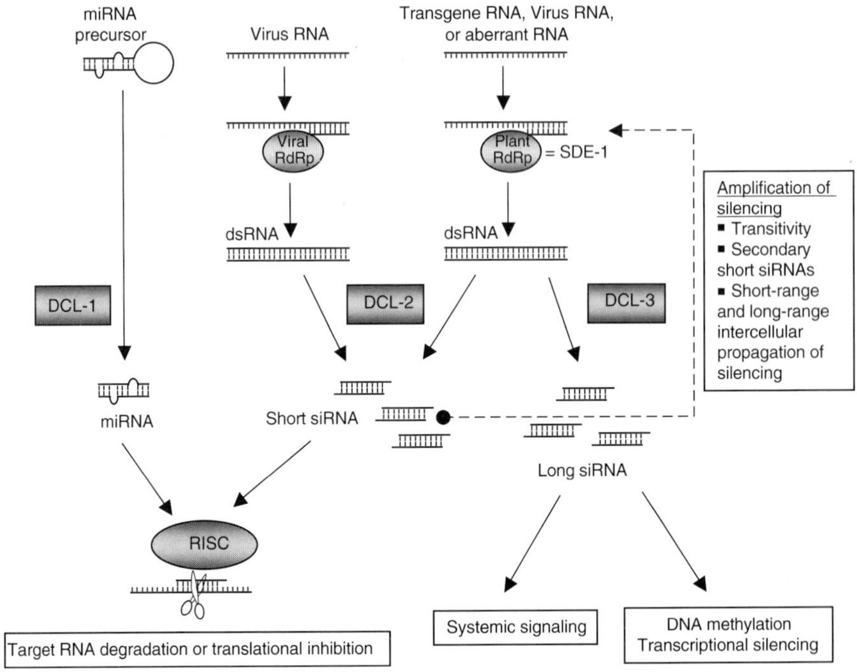


**Figure 10.1** Model of the RNA silencing pathway. Plant RNA silencing is divided into three branches. The first branch is involved in developmental regulation. Here, nuclear gene-encoded miRNA precursors are processed by DCL-1 into miRNAs that upon association with RISC cause target RNA cleavage or translational arrest. The second branch functions in antiviral defense and is initiated by viral dsRNA produced by virus-encoded RNA-dependent RNA polymerase (RdRp) during virus replication. The dsRNA is recognized and cleaved by DCL-2, leading to the accumulation of short virus-specific (21 nt) siRNAs. These siRNAs associate with RISC causing the degradation of viral (target) RNA. The third branch of the RNA silencing pathway involves a plant RdRp (SDE-1), which produces dsRNA from either foreign or aberrant RNA. The dsRNA is cleaved by DCL-2 and DCL-3 leading to the accumulation of short (21 nt) and long (25 nt) siRNAs, respectively. The long siRNAs are involved in systemic silencing signaling as well as in DNA methylation and transcriptional silencing. The short siRNAs produced by the second and third branches may associate with RISC for target RNA degradation, or serve as primers for SDE-1. SDE-1 activity primed by siRNAs generates new dsRNA and, therefore, new 'secondary' siRNAs. Because the secondary siRNAs correspond to sequences located outside the primary targeted regions of the RNA, and again can serve as primers for dsRNA synthesis and siRNA production, this process leads to transitive silencing, i.e. the spread of siRNA-targeted regions along the gene (or viral RNA). SDE-1-mediated amplification of silencing enhances the silencing pathway in cells in which silencing is triggered. Since silencing spreads between cells, this amplification is an essential step for maintaining silencing in cells in which the 'primary' siRNAs are no longer present (see Figure 10.3).

Humans and *C. elegans* indeed encode only one Dicer enzyme, which can process both dsRNA and miRNA precursors. However, human and mouse genomes encode three RNaseIII family members (Fortin *et al.*, 2002), of which one, 'Drosha', is able to cleave miRNA precursors and leads to accumulation of precursors if deficient in HeLa cells (Lee *et al.*, 2003b). In *Drosophila* and plants, the short RNAs are processed by distinct Dicers or Dicer-related enzymes (Schauer *et al.*, 2002; Finnegan *et al.*, 2003; Lee *et al.*, 2004; Tijsterman and Plasterk, 2004). *Drosophila* has two Dicers, Dicer-1 and Dicer-2, which are specifically required for miRNA and siRNA production, respectively (Lee *et al.*, 2004; Tijsterman and Plasterk, 2004). *Arabidopsis* contains four Dicer-like activities (DCL1 to DCL4), of which one (DICER-LIKE1 = DCL1) has been known for some time to have a definitive function in small RNA biogenesis (Park *et al.*, 2002; Reinhart *et al.*, 2002; Schauer *et al.*, 2002) and to be required for miRNA production but not for siRNA production (Boutet *et al.*, 2003; Finnegan *et al.*, 2003). Recent studies in *Arabidopsis* have confirmed the role of DCL1 in miRNA production and indicated a role of DCL2 and DCL3 in siRNA production (Xie *et al.*, 2004). Plants process long dsRNAs into two functionally different classes of short (21–22 nt) siRNAs and long (24–26 nt) siRNAs (Hamilton *et al.*, 2002). The long and short siRNAs appear to be produced by DCL3 and DCL2, respectively (Xie *et al.*, 2004); this is consistent with previous biochemical data derived from experiments using wheat germ extracts, which indicated that the long and short siRNAs are formed by two separate Dicer homologs (Tang *et al.*, 2003).

miRNAs are numerous and widespread among eukaryotes and have been conserved throughout evolution. Their major role is in developmental regulation. For example, plant miRNAs regulate the transcript level of different transcription factors, hormone response genes and genes involved in silencing (Llave *et al.*, 2002b; Rhoades *et al.*, 2002; Aukerman and Sakai, 2003; Bartel and Bartel, 2003; Carrington and Ambros, 2003; Kasschau *et al.*, 2003; Palatnik *et al.*, 2003; Xie *et al.*, 2003). It is therefore not surprising that plants mutated in *DCL-1* (*CAF*) and *HEN1* – genes known to influence miRNA accumulation – and *AGO1*, a gene with putative roles in miRNA function, have strong pleiotropic developmental phenotypes (Bohmert *et al.*, 1998; Jacobsen *et al.*, 1999; Park *et al.*, 2002; Reinhart *et al.*, 2002; Schauer *et al.*, 2002; Kidner and Martienssen, 2003). In contrast to miRNAs, siRNAs are mainly involved in the suppression of molecular parasites (Baulcombe, 2002; Plasterk, 2002; Voinnet, 2002; Carrington and Ambros, 2003), such as transposons, transgenes and viruses (Voinnet, 2001). Although there is evidence suggesting that siRNAs can play a role in developmental gene control in *Drosophila* (Aravin *et al.*, 2001) there is little evidence yet for such a role in plants.

Although miRNAs and siRNAs appear to have different roles, the effector steps involved in miRNA- and siRNA-mediated pathways share components. Both types of short RNAs incorporate and guide a multicomponent nuclease complex (RNA-Induced Silencing Complex = RISC) to inactivate homologous target RNAs (Hammond *et al.*, 2000). Upon association with RISC, the short RNAs are unwound and activated in an ATP-dependent manner (Nykanen *et al.*, 2001). Only as single-stranded species are they competent to bind complementary mRNA targets for RISC-mediated destruction (Marathe *et al.*, 2000).

The Dicer/RISC process ensures that RNA species are degraded if they share sequence similarity with the dsRNA. Indeed, the Dicer and RISC components are thought to directly interact in order to direct sequence-specific RNA processing. Thus, several recent *in vitro* and *in vivo* studies have established that Dicers and associated factors from diverse organisms actively participate in RISC assembly and are components of RISC (Hammond *et al.*, 2001a; Tabara *et al.*, 2002; Doi *et al.*, 2003; Liu *et al.*, 2003; Tang *et al.*, 2003; Lee *et al.*, 2004; Pham *et al.*, 2004). The two different Dicers in *Drosophila* apparently participate in distinct steps of RISC assembly, leading to siRNA- and miRNA-associated RISCs that differ with respect to the associated Dicer (Lee *et al.*, 2004). The set of proteins associated with RISCs also contains ARGONAUTE (AGO) family members that likely bind siRNAs or target sequences (Carmell *et al.*, 2002; Lingel *et al.*, 2003; Song *et al.*, 2003; Yan *et al.*, 2003) and appear to contribute the mRNA-cleaving 'slicer' activity of RISC (Liu *et al.*, 2004; Song *et al.*, 2004). RISC complexes cleave targeted mRNA in cases of perfect or near-perfect base pairing between mRNA and short-guide RNA, whereas they mediate translational repression in cases of partial complementarity (Hutvagner and Zamore, 2002; Aukerman and Sakai, 2003; Doench *et al.*, 2003; Tang *et al.*, 2003; Chen, 2004; Doench and Sharp, 2004). Most plant miRNAs have high sequence complementarity with their targets and thus act as negative regulators by a cleavage-type mechanism (Llave *et al.*, 2002b; Rhoades *et al.*, 2002; Emery *et al.*, 2003; Kasschau *et al.*, 2003; Palatnik *et al.*, 2003; Tang *et al.*, 2003; Xie *et al.*, 2003), although there are counterexamples in which miRNA-mediated translation inhibition has been described (Aukerman and Sakai, 2003; Chen, 2004). Animal miRNAs, on the other hand, generally have lower sequence complementarity with their targets and usually suppress translation (Olsen and Ambros, 1999; Reinhart *et al.*, 2000), although, again, exceptions exist in which some human and animal miRNAs act through target cleavage (Zeng *et al.*, 2003; Yekta *et al.*, 2004). These considerations, however, should not obscure the fundamental difference between miRNAs and siRNAs: siRNAs mediate the silencing of the same or similar genes from which they originate, whereas miRNAs are encoded by their own genes and regulate different genes (Bartel and Bartel, 2003; Bartel, 2004).

siRNAs also regulate other silencing effector complexes. For example, in plants and *C. elegans*, they act as primers to guide an RNA-dependent RNA polymerase (RdRp) to transform target RNAs into dsRNAs (Figure 10.1). Since the newly created dsRNAs again serve as substrate for Dicer, and thus lead to the generation of new 'secondary' siRNAs, this interaction increases the initial pool of siRNAs and thus amplifies silencing (Dalmay *et al.*, 2000; Mourrain *et al.*, 2000; Sijen *et al.*, 2001). Because the secondary siRNAs correspond to sequences located outside the primary targeted regions of the RNA, and again can serve as primers for dsRNA synthesis and siRNA production, this process leads to transitive silencing, i.e. the spread of siRNA-targeted regions along the gene (Vaistij *et al.*, 2002; Himber *et al.*, 2003). Whereas transitivity in *C. elegans* proceeds mainly in the $3' \rightarrow 5'$ direction along the target transgene, the spread of silencing in *Arabidopsis* occurs in both the $5'$ and $3'$ direction of the target transgene (Vaistij *et al.*, 2002). The spread of silencing in the $5'$ direction is consistent with the hypothesis that siRNAs prime

$5' \rightarrow 3'$ dsRNA synthesis on sense mRNA. The mechanism by which secondary siRNAs located 3′ to the triggering dsRNA are generated is not entirely understood.

RdRp-dependent synthesis is generally required for the formation of siRNAs involved in the silencing of endogenous genes and transgenes, as was shown genetically in screens for RNA-silencing-defective mutants (Cogoni and Macino, 1999; Dalmay *et al.*, 2000; Mourrain *et al.*, 2000; Smardon *et al.*, 2000; Volpe *et al.*, 2002). *Arabidopsis* plants contain at least three active RdRp genes, termed RDR1, RDR2 and RDR6. The RDR6 gene is also known as SDE1/SDE2 (Dalmay *et al.*, 2000; Mourrain *et al.*, 2000; Yu *et al.*, 2003) and, for simplicity, here the gene is referred to as SDE1. Transgene silencing in *Arabidopsis* also depends on SDE3. This protein has sequences characteristic of an RNA helicase and, like SDE1, has been positioned by genetic experiments upstream of, or at, the step of dsRNA synthesis (Dalmay *et al.*, 2001). However, transgenes that generate dsRNA, and viruses that produce dsRNA during replication, bypass the requirement for RdRp-dependent dsRNA synthesis in silencing (Dalmay *et al.*, 2000, 2001; Beclin *et al.*, 2002).

siRNAs could also play a role in transcriptional silencing by directing DNA and/or histone methylation of homologous DNAs (Mlotshwa *et al.*, 2002; Grewal and Moazed, 2003; Pickford and Cogoni, 2003; Zilberman *et al.*, 2003; Ekwall, 2004; Verdel *et al.*, 2004). Such siRNA-mediated chromatin-based silencing may serve to suppress mobile genetic elements or invasive DNA (Schramke and Allshire, 2003). Evidence for such a role is provided by the high abundance of siRNAs derived from repeated sequences corresponding to transposons, retroelements and centromeres (Hamilton *et al.*, 2002; Llave *et al.*, 2002a; Mette *et al.*, 2002; Reinhart and Bartel, 2002), and also by the requirement of RNA-silencing-related factors for *de novo* DNA methylation and initiation of transcriptional silencing in *Arabidopsis* (Zilberman *et al.*, 2003; Chan *et al.*, 2004; Xie *et al.*, 2004) as well as for maintenance of transcriptionally silenced or heterochromatic states (Hall *et al.*, 2002; Volpe *et al.*, 2002, 2003) in yeast. Moreover, heterochromatin formation can be triggered, in a sequence-dependent manner, by cytoplasmic RNA silencing (Jones *et al.*, 1999; Aufsatz *et al.*, 2002; Schramke and Allshire, 2003; Volpe *et al.*, 2003; Pal-Bhadra *et al.*, 2004; Verdel *et al.*, 2004). Recent studies in human cells indicate that siRNAs targeted to CpG islands within the promoter of a specific gene can indeed induce transcriptional gene silencing by DNA-methyltransferase-dependent methylation (Kawasaki and Taira, 2004).

RNA viruses and inverted-repeat transgenes that are prone to RNA silencing have been associated with methylation of homologous genomic DNA sequences (Matzke *et al.*, 2001). The functional link between cytoplasmic silencing and nuclear silencing that depends on RNA-silencing-related genes, as well as the association of DNA elements known to be controlled by silencing with high levels of siRNAs, indicate that siRNA may function as the RNA-directed DNA methylation and silencing signal that is transmitted from the cytoplasm to the nucleus. Indeed, siRNA may recruit chromatin modification complexes to specific loci (Grewal and Moazed, 2003).

## 10.3 Virus-induced RNA silencing

RNA silencing has been implicated in anti-virus defense in plants (Voinnet, 2001; Waterhouse *et al.*, 2001), insects (Li *et al.*, 2002) and perhaps other eukaryotes (Gitlin and Andino, 2003). The first indications that RNA-mediated responses play an important antiviral role came from observations that the transgenic expression of viral sequences protected plants from homologous viruses by conferring sequence-specific degradation of viral RNAs (Lindbo and Dougherty, 1992; Dougherty *et al.*, 1994; Baulcombe, 1996). In addition, non-viral transgenes are able to suppress virus infection if the virus is modified by insertion of the transgene sequence into the viral genome (English *et al.*, 1996). Viruses are also able to induce the silencing of host genes if they carry host-related inserts or if a transgene possesses homology to the sequence of the inoculated virus (Lindbo and Dougherty, 1992; Lindbo *et al.*, 1993; Kumagai *et al.*, 1995; Ruiz *et al.*, 1998; Liu *et al.*, 2002; Dinesh-Kumar *et al.*, 2003; Lu *et al.*, 2003; Brigneti *et al.*, 2004; Burch-Smith *et al.*, 2004). Virus-induced gene silencing (VIGS) of host or transgenes depends on successful virus infection and therefore is usually observed once the virus has systemically infected the plant. Once triggered, virus-induced silencing may affect the virus as well as the homologous plant gene resulting in 'recovery' from disease (Lindbo and Dougherty, 1992; Lindbo *et al.*, 1993). The observation that infection of plants by nepoviruses and caulimoviruses results in RNA-silencing-mediated recovery, even in the absence of a transgene (Covey *et al.*, 1997; Ratcliff *et al.*, 1997), indicated that sequence-specific RNA degradation acts as a natural plant antiviral response (Ratcliff *et al.*, 1997). This was confirmed by showing that virus-induced RNA silencing can also occur in the absence of a recovery phenotype, as is shown by resistance of plants against secondary infections with the same virus or with a virus carrying homology to the silencing inducing virus or a co-silenced transgene (Ratcliff *et al.*, 1999).

Viral proteins have been identified that suppress RNA-mediated defense (Anandalakshmi *et al.*, 1998; Beclin *et al.*, 1998; Brigneti *et al.*, 1998; Kasschau and Carrington, 1998) (Table 10.1), and siRNAs known to act as regulators of the previously described homology-dependent gene inactivation phenomena were also recognized as specificity determinants of RNA-mediated defense (Hamilton and Baulcombe, 1999). These findings led to the current model in which, in infected plants, the accumulation of viral RNAs induces RNA silencing as a plant defense response, which in turn generates virus-specific siRNAs (viral siRNAs). These siRNAs then program silencing effector complexes to carry out antiviral responses (Voinnet, 2001). Further studies are required to determine the mechanisms by which plant viruses trigger RNA silencing, and which silencing pathways are involved in antiviral responses. It has been proposed that silencing is triggered by dsRNA replicative intermediates (Ahlquist, 2002) which is consistent with the finding that RNA-virus-induced silencing is RdRp-independent (Dalmay *et al.*, 2000). These dsRNA replication intermediates could serve as dsRNA substrates for Dicer to produce viral siRNAs. However, viral siRNAs could also be generated from structured regions of viral ssRNAs, as is suggested by the finding that plus-strand-specific

**Table 10.1** Examples of virus-encoded silencing suppressor proteins

| Virus | Virus genus | Suppressor | References |
|---|---|---|---|
| *Positive-stranded RNA viruses* | | | |
| Tobacco mosaic virus | Tobamovirus | P126 | Ding *et al.*, 2004 |
| Tomato mosaic virus | | P130 | Kubota *et al.*, 2003 |
| Tomato bushy stunt virus | Tombusvirus | P19 | Ye *et al.*, 2003 |
| Cymbidium ringspot virus | | P19 | Silhavy *et al.*, 2002 |
| Carnation italian ringspot virus | | P19 | Vargason *et al.*, 2003 |
| Turnip crinkle virus | Carmovirus | P38 | Thomas *et al.*, 2003 |
| Cucumber mosaic virus | Cucumovirus | 2b | Lucy *et al.*, 2000 |
| Beet yellows virus | Closterovirus | P21 | Reed *et al.*, 2003 |
| Tobacco etch virus | Potyvirus | HC-Pro | Kasschau and Carrington, 1998 |
| Barley yellow mosaic virus | Hordeivirus | $\gamma$b | Yelina *et al.*, 2002 |
| Beet western yellows virus | Polerovirus | P0 | Pfeffer *et al.*, 2002 |
| Peanut clump virus | Pecluvirus | P15 | Dunoyer *et al.*, 2002 |
| Potato virus X | Potexvirus | P25 | Voinnet *et al.*, 2000 |
| *Negative-stranded RNA viruses* | | | |
| Tomato spotted wilt virus | Tospovirus | NSs | Takeda *et al.*, 2002 |
| Rice hoja blanca virus | Tenuivirus | NS3 | Bucher *et al.*, 2003 |
| *DNA viruses* | | | |
| African cassava mosaic virus | Begomovirus | AC2 | Voinnet *et al.*, 1999 |
| Tomato yellow leaf curl virus | | C2 | Dong *et al.*, 2003 |

siRNAs are dramatically overrepresented in an RNA-virus-infected plant (Szittya *et al.*, 2002). Interestingly, viral siRNAs appear to be restricted to the short (21–22 nt) class of siRNAs (Hamilton *et al.*, 2002; Silhavy *et al.*, 2002). This finding is consistent with recent evidence showing that accumulation of viral siRNAs depends on a specific Dicer enzyme, DCL2 (Xie *et al.*, 2004).

## 10.4 Systemic silencing

A remarkable feature of RNA silencing is its ability to act beyond the cells in which it is initiated. First indications for the spread of silencing came from observations with transgenic plants, in which a suppressed phenotype first occurred in lower leaves and then propagated into upper leaves (Boerjan *et al.*, 1994; Palauqui *et al.*, 1996). Subsequent grafting experiments showed that the spread of silencing can propagate from silenced stocks into non-silenced scions and thus demonstrated the existence of a diffusible signal molecule. The spread of silencing from stock to scion was not observed if the transgene in the scion was different from the gene silenced in the stock, indicating that the signal carries sequence-specific information, likely in the form of RNA (Palauqui *et al.*, 1997). Systemic silencing is also evident in plants carrying an expressed GFP transgene, where RNA silencing can be initiated by localized introduction of an additional ectopic GFP transgene (Plate 10.1). Following local transfection of such GFP-transgenic plants (e.g. by agroinfiltration or microparticle bombardment of leaves), RNA silencing is initially manifested locally in the tissues containing the ectopic DNA but eventually becomes systemic, illustrating that a silencing signal moves between cells and in the vascular system of the plant (Voinnet and Baulcombe, 1997; Voinnet *et al.*, 1998). The production and movement of this signal persists throughout the lifetime of the plant even if the tissues containing the ectopic initiator are removed. The ectopic DNA thus cannot be the only source of silencing signal. Upon exit from transfected cells, the signal must depend upon a different source.

Insights into the nature of silencing signaling were provided by the group of Voinnet (Himber *et al.*, 2003). These authors performed elegant assays combining the use of diverse silencing triggers, viral-derived silencing suppressors and silencing-deficient mutants of *Arabidopsis*, and found evidence for the existence of two phases in local silencing signaling. The first phase results in silencing of cells up to 10–15 cells away from the cells in which silencing was initially triggered (Figures 10.2 and 10.3). This short-range spread of silencing is independent of SDE1 and SDE3 and is not influenced by homologous transcripts in the recipient cells (i.e. it is independent of a transgene), thus excluding the participation of secondary siRNAs produced by transitivity. In contrast, further spread of silencing is

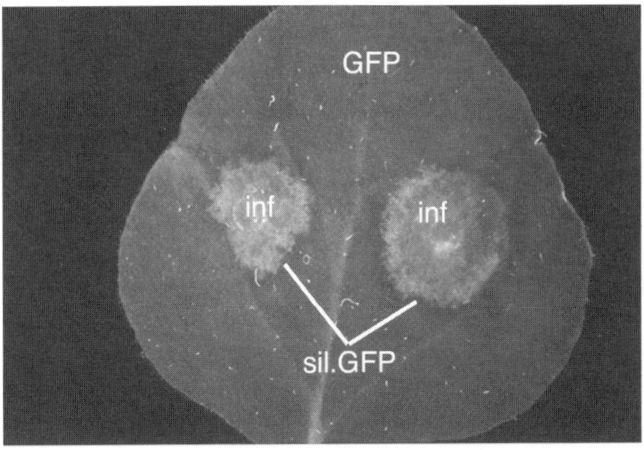


**Figure 10.2** Local induction of silencing. Leaf of *Nicotiana benthamiana* line 16c (Ruiz *et al.*, 1998), which carries a highly expressed green fluorescent protein (GFP) transgene. The leaf was locally infiltrated with an *Agrobacterium* culture (agro-infiltration) in which the T-DNA carries a copy of the GFP transgene. The infiltrated tissue (inf) is strongly green fluorescent (seen as light grey-white) due to superimposed expression of the ectopic and stably integrated GFP transgenes. However, silencing has already spread to adjacent, non-infiltrated, cells, as is evident from the rim of cells (sil. GFP) that turned red (chlorophyll autofluorescence, seen as dark grey) due to the silencing of the stably integrated plant gene. This red (dark grey) rim reflects the short-range spread of primary siRNAs into 10–15 cells adjacent to infiltrated cells. This short-range spread of silencing is SDE-1 independent.

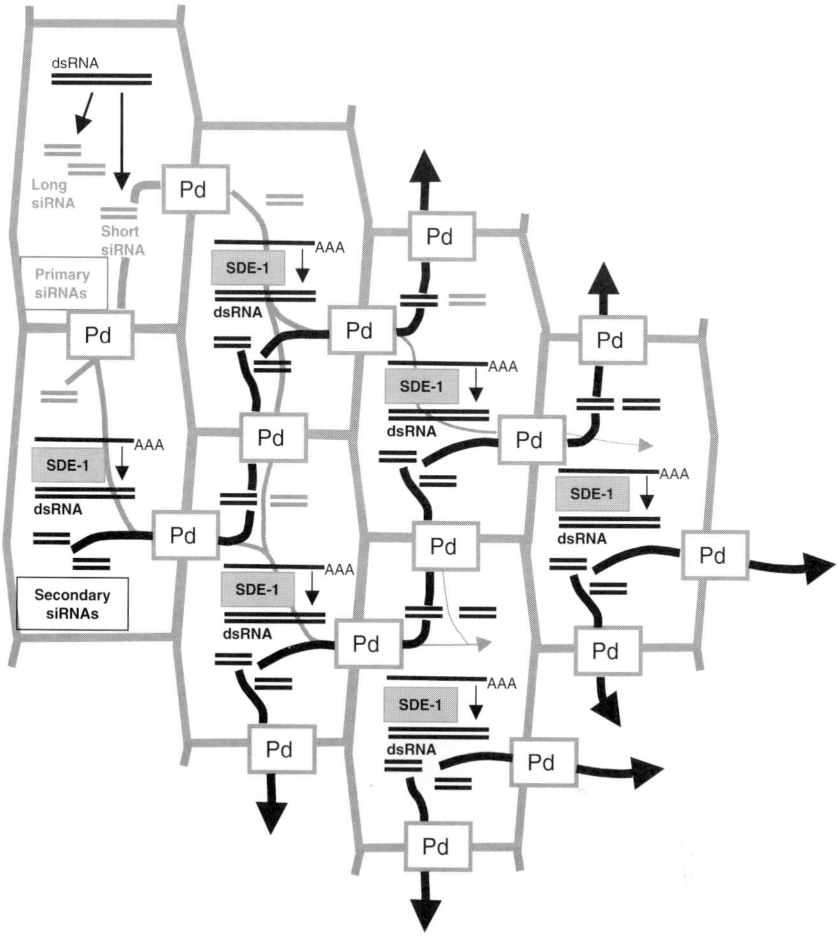


**Figure 10.3** Model for short-range and long-range cell-to-cell spread of silencing. Cells in which silencing is initially triggered (e.g. the cell shown in the top left corner) accumulate long and 'primary' short siRNAs (grey). The function of long siRNAs is not considered here (long siRNAs play an unknown role in phloem-mediated spread of silencing into other plant organs). The local cell-to-cell propagation of silencing is mediated by the short siRNAs. The short siRNAs can associate with RISC to degrade target RNA, move cell-to-cell via plasmodesmata (Pd) or serve as primers for SDE-1, thus leading to the production of 'secondary' siRNAs (black) in the cell in which silencing is initially triggered, as well as in adjacent cells. Individual siRNAs can trigger silencing only in a small halo of 10–15 cells into which they can diffuse. Thus, the 'primary' siRNAs (grey) become more and more diluted (thinner grey lines) until their concentration is below a threshold required for efficient target RNA degradation. This short-range silencing is evident in *sde-1* mutants, in which secondary siRNA production is inhibited. In the presence of SDE-1, however, secondary siRNAs are produced in each new cell in which silencing is triggered. By creating high levels of secondary siRNAs, this mechanism amplifies silencing and maintains its efficient long-range cell-to-cell spread (thick black lines).

SDE1- and SDE3-dependent and involves the presence and participation of secondary products produced by transitivity using a homologous template. In contrast to GFP mRNA silencing, the silencing of Rubisco small subunit (RbcS) mRNA did not show capacity for extensive cell-to-cell spread beyond the initial 10–15 cells (Himber *et al.*, 2003), which is in agreement with the finding that this mRNA does not support transitivity (Vaistij *et al.*, 2002) and thus is probably unable to support signal relay amplification. Indeed, in *Drosophila*, where RNA silencing is a cell-autonomous process, the lack of systemic silencing is correlated with the lack of transitivity and secondary siRNA production (Roignant *et al.*, 2003). Taken together, these findings indicate that the silencing signal is able to spread short distances from cells undergoing silencing during the period when the signal was generated. This short-distance movement is independent of silencing in recipient cells. However, any further spread of silencing depends on the production of new signal molecules generated by RdRp-mediated dsRNA synthesis and Dicer-mediated dsRNA degradation. Thus, the silencing signal is a molecule that moves short distances covering 10–15 cells. The long-range cell-to-cell propagation of silencing can only occur if a current signal is able to transform a non-silenced recipient cell into a silenced cell so that this cell can act as a new source of signal (Figure 10.3).

Intercellular movement and amplification of silencing signals could act as an effective means for the systemic propagation of gene regulation in response to environmental cues. Well-known examples for such systemic reactions are systemic acquired resistance (SAR) (Dempsey *et al.*, 1999; Dong, 2001), systemic wound signaling (Pearce *et al.*, 1991), systemic acclimation to light (Karpinski *et al.*, 1999) and the systemic induction of recombination upon virus infection (Kovalchuk *et al.*, 2003). In fact, the SAR pathway appears to be strongly influenced by the expression of viral silencing suppressors (Ji and Ding, 2001; Pruss *et al.*, 2004).

The finding that viruses act as triggers and targets of silencing and also encode suppressors of silencing has led to the view that the primary role of systemic silencing is defense against invading viruses (Voinnet, 2001). The ability of the silencing signal to move 10–15 cells ahead of the cells in which it is produced may ensure that an antiviral response is mounted in cells that are about to be infected. Such a defense would depend on the ability of the signal to move ahead of the infection and would be much enhanced if the replicating virus displayed homology to transcribed plant nuclear sequences, allowing relay amplification and systemic spread of the antiviral state much further ahead of virus infection.

Limited silencing movement could also participate in non-cell-autonomous regulation of gene activity through miRNA trafficking. miRNAs and siRNAs are of similar size and miRNAs in plants mostly function as siRNA by mediating target RNA cleavage (Bartel and Bartel, 2003). For example, limited cell-to-cell movement of miRNAs could establish gradients of gene expression in meristems and leaf primordia. Indeed, polarized gradients of miRNA have been correlated with complementary target gene expression and resulting leaf asymmetry along the adaxial/abaxial axis in *Arabidopsis* (Juarez *et al.*, 2004; Kidner and Martienssen, 2004).

## 10.5 Nature of the systemic silencing signal

Local and systemic silencing of GFP in *N. benthamiana* involves two functionally distinct siRNAs (Hamilton *et al.*, 2002) (Figure 10.1) that likely arise from separate Dicer activities (Tang *et al.*, 2003; Xie *et al.*, 2004) and both are among the sequence-specific candidate molecules that may serve as a silencing signal. In extracts of local silencing tissue, the longer siRNAs (25 nt) are as abundant as the shorter species, whereas in systemic silencing tissue, the shorter siRNAs (21–22 nt) are by far the more abundant species. To correlate the spread of silencing with the presence of either long or short siRNA, a set of different viral silencing suppressor proteins was tested for their effects on the abundance of siRNA species in local tissues co-agroinfiltrated with a silencing-inducing construct (Hamilton *et al.*, 2002). Although all tested suppressor proteins [P1 of *Rice yellow mottle virus* (RYMV), P19 of *Tomato bushy stunt virus* (TBSV), the Hc-Pro protein of *Potato virus Y* (PVY), the 2b protein of *Cucumber mosaic virus* (CMV) and the AC2 protein of *African cassava mosaic virus* (ACMV)] caused an overall reduction in the abundance of siRNAs in the infiltrated tissue, the degree of reduction in the abundance of short and long siRNAs differed between suppressors. Importantly, suppression of GFP mRNA accumulation was correlated with the presence of short siRNAs and not with the presence of long siRNAs, indicating that short siRNAs, but not long siRNAs, have a role in local silencing. Moreover, systemic, long-distance, silencing of GFP was suppressed in the presence of silencing suppressors (P1, Hc-Pro, P19) that affect the production of long siRNAs in the infiltrated cells. In contrast, those suppressors (AC2, 2b) that only had slight effects on the abundance of long siRNAs did not interfere with the systemic propagation of GFP silencing. These findings were confirmed by using GFP-expressing PVX constructs. Such constructs only caused single-cell infections since they lacked coat protein, which is required for the cell-to-cell spread of the virus. Previous studies had established that the P25 protein of this virus inhibits systemic but not local silencing (Voinnet *et al.*, 2000). Inoculation of GFP-transgenic plants with CP-deficient PVX-GFP that either did, or did not, encode P25 revealed that there was systemic silencing and long siRNA only if the virus had a deletion that included the P25 gene. However, if the P25 was encoded by the viral genome, or was provided *in trans*, there was suppression of systemic silencing and the longer siRNA was absent. These observations thus provided strong indications that systemic silencing is associated with long siRNA (Hamilton *et al.*, 2002).

Further studies have established an SDE4-dependent correlation between retrotransposon methylation and the occurrence of retrotransposon-specific long siRNAs. Thus, long siRNAs are dispensable for sequence-specific mRNA degradation, but correlate with systemic silencing and methylation of homologous DNA. Conversely, the short siRNA class correlates with mRNA degradation but not with systemic silencing or methylation (Hamilton *et al.*, 2002). It appears that the short siRNA, but not the long siRNA, represents the siRNA that guides RISC ribonuclease to the target of RNA silencing (Zamore *et al.*, 2000; Elbashir *et al.*, 2001a). Indeed, synthetic siRNAs are inactive in RISC assays if they are longer than 23 nt (Elbashir *et al.*, 2001b). The correlation of long siRNAs with systemic silencing could suggest that these molecules are directly involved in this process by acting as mobile long-distance signal molecules. However, it is also possible that systemic silencing is mediated by a derivative or precursor of the long siRNAs (Hamilton *et al.*, 2002), or that systemic silencing is mediated by yet undetected nucleic acid molecules that are produced in parallel and independently of long siRNAs.

The coinfiltration of various viral silencing suppressors together with a GFP-silencing inducer was also used to further investigate the nature of the short-range spread of local silencing that leads to the production of a rim of silenced cells surrounding the patch of infiltrated cells (Himber *et al.*, 2003) (Figure 10.2). As already reported by Hamilton *et al.* (2002), the infiltration with P1 abolished accumulation of long siRNAs but not of short siRNAs in infiltrated tissue. However, the suppressor did not interfere with short-range spread of silencing, indicating that the long siRNAs that previously had been shown to be correlated with long-distance systemic silencing (Hamilton *et al.*, 2002) are dispensable for short-range local silencing. The rim of silenced cells did not occur if leaves were infiltrated with P19, which abolished the accumulation of both short and long siRNAs, thus correlating the presence of siRNAs with short-range silencing signaling. However, the production of short siRNAs *per se* is insufficient for the short-range silencing signaling, since patches infiltrated with AC2 accumulated long and short siRNAs but were not surrounded by silenced cells. Thus, the correlation between the presence of short siRNAs and the intercellular silencing signaling does not hold true for all cases. Nevertheless, small siRNAs are likely candidates to serve as a signal. Transfection of cells with small siRNA duplexes induces silencing *in vivo* (Klahre *et al.*, 2002) and *in vitro* (Elbashir *et al.*, 2001a,b). Moreover, secondary siRNAs produced by transitivity that are required for extensive spread of silencing (beyond the initial short range of 10–15 cells) are of the 21-nt size class (Himber *et al.*, 2003). This latter finding is also consistent with the vast overrepresentation of short siRNAs in systemic tissues (Hamilton *et al.*, 2002). Because of the recurrent production of small secondary siRNAs, the 25-nt class siRNAs, which are produced along with small siRNAs in cells in which silencing was originally triggered, are progressively diluted away as silencing spreads from cell to cell.

The correlation between long siRNAs in agro-infiltrated leaves and the onset of silencing in systemic leaves (Hamilton *et al.*, 2002) suggests that long siRNAs, which are abundantly produced in agro-infiltrated leaves, could act as signaling molecules for systemic silencing. Long-distance transport of such molecules could induce the *de novo* production of 21-nt siRNAs in and around the vasculature. Extensive cell-to-cell movement of silencing would then proceed from the veins independently of long siRNA (Himber *et al.*, 2003). A role of long siRNAs as systemic signaling molecules would depend on the lack of any relay amplification in the phloem, since this process would result in short, secondary siRNAs. Indeed, it has been demonstrated by triple grafting experiments that silencing can spread through sections of non-transgenic plants in which the likelihood of amplification is very low (Palauqui *et al.*, 1997).

The existence of separate long-siRNA-mediated systemic and short-siRNA-mediated local pathways for the spread of silencing is supported by the specific effects on the systemic, but not on the cell-to-cell propagation of silencing observed upon treatments of plants with low concentrations of cadmium (Ueki and Citovsky, 2001). However, although all these studies correlate the capacity for the occurrence and spread of silencing with the accumulation of siRNAs, proof that siRNAs indeed represent the mobile signal is lacking. The apparent role of long-siRNA accumulation in systemic silencing seen in the *Agrobacterium*-infiltration experiments (Hamilton *et al.*, 2002) and in DNA methylation in *Arabidopsis* (Zilberman *et al.*, 2003) seems not to hold for certain stably transformed tobacco lines (Mallory *et al.*, 2001). Thus, expression of the potyviral silencing suppressor HC-Pro, which is likely to inhibit RISC effector complexes (Kasschau *et al.*, 2003; Xie *et al.*, 2003; Dunoyer *et al.*, 2004), eliminates detectable levels of siRNAs (Anandalakshmi *et al.*, 1998; Mallory *et al.*, 2001, 2002a,b) but does not abolish systemic silencing in GUS-transgenic tobacco (Mallory *et al.*, 2001). Moreover, by probing HC-Pro function in plant lines carrying different types of transgene, no consistent correlation was observed between the capacity to trigger systemic RNA silencing or cognate transgene methylation and the accumulation of long and short siRNAs, or with that of full-length dsRNAs (Mallory *et al.*, 2003).

Although it is possible that undetectable amounts of long siRNA are sufficient to function as a signal, a role for other potential candidate molecules cannot be ruled out. Despite the fact that siRNAs would theoretically be expected to offer an advantage for cell-to-cell spread due to their small size, plants are known to have the capacity to support intercellular and systemic spread of quite long RNA molecules, such as gene transcripts, viroids and viruses (Jørgensen *et al.*, 1998; Ruiz-Medrano *et al.*, 1999; Citovsky and Zambryski, 2000; Kim *et al.*, 2001; Lucas *et al.*, 2001; Heinlein, 2002a). This raises the possibility that the mobile silencing signal is an mRNA or mRNA/protein complex that moves via the normal pathways for RNA trafficking. Upon entry into a cell, the translocated RNA could act as a template for RdRp, leading to production of dsRNA and thus the initiation of silencing. The signal might also be the dsRNA itself. The RNA genomes of viruses and viroids are highly structured and can fold into double-stranded regions; they could serve as a model for dsRNA trafficking. Most viral siRNAs in infected plants are derived from the plus strand (Szittya *et al.*, 2002), suggesting that silencing in recipient cells is triggered from structured, double-stranded regions of the genome before a viral or cellular RdRp creates RNA duplex molecules through minus-strand synthesis. Thus, viral RNA may indeed exemplify the movement of dsRNA that efficiently triggers silencing.

More evidence for the role of dsRNA in systemic silencing comes from studies in *C. elegans*. Injection, feeding and soaking with dsRNA can result in RNA silencing in all cells of this animal and its progeny, indicating the existence of a mobile signal that can be taken up by different tissues (Fire *et al.*, 1998). The induction of silencing requires only a few molecules of dsRNA per cell (Fire *et al.*, 1998; Kennerdell and Carthew, 1998). Evidence that dsRNA may serve as a systemic signaling molecule in *C. elegans* came from two types of studies. The first study made use of the *rde-4* mutation which interferes with the accumulation of siRNAs, and therefore with RNA silencing (Parrish and Fire, 2001; Tabara *et al.*, 2002). Although *rde-4/rde-4* homozygotes are not able to process large dsRNAs into siRNAs, RNA silencing was observed in heterozygous *rde*-4/+ progeny, suggesting that unprocessed dsRNAs can elicit systemic RNA silencing (Timmons *et al.*, 2003). The other line of evidence for the role of dsRNA as a non-cell-autonomous silencing signal comes from genetic screens aimed at isolating mutants affected in systemic silencing. One identified gene (*RSD8/SID-1*) encodes a transmembrane dsRNA receptor or channel that is essential for systemic but not cell-autonomous RNA silencing (Winston *et al.*, 2002; Tijsterman *et al.*, 2004). Upon expression, this protein enables passive cellular uptake of dsRNA and sensitizes cells to soaking-induced silencing with a potency that is positively correlated with the dsRNA length. These data provide strong indications that systemic RNA silencing in *C. elegans* involves SID-1-mediated intercellular transport of dsRNA (Feinberg and Hunter, 2003).

## 10.6 Pathway of the silencing signal in plants – plasmodesmata

The cell-to-cell and systemic spread of silencing in plants appears to follow the symplasmic route. The silencing of GFP in transgenic *N. benthamiana* plants does not spread into guard cells (Plate 10.1), which are known to have lost their plasmodesmal connections to neighboring cells during differentiation (Willmer and Sexton, 1979; Wille and Lucas, 1984; Palevitz and Hepler, 1985). This observation provides circumstantial evidence that the silencing signal moves from cell to cell through plasmodesmata (Voinnet *et al.*, 1998). Systemic movement via the phloem is implicated by the observation that silencing in systemic tissues is first established along the veins before it further progresses into the mesophyll (Voinnet *et al.*, 1998). Systemic silencing also tends to spread first into leaves that are located on the same side of the plant as the leaf in which silencing was initiated (Voinnet *et al.*, 1998). This pattern of systemic silencing is reminiscent of the patterns of phloem transport of dyes and virus infection (Roberts *et al.*, 1997). Both the systemic spread of viruses and the systemic spread of silencing are sensitive to nontoxic concentrations of cadmium (Citovsky *et al.*, 1998; Goshroy *et al.*, 1998; Ueki and Citovsky, 2001, 2002; Carr and Murphy, 2002). Recent studies indicate that cadmium treatment as well as expression of a cadmium-induced protein cdiGRP, induces the deposition of callose in the vascular bundles, which is consistent with the trafficking of both viruses and silencing signal via the phloem (Ueki and Citovsky, 2002).

Since the silencing signal is likely to be an RNA molecule, the aforesaid findings suggest that the signal joins viral RNA genomes and mRNAs as a third RNA species that moves via plasmodesmata and phloem throughout the plant (Citovsky and Zambryski, 2000; Lucas *et al.*, 2001; Lucas and Lee, 2004). However, caution is still in order since the evidence for symplasmic movement of the silencing signal is not yet conclusive. For example, plasmodesmatal isolation is only one possible explanation for the lack of GFP silencing in guard cells. Alternatively, the highly specialized guard cells may have lost the ability to establish silencing following differentiation, or may be characterized by a slow turnover of GFP. In this scenario, the GFP silencing signal could spread into guard cells by a pathway other than the plasmodesmata without producing a GFP-silenced phenotype. Moreover, since systemic silencing also occurs in *C. elegans*, pathways for the spread of the silencing signal other than through direct cytoplasmic bridges appear to exist. As noted earlier, in *C. elegans*, the propagation of silencing signal from one tissue to another depends on a transmembrane protein, SID-1 (Winston *et al.*, 2002) capable of mediating the transport of nucleic acids (Feinberg and Hunter, 2003). The role of this protein in the spread of silencing is supported by the correlation between cell-autonomous silencing and the lack of an SID-1 homolog in *Drosophila* (Winston *et al.*, 2002; Roignant *et al.*, 2003). Moreover, although silencing in plants usually spreads into leaves above the originally silenced leaf, other patterns of silencing have also been observed (Boerjan *et al.*, 1994; Palauqui *et al.*, 1996; Kunz *et al.*, 2001).

However, in spite of these reservations, movement of the silencing signal through plasmodesmata has received fresh support from a recent report showing the presence of authentic siRNA and miRNA species in the phloem sap of cucurbita species. The sap also contains small RNA binding proteins that in microinjection assays increase the size exclusion limit of plasmodesmata and mediate the selective cell-to-cell trafficking of 25-nt ssRNA (Yoo *et al.*, 2004). Additional support for a role of plasmodesmata also comes from recent studies on the spread of silencing in the presence of transgenes expressing viral movement proteins. Expression of the 25-kDa 'triple-gene-block' movement protein (MP; TGBp1) of *White clover mosaic virus* (WClMV) in transgenic *N. benthamiana* allows the spread of silencing signal into meristem tissue (Foster *et al.*, 2002) – a tissue that normally does not respond to the signal (Beclin *et al.*, 1998; Voinnet *et al.*, 1998). Although expression of this MP may interfere with a 'meristem surveillance mechanism' proposed to control entry of RNA molecules (Foster *et al.*, 2002), this protein may also contribute to the spread of silencing signal as a consequence of its movement functions. Recent studies in our laboratory indicate that expression of the MP of *Tobacco mosaic virus* (TMV) enhances the cell-to-cell spread of GFP silencing in infiltrated and non-infiltrated, systemic leaves (Kwon, M.-O., Karsies, A., Vogler, H. and Heinlein, M. unpublished results), indicating that the ability of MPs to facilitate the trafficking of RNA virus genomes and silencing signals is not restricted to meristematic cells. Although at present it is unknown whether MP influences the silencing pathway (i.e. it may influence siRNA production or the ratio between long and short siRNAs), the observations are consistent with the facilitated spread of the silencing signal through plasmodesmata manipulated by MP (Wolf *et al.*, 1989; Waigmann *et al.*, 1994). Although the small size of siRNAs may be consistent with movement through plasmodesmata (Voinnet *et al.*, 1998; Zambryski and Crawford, 2000), recent studies by Yoo *et al.* (2004) indicate that siRNAs do not move cell-to-cell without a helping small RNA binding protein. The enhanced cell-to-cell progression of silencing in the presence of

MP may thus suggest that viral MPs complement the pool of such proteins. However, the ability of MP to support the spread of silencing may also indicate that the true silencing signal is not an siRNA molecule but a larger molecule, with a possible size at least up to that of a viral genome (e.g. 10 kb).

Based on studies employing Fluorescence Recovery after Photobleaching (FRAP) the diffusion of nucleic acids in the cytosol (compared to water) is strongly inhibited (Verkman, 2002). Therefore, the spread of the silencing signal and of other RNA molecules is likely to depend on active transport. Studies of viral MPs suggest a role of the cytoskeleton and ER membranes in the spread of viral RNA (Boyko *et al.*, 2000a,b,c, 2002; Gillespie *et al.*, 2002; Heinlein, 2002b; Heinlein and Epel, 2004; Kawakami *et al.*, 2004), and further research is likely to reveal the particular mechanical roles of these cellular components in the transport of nucleic acids. Moreover, proteins that interact with MPs, or with plasmodesmata, as well as proteins that facilitate macromolecular trafficking through plasmodesmata, are being identified (Medina-Escobar *et al.*, 2003; Heinlein and Epel, 2004) and it will be interesting to know whether such proteins intervene in a mechanism common to the intercellular trafficking of the viral RNA and silencing signal. It should be noted that silencing signal trafficking also involves transport into and out of the nucleus (Cerutti, 2003), as indicated, e.g. by evidence that dsRNA and replicating virus can direct methylation of homologous DNA sequences (Matzke *et al.*, 2001; Mette *et al.*, 2001; Vance and Vaucheret, 2001; Waterhouse *et al.*, 2001). Possibly, the pathway involved may again be similar to the mechanism harnessed by RNA virus genomes (Whittaker and Helenius, 1998; Cros and Palese, 2003).

## 10.7 Concluding remarks

Systemic RNA silencing in plants is a novel phenomenon that relies on the cell-to-cell and systemic movement of a sequence-specific signal. However, apart from first insights provided by Yoo *et al.* (2004), neither the nature of the signal nor its pathway and mechanism of intercellular transport is known. Although current evidence suggests that the signal involves siRNA molecules, a role of large RNA molecules in silencing signaling has not been ruled out and is indeed supported by the ability of plants to transport large RNA molecules, and also by observations that the spread of silencing is enhanced by viral MPs (Foster *et al.*, 2002; our unpublished observations). Conclusive evidence will ultimately depend on the cloning of the silencing signal and the demonstration that the molecule spreads between cells and triggers silencing. One possible approach might be to graft non-silenced scions onto silenced root-stocks, and to infect the scions with a phloem-limited virus. Viruses could then be isolated with phloem sap through aphid stylets, and viruses that have packaged the silencing signal could be identified through their capacity to induce specific silencing on test plants.

The role of plasmodesmata in silencing signaling is supported by the lack of movement of systemic silencing into symplasmically isolated guard cells and the effects of viral MP on the spread of silencing. Moreover, small RNA binding proteins have been identified that modify the size exclusion limit of plasmodesmata and facilitate the cell-to-cell spread of synthetic 25-nt ssRNA molecules in microinjection assays (Yoo *et al.*, 2004). However, this evidence for the spread of silencing signal through plasmodesmata requires confirmation by more direct *in vivo* observations. For example, genetic screening could lead to the identification of genes that are required for the spread of the silencing signal and which encode proteins that localize to the channel. Following the translocation of fluorescence-labeled silencing signal candidates across plasmodesmata by real time microscopy would be an even more direct approach.

Another important question to solve in the future is the mechanism by which silencing signals are transported within the cell to reach the plasmodesmata, or to enter the nucleus. Since the signal is proposed to be an RNA molecule, ongoing studies in several laboratories on viral RNA and mRNA trafficking may well provide insights. Enhancement of silencing signaling by transgenes expressing viral MP (Foster *et al.*, 2002; our unpublished observations) suggests that the mechanisms for the spread of the silencing signal and viral RNA may indeed overlap.

As noted above, RNA silencing is believed to have evolved as an anti-viral plant defense response (Voinnet, 2001). However, the *in vivo* interplay between a virus and silencing during infection is not well characterized. Whereas the observation of the phenomenon of recovery from virus infection depends on systemic infection and subsequent attenuation of the disease by silencing, other outcomes of anti-viral silencing could also occur but would be less obvious. For example, subliminal infections, in which the virus infects one cell and fails to infect adjacent cells, could be the result of the spread of silencing signal from the infected cell into adjacent cells where it would prevent virus accumulation. Similarly, plant mutations that interfere with systemic virus movement but not with local movement could interfere with the ability of the virus to prevent the spread of silencing signal (Chisholm *et al.*, 2000). The requirement for coat protein for systemic movement of viruses (Hilf and Dawson, 1993; Taliansky and Garcia-Arenal, 1995; Vaewhongs and Lommel, 1995; Gilbertson and Lucas, 1996) may be explained by protection of the viral RNA against RNA-silencing-mediated degradation. Thus, to include analyses of plant silencing responses in virus-infected tissues in the analysis of virus : host interactions will likely enhance our understanding of how viral host range is regulated. Since the pathways for the spread of the silencing signal and viral RNA may overlap, the host range of a given virus might be determined by viral and host factors that control the delicate balance between the trafficking of viral RNA versus the trafficking of the silencing signal. Indeed, MP : host interactions in a given cell may be regulated specifically to manipulate the RNA transport pathway so as to permit RNA transport, and thus viral RNA spread during early infection but to inhibit RNA transport, and hence the spread of the silencing signal, during late infection (Heinlein and Epel, 2004).

Similarly, analysis of the ability of miRNA to move systemically could shed light on developmental processes. The occurrence of miRNA gradients in polarized plant tissues (such as the adaxial and abaxial sides of the leaf) suggests that miRNAs may control polarized tissue differentiation. The miRNA gradients may be generated either by cell-specific transcription levels or by the capacity of miRNAs to move between cells similarly as proposed for siRNAs. However, recent *in planta* studies by the group of O. Voinnet (IBMP, Strasbourg), argue against the cell-to-cell movement of miRNAs in *Arabidopsis*. This group found that miR171 and other miRNAs triggered transitive and non-cell-autonomous silencing of their targets, if tested in a wild-type background. However, if tested in the mutant *sde1* background, which interferes with secondary siRNA formation, a near-perfect spatial overlap between the transcription and endonucleolytic activity of miR171 was observed, providing compelling evidence that miRNAs act in a cell-autonomous manner (Parizotto *et al*., 2004).

In closing, it should be clear to the reader that current knowledge of both siRNA and miRNA pathways is growing with an increasing pace. New classes of small RNA species are continuously being identified (Ambros *et al*., 2003; Dunoyer *et al*., 2004; Xie *et al*., 2004). Some of these are not affected by expression of the silencing suppressor HC-Pro (Dunoyer *et al*., 2004), which interferes with siRNA accumulation (Mallory *et al*., 2001), thus suggesting the existence of new candidate molecules for signaling in systemic silencing and DNA methylation. The now apparent complex relationships between different small RNA-mediated pathways and the pleiotropic effects of mutations and silencing suppressors (Kasschau *et al*., 2003; Cook *et al*., 2004; Dunoyer *et al*., 2004; Pruss *et al*., 2004) will present significant hurdles before final conclusions with respect to the nature of systemic silencing signaling can be drawn. Indeed, it is likely that cellular and systemic silencing signaling play more diverse roles in organismal biology than we presently are able to imagine.

## Acknowledgments

I am grateful to Ken Richards for his support in preparing this manuscript and I thank the Swiss National Science Foundation (SNF 631-65953.01), the Swiss Bundesanstalt für Bildung und Wissenschaft (BBW; EC contract G2-CT-2002-01673) and the French CNRS for financial support.

## References

Ahlquist, P. (2002) 'RNA-dependent RNA polymerases, viruses, and RNA silencing', *Science* **296**, 1270–1273.

Ambros, V., Lee, R.C., Lavanway, A., Williams, P.T. and Jewell, D. (2003) 'MicroRNAs and other tiny endogenous RNAs in *C. elegans*', *Current Biology* **13**, 807–818.

Anandalakshmi, R., Pruss, G.J., Xin, G. *et al*. (1998) 'A viral suppressor of gene silencing in plants', *Proceedings of the National Academy of Sciences USA* **95**, 13079–13084.

Aravin, A.A., Naumova, N.M., Tulin, A.V., Vagin, V.V., Rozovsky, Y.M. and Gvozdev, V.A. (2001) 'Double-stranded RNA-mediated silencing of genomic tandem repeats and transposable elements in the *D. melanogaster* germline', *Current Biology* **11**, 1017–1027.

Aufsatz, W., Mette, M.F., Van Der Winden, J., Matzke, A.J. and Matzke, M. (2002) 'RNA-directed DNA methylation in *Arabidopsis*', *Proceedings of the National Academy of Sciences USA* **99** (Suppl. 4), 16499–16506.

Aukerman, M.J. and Sakai, H. (2003) 'Regulation of flowering time and floral organ identity by a MicroRNA and its APETALA2-like target genes', *Plant Cell* **15**, 2730–2741.

Bartel, B. and Bartel, D.P. (2003) 'MicroRNAs: at the root of plant development?', *Plant Physiology* **132**, 709–717.

Bartel, D.P. (2004) 'MicroRNAs: genomics, biogenesis, mechanism, and function', *Cell* **116**, 281–297.

Baulcombe, D. (2002) 'RNA silencing', *Current Biology* **12**, R82–R84.

Baulcombe, D.C. (1996) 'Mechanisms of pathogen-derived resistance to viruses in transgenic plants', *Plant Cell* **8**, 1833–1844.

Beclin, C., Berthome, R., Palauqui, J.C., Tepfer, M. and Vaucheret, H. (1998) 'Infection of tobacco or *Arabidopsis* plants by CMV counteracts systemic post-transcriptional silencing of nonviral (trans) genes', *Virology* **252**, 313–317.

Beclin, C., Boutet, S., Waterhouse, P. and Vaucheret, H. (2002) 'A branched pathway for transgene-induced RNA silencing in plants', *Current Biology* **12**, 684–688.

Bernstein, E., Caudy, A.A., Hammond, S.M. and Hannon, G.J. (2001) 'Role of a bidentate ribonuclease in the initiation step of RNA interference', *Nature* **409**, 295–296.

Boerjan, W., Bauw, G., Van Montagu, M. and Inze, D. (1994) 'Distinct phenotypes generated by over-expression and suppression of S-adenosyl-L-methionine synthetase reveal developmental patterns of gene silencing in tobacco', *Plant Cell* **6**, 1401–1414.

Bohmert, K., Camus, I., Bellini, C., Bouchez, D., Caboche, M. and Benning, C. (1998) 'AGO1 defines a novel locus of *Arabidopsis* controlling leaf development', *EMBO Journal* **17**, 170–180.

Bosher, J.M. and Labouesse, M. (2000) 'RNA interference: genetic wand and genetic watchdog', *Nature Cell Biology* **2**, E31–E36.

Boutet, S., Vazquez, F., Liu, J. *et al.* (2003) '*Arabidopsis* HEN1: a genetic link between endogenous miRNA controlling development and siRNA controlling transgene silencing and virus resistance', *Current Biology* **13**, 843–848.

Boyko, V., Ashby, J.A., Suslova, E. *et al.* (2002) 'Intramolecular complementing mutations in *Tobacco mosaic virus* movement protein confirm a role for microtubule association in viral RNA transport', *Journal of Virology* **76**, 3974–3980.

Boyko, V., Ferralli, J., Ashby, J., Schellenbaum, P. and Heinlein, M. (2000a) 'Function of microtubules in intercellular transport of plant virus RNA', *Nature Cell Biology* **2**, 826–832.

Boyko, V., Ferralli, J. and Heinlein, M. (2000b) 'Cell-to-cell movement of TMV RNA is temperature-dependent and corresponds to the association of movement protein with microtubules', *Plant Journal* **22**, 315–325.

Boyko, V., Van Der Laak, J., Ferralli, J., Suslova, E., Kwon, M.-O. and Heinlein, M. (2000c) 'Cellular targets of functional and dysfunctional mutants of *Tobacco mosaic virus* movement protein fused to GFP', *Journal of Virology* **74**, 11339–11346.

Brigneti, G., Martin-Hernandez, A.M., Jin, H. *et al.* (2004) 'Virus-induced gene silencing in *Solanum* species', *Plant Journal* **39**, 264–272.

Brigneti, G., Voinnet, O., Li, W.X., Ji, L.H., Ding, S.W. and Baulcombe, D.C. (1998) 'Viral pathogenicity determinants are suppressors of transgene silencing in *Nicotiana benthamiana*', *EMBO Journal* **17**, 6739–6746.

Bucher, E., Sijen, T., De Haan, P., Goldbach, R. and Prins, M. (2003) 'Negative-strand tospoviruses and tenuiviruses carry a gene for a suppressor of gene silencing at analogous genomic positions', *Journal of Virology* **77**, 1329–1336.

Burch-Smith, T.M., Anderson, J.C., Martin, G.B. and Dinesh-Kumar, S.P. (2004) 'Applications and advantages of virus-induced gene silencing for gene function studies in plants', *Plant Journal* **39**, 734–746.

Caplen, N.J., Parrish, S., Imani, F., Fire, A. and Morgan, R.A. (2001) 'Specific inhibition of gene expression by small double-stranded RNAs in invertebrate and vertebrate systems', *Proceedings of the National Academy of Sciences USA* **98**, 9742–9747.

Carmell, M.A., Xuan, Z., Zhang, M.Q. and Hannon, G.J. (2002) 'The Argonaute family: tentacles that reach into RNAi, developmental control, stem cell maintenance, and tumorigenesis', *Genes and Development* **16**, 2733–2742.

Carpenter, A.E. and Sabatini, D.M. (2004) 'Systematic genome-wide screens of gene function', *Nature Reviews Genetics* **5**, 11–22.

Carr, J.P. and Murphy, A.M. (2002) 'Cadmium blocks viral invasion in plants', *Nature Cell Biology* **4**, E167–E168.

Carrington, J.C. and Ambros, V. (2003) 'Role of microRNAs in plant and animal development', *Science* **301**, 336–338.

Cerutti, H. (2003) 'RNA interference: traveling in the cell and gaining functions?', *Trends in Genetics* **19**, 39–46.

Chan, S.W., Zilberman, D., Xie, Z., Johansen, L.K., Carrington, J.C. and Jacobsen, S.E. (2004) 'RNA silencing genes control *de novo* DNA methylation', *Science* **303**, 1336.

Chen, X. (2004) 'A microRNA as a translational repressor of APETALA2 in *Arabidopsis* flower development', *Science* **303**(5666), 2022–2025.

Chisholm, S.T., Mahajan, S.K., Whitham, S.A., Yamamoto, M.L. and Carrington, J.C. (2000) 'Cloning of the *Arabidopsis RTM1* gene, which controls restriction of long-distance movement of tobacco etch virus', *EMBO Journal* **97**, 489–494.

Citovsky, V. and Zambryski, P. (2000) 'Systemic transport of RNA in plants', *Trends in Plant Science* **5**, 52–54.

Citovsky, V., Ghoshroy, S., Tsui, F. and Klessig, D. (1998) 'Non-toxic concentrations of cadmium inhibit systemic movement of turnip vein clearing virus by a salicylic acid-dependent mechanism', *Plant Journal* **16**, 13–20.

Cogoni, C. and Macino, G. (1999) 'Gene silencing in *Neurospora crassa* requires a protein homologous to RNA-dependent RNA polymerase', *Nature* **399**, 166–169.

Cook, H.A., Koppetsch, B.S., Wu, J. and Theurkauf, W.E. (2004) 'The Drosophila SDE3 homolog armitage is required for oskar mRNA silencing and embryonic axis specification', *Cell* **116**, 817–829.

Covey, S.N., Al-Kaff, N.S., Langara, A. and Turner, D.S. (1997) 'Plants combat infection by gene silencing', *Nature* **385**, 781–782.

Cros, J.F. and Palese, P. (2003) 'Trafficking of viral genomic RNA into and out of the nucleus: influenza, Thogoto and Borna disease viruses', *Virus Research* **95**, 3–12.

Dalmay, T., Hamilton, A., Rudd, S., Angell, S. and Baulcombe, D.C. (2000) 'An RNA-dependent RNA polymerase gene in *Arabidopsis* is required for post-transcriptional gene silencing mediated by a transgene but not by a virus', *Cell* **101**, 543–553.

Dalmay, T., Horsefield, R., Braunstein, T.H. and Baulcombe, D.C. (2001) 'SDE3 encodes an RNA helicase required for post-transcriptional gene silencing in *Arabidopsis*', *EMBO Journal* **20**, 2069–2078.

Dempsey, D.A., Shah, J. and Klessig, D. (1999) 'Salicylic acid and disease resistance in plants', *Critical Reviews in Plant Sciences* **18**, 547–575.

Denli, A.M. and Hannon, G.J. (2003) 'RNAi: an ever-growing puzzle', *Trends in Biochemical Sciences* **28**, 196–201.

Depicker, A. and van Montagu, M. (1997) 'Post-transcriptional gene silencing in plants', *Current Opinion in Cell Biology* **9**, 373–382.

Dinesh-Kumar, S.P., Anandalakshmi, R., Marathe, R., Schiff, M. and Liu, Y. (2003) 'Virus-induced gene silencing', *Methods in Molecular Biology* **236**, 287–294.

Ding, X.S., Liu, J., Cheng, N.H. *et al.* (2004) 'The Tobacco mosaic virus 126-kDa protein associated with virus replication and movement suppresses RNA silencing', *Molecular Plant–Microbe Interactions* **17**, 583–592.

Doench, J.G. and Sharp, P.A. (2004) 'Specificity of microRNA target selection in translational repression', *Genes and Development* **18**(5), 504–511.

Doench, J.G., Petersen, C.P. and Sharp, P.A. (2003) 'siRNAs can function as miRNAs', *Genes and Development* **17**, 438–442.

Doi, N., Zenno, S., Ueda, R., Ohki-Hamazaki, H., Ui-Tei, K. and Saigo, K. (2003) 'Short-interfering-RNA-mediated gene silencing in mammalian cells requires Dicer and eIF2C translation initiation factors', *Current Biology* **13**, 41–46.

Dong, X. (2001) 'Genetic dissection of systemic acquired resistance', *Current Opinion in Plant Biology* **4**, 309–314.

Dong, X., Van Wezel, R., Stanley, J. and Hong, Y. (2003) 'Functional characterization of the nuclear localization signal for a suppressor of post-transcriptional gene silencing', *Journal of Virology* **77**, 7026–7033.

Dougherty, W.G., Lindbo, J.A., Smith, H.A., Parks, T.D., Swaney, S. and Proebsting, W.M. (1994) 'RNA-mediated virus resistance in transgenic plants: exploitation of a cellular pathway possibly involved in RNA degradation', *Molecular Plant–Microbe Interactions* **7**, 544–552.

Dunoyer, P., Lecellier, C.H., Parizotto, E.A., Himber, C. and Voinnet, O. (2004) 'Probing the microRNA and small interfering RNA pathways with virus-encoded suppressors of RNA silencing', *The Plant Cell* **16**, 1235–1250.

Dunoyer, P., Pfeffer, S., Fritsch, C., Hemmer, O., Voinnet, O. and Richards, K.E. (2002) 'Identification, subcellular localization and some properties of a cysteine-rich suppressor of gene silencing encoded by peanut clump virus', *Plant Journal* **29**, 555–567.

Dykxhoorn, D.M., Novina, C.D. and Sharp, P.A. (2003) 'Killing the messenger: short RNAs that silence gene expression', *Nature Reviews Molecular Cell Biology* **4**, 457–467.

Ekwall, K. (2004) 'The RITS complex – a direct link between small RNA and heterochromatin', *Molecular Cell* **13**, 304–305.

Elbashir, S.M., Lendeckel, W. and Tuschl, T. (2001a) 'RNA interference is mediated by 21- and 22-nucleotide RNAs', *Genes and Development* **15**, 188–200.

Elbashir, S.M., Martinez, J., Patkaniowska, A., Lendeckel, W. and Tuschl, T. (2001b) 'Functional anatomy of siRNAs for mediating efficient RNAi in *Drosophila melanogaster* embryo lysate', *EMBO Journal* **20**, 6877–6888.

Emery, J.F., Floyd, S.K., Alvarez, J. *et al.* (2003) 'Radial patterning of *Arabidopsis* shoots by class III HD-ZIP and KANADI genes', *Current Biology* **13**, 1768–1774.

English, J.J., Mueller, E. and Baulcombe, D.C. (1996) 'Suppression of virus accumulation in transgenic plants exhibiting silencing of nuclear genes', *The Plant Cell* **8**, 179–188.

Feinberg, E.H. and Hunter, C.P. (2003) 'Transport of dsRNA into cells by the transmembrane protein SID-1', *Science* **301**, 1545–1547.

Finnegan, E.J. and Matzke, M.A. (2003) 'The small RNA world', *Journal of Cell Science* **116**, 4689–4693.

Finnegan, E.J., Margis, R. and Waterhouse, P.M. (2003) 'Posttranscriptional gene silencing is not compromised in the *Arabidopsis* CARPEL FACTORY (DICER-LIKE1) mutant, a homolog of Dicer-1 from *Drosophila*', *Current Biology* **13**, 236–240.

Fire, A., Xu, S., Montgomery, M.K., Kostas, S.A., Driver, S.E. and Mello, C.C. (1998) 'Potent and specific genetic interference by double-stranded RNA in *Caenorhabditis elegans*', *Nature* **391**, 806–811.

Fortin, K.R., Nicholson, R.H. and Nicholson, A.W. (2002) 'Mouse ribonuclease III. cDNA structure, expression analysis, and chromosomal location', *BMC Genomics* **3**, 26.

Foster, T.M., Lough, T.J., Emerson, S.J. *et al.* (2002) 'A surveillance system regulates selective entry of RNA into the shoot apex', *The Plant Cell* **14**, 1497–1508.

Gilbertson, R.L. and Lucas, W.J. (1996) 'How do viruses traffic on the "vascular highway"?', *Trends in Plant Science* **1**, 260–268.

Gillespie, T., Boevink, P., Haupt, S. *et al.* (2002) 'Functional analysis of a DNA shuffled movement protein reveals that microtubules are dispensable for the cell-to-cell movement of tobacco mosaic virus', *The Plant Cell* **14**, 1207–1222.

Gitlin, L. and Andino, R. (2003) 'Nucleic acid-based immune system: the antiviral potential of mammalian RNA silencing', *Journal of Virology* **77**, 7159–7165.

Goshroy, S., Freedman, K., Lartey, R. and Citovsky, V. (1998) 'Inhibition of plant viral systemic infection by non-toxic concentrations of cadmium', *Plant Journal* **13**, 591–602.

Grewal, S.I. and Moazed, D. (2003) 'Heterochromatin and epigenetic control of gene expression', *Science* **301**, 798–802.

Griffiths-Jones, S. (2004) 'The microRNA registry', *Nucleic Acids Research* **32** (Database issue), D109–D111.

Grishok, A., Pasquinelli, A.E., Conte, D. *et al.* (2001) 'Genes and mechanisms related to RNA interference regulate expression of the small temporal RNAs that control *C. elegans* developmental timing', *Cell* **106**, 23–34.

Gupta, S., Schoer, R.A., Egan, J.E., Hannon, G.J. and Mittal, V. (2004) 'Inducible, reversible, and stable RNA interference in mammalian cells', *Proceedings of the National Academy of Sciences USA* **101**, 1927–1932.

Hall, I.M., Shankaranarayana, G.D., Noma, K., Ayoub, N., Cohen, A. and Grewal, S.I. (2002) 'Establishment and maintenance of a heterochromatin domain', *Science* **297**, 2232–2237.

Hamilton, A., Voinnet, O., Chappel, L. and Baulcombe, D. (2002) 'Two classes of short interfering RNA in RNA silencing', *EMBO Journal* **21**, 4671–4679.

Hamilton, A.J. and Baulcombe, D.C. (1999) 'A species of small antisense RNA in posttranscriptional gene silencing in plants', *Science* **286**, 950–952.

Hammond, S.M., Bernstein, E., Beach, D. and Hannon, G.J. (2000) 'An RNA-directed muclease mediates post-transcriptional gene silencing in *Drosophila* cells', *Nature* **404**, 293–296.

Hammond, S.M., Boettcher, S., Caudy, A.A., Kobayashi, R. and Hannon, G.J. (2001a) 'Argonaute2, a link between genetic and biochemical analyses of RNAi', *Science* **293**, 1146–1150.

Hammond, S.M., Caudy, A.A. and Hannon, G.J. (2001b) 'Post-transcriptional gene silencing by double-stranded RNA', *Nature Reviews Genetics* **2**, 110–119.

Hannon, G.J. (2002) 'RNA interference', *Nature* **418**, 244–251.

Hannon, G.J. and Conklin, D.S. (2004) 'RNA interference by short hairpin RNAs expressed in vertebrate cells', *Methods in Molecular Biology* **257**, 255–266.

Heinlein, M. (2002a) 'Plasmodesmata: dynamic regulation and role in macromolecular cell-to-cell signalling', *Current Opinion in Plant Biology* **5**, 543–552.

Heinlein, M. (2002b) 'The spread of Tobacco mosaic virus infection: insights into the cellular mechanism of RNA transport', *Cellular and Molecular Life Sciences* **59**, 58–82.

Heinlein, M. and Epel, B.L. (2004) 'Macromolecular transport and signaling through plasmodesmata', *International Review of Cytology* **235**, 93–164.

Hilf, M.E. and Dawson, W.O. (1993) 'The tobacco virus capsid protein functions as a host-specific determinant of long-distance movement', *Virology* **193**, 106–114.

Himber, C., Dunoyer, P., Moissiard, G., Ritzenthaler, C. and Voinnet, O. (2003) 'Transitivity-dependent and -independent cell-to-cell movement of RNA silencing', *EMBO Journal* **22**, 4523–4533.

Hutvagner, G. and Zamore, P.D. (2002) 'RNAi: nature abhors a double-strand', *Current Opinion in Genetics and Development* **12**, 225–232.

Hutvagner, G., Mclachlan, J., Pasquinelli, A.E., Balint, E., Tuschl, T. and Zamore, P.D. (2001) 'A cellular function for the RNA-interference enzyme Dicer in the maturation of the let-7 small temporal RNA', *Science* **293**, 834–838.

Jacobsen, S.E., Running, M.P. and Meyerowitz, E.M. (1999) 'Disruption of an RNA helicase/RNAse III gene in *Arabidopsis* causes unregulated cell division in floral meristems', *Development* **126**, 5231–5243.

Ji, L.H. and Ding, S.W. (2001) 'The suppressor of transgene RNA silencing encoded by Cucumber mosaic virus interferes with salicylic acid-mediated virus resistance', *Molecular Plant–Microbe Interactions* **14**, 715–724.

Jones, L., Hamilton, A.J., Voinnet, O., Thomas, C.L., Maule, A.J. and Baulcombe, D.C. (1999) 'RNA–DNA interactions and DNA methylation in post-transcriptional gene silencing', *The Plant Cell* **11**, 2291–2301.

Jones, S.W., Souza, P.M. and Lindsay, M.A. (2004) 'siRNA for gene silencing: a route to drug target discovery', *Current Opinion in Pharmacology* **4**, 522–527.

Jørgensen, R.A., Atkinson, R.G., Forster, R.L. and Lucas, W.J. (1998) 'An RNA-based information superhighway in plants', *Science* **279**, 1486–1487.

Juarez, M.T., Kui, J.S., Thomas, J., Heller, B.A. and Timmermans, M.C. (2004) 'MicroRNA-mediated repression of rolled leaf1 specifies maize leaf polarity', *Nature* **428**, 84–88.

Karpinski, S., Reynolds, H., Karpinska, B., Wingsle, G., Creissen, G. and Mullineaux, P. (1999) 'Systemic signaling and acclimation in response to excess excitation energy in *Arabidopsis*', *Science* **284**, 654–657.

Kasschau, K.D. and Carrington, J.C. (1998) 'A counterdefensive strategy of plant viruses: suppression of post-transcriptional gene silencing', *Cell* **95**, 461–470.

Kasschau, K.D., Xie, Z., Allen, E. *et al.* (2003) 'P1/HC-pro, a viral suppressor of RNA silencing, interferes with *Arabidopsis* development and miRNA function', *Developmental Cell* **4**, 205–217.

Kawakami, S., Watanabe, Y. and Beachy, R.N. (2004) 'Tobacco mosaic virus infection spreads cell to cell as intact replication complexes', *Proceedings of the National Academy of Sciences USA* **101**, 6291–6296.

Kawasaki, H. and Taira, K. (2004) 'Induction of DNA methylation and gene silencing by short interfering RNAs in human cells', *Nature* **431**, 211–217.

Kennerdell, J.R. and Carthew, R.W. (1998) 'Use of dsRNA-mediated genetic interference to demonstrate that frizzled and frizzled 2 act in the wingless pathway', *Cell* **95**, 1017–1026.

Kessler, A., Halitschke, R. and Baldwin, I.T. (2004) 'Silencing the jasmonate cascade: induced plant defenses and insect populations', *Science* **305**, 665–668.

Ketting, R.F., Fischer, S.E.J., Bernstein, E., Sijen, T., Hannon, G.J. and Plasterk, R.H.A. (2001) 'Dicer functions in RNA interference and in synthesis of small RNA involved in developmental timing in *C. elegans*', *Genes and Development* **15**, 2654–2659.

Kidner, C.A. and Martienssen, R.A. (2003) 'Macro effects of microRNAs in plants', *Trends in Genetics* **19**, 13–16.

Kidner, C.A. and Martienssen, R.A. (2004) 'Spatially restricted microRNA directs leaf polarity through ARGONAUTE1', *Nature* **428**, 81–84.

Kim, M., Canio, W., Kessler, S. and Sinha, N. (2001) 'Developmental changes due to long-distance movement of a homeobox fusion transcript in tomato', *Science* **293**, 287–289.

Klahre, U., Crete, P., Leuenberger, S.A., Iglesias, V.A. and Meins, F. Jr. (2002) 'High molecular weight RNAs and small interfering RNAs induce systemic post-transcriptional gene silencing in plants', *EMBO Journal* **99**, 11981–11986.

Knight, S.W. and Bass, B.L. (2001) 'A role for the RNase III enzyme DCR-1 in RNA interference and germ line development in *Caenorhabditis elegans*', *Science* **293**, 2269–2271.

Kovalchuk, I., Kovalchuk, O., Kalck, V., Boyko, V., Filkowski, J., Heinlein, M. and Hohn, B. (2003) 'Pathogen-induced systemic signal triggers genome instability', *Nature* **423**, 760–762.

Kubota, K., Tsuda, S., Tamai, A. and Meshi, T. (2003) 'Tomato mosaic virus replication protein suppresses virus-targeted post-transcriptional gene silencing', *Journal of Virology* **77**, 11016–11026.

Kumagai, M.H., Donson, J., Della-Cioppa, G., Harvey, D., Hanley, K. and Grill, L.K. (1995) 'Cytoplasmic inhibition of carotenoid biosynthesis with virus-derived RNA', *Proceedings of the National Academy of Sciences USA* **92**, 1679–1683.

Kunz, C., Schob, H., Leubner-Metzger, G., Glazov, E. and Meins, F. Jr. (2001) 'Beta-1,3-glucanase and chitinase transgenes in hybrids show distinctive and independent patterns of post-transcriptional gene silencing', *Planta* **212**, 243–249.

Lai, E.C. (2003) 'MicroRNAs: runts of the genome assert themselves', *Current Biology* **13**, R925–R936.

Lee, J.-Y., Yoo, B.-C., Rojas, M.R., Gomez-Ospina, N., Staehelin, L.A. and Lucas, W.J. (2003a) 'Selective trafficking of non-cell-autonomous proteins mediated by NtNCAPP1', *Science* **299**, 392–396.

Lee, Y., Ahn, C., Han, J. *et al.* (2003b) 'The nuclear RNase III Drosha initiates microRNA processing', *Nature* **425**, 415–419.

Lee, Y.S., Nakahara, K., Pham, J.W. *et al.* (2004) 'Distinct roles for Drosophila Dicer-1 and Dicer-2 in the siRNA/miRNA silencing pathways', *Cell* **117**, 69–81.

Li, H., Li, W.X. and Ding, S.W. (2002) 'Induction and suppression of RNA silencing by an animal virus', *Science* **296**, 1319–1321.

Lindbo, J.A. and Dougherty, W.G. (1992) 'Pathogen-derived resistance to a potyvirus: immune and resistant phenotypes in transgenic tobacco expressing altered forms of a potyvirus coat protein nucleotide sequence', *Molecular Plant–Microbe Interactions* **5**, 144–153.

Lindbo, J.A., Silva-Rosales, L., Proebsting, W.M. and Dougherty, W.G. (1993) 'Induction of a highly specific antiviral state in transgenic plants: implications for regulation of gene expression and virus resistance', *The Plant Cell* **5**, 1749–1759.

Lingel, A., Simon, B., Izaurralde, E. and Sattler, M. (2003) 'Structure and nucleic-acid binding of the *Drosophila* Argonaute 2 PAZ domain', *Nature* **426**, 465–469.

Liu, J., Carmell, M.A., Rivas, F.V. *et al.* (2004) 'Argonaute2 is the catalytic engine of mammalian RNAi', *Science* **305**, 1437–1441.

Liu, Q., Rand, T.A., Kalidas, S. *et al.* (2003) 'R2D2, a bridge between the initiation and effector steps of the *Drosophila* RNAi pathway', *Science* **301**, 1921–1925.

Liu, Y., Schiff, M. and Dinesh-Kumar, S.P. (2002) 'Virus-induced gene silencing in tomato', *Plant Journal* **31**, 777–786.

Llave, C., Kasschau, K.D., Rector, M.A. and Carrington, J.C. (2002a) 'Endogenous and silencing-associated small RNAs in plants', *The Plant Cell* **14**, 1605–1619.

Llave, C., Xie, Z., Kasschau, K.D. and Carrington, J.C. (2002b) 'Cleavage of *Scarecrow-like* mRNA targets directed by a class of *Arabidopsis* miRNA', *Science* **297**, 2053–2056.

Love, R. (2004) 'First disease model created by RNA interference', *The Lancet Neurology* **3**, 7.

Lu, R., Martin-Hernandez, A.M., Peart, J.R., Malcuit, I. and Baulcombe, D.C. (2003) 'Virus-induced gene silencing in plants', *Methods* **30**, 296–303.

Lucas, W.J. and Lee, J.Y. (2004) 'Plasmodesmata as a supracellular control network in plants', *Nature Reviews Molecular Cell Biology* **5**, 712–726.

Lucas, W.J., Yoo, B.-C. and Kragler, F. (2001) 'RNA as a long-distance information macromolecule in plants', *Nature Reviews Molecular Cell Biology* **2**, 849–857.

Lucy, A.P., Guo, H.-S., Li, W.-X. and Ding, S.-W. (2000) 'Suppression of post-transcriptional gene silencing by a plant viral protein localized in the nucleus', *EMBO Journal* **19**, 1672–1680.

Mallory, A.C., Ely, L., Smith, T.H. *et al.* (2001) 'HC-Pro suppression of transgene silencing eliminates the small RNAs but not transgene methylation or the mobile signal', *The Plant Cell* **13**, 571–583.

Mallory, A.C., Mlotshwa, S., Bowman, L.H. and Vance, V.B. (2003) 'The capacity of transgenic tobacco to send a systemic RNA silencing signal depends on the nature of the inducing transgene locus', *Plant Journal* **35**, 82–92.

Mallory, A.C., Parks, G., Endres, M.W. *et al.* (2002a) 'The amplicon-plus system for high-level expression of transgenes in plants', *Nature Biotechnology* **20**, 622–625.

Mallory, A.C., Reinhard, B.J., Bartel, D.B., Vance, V.B. and Bowman, L.H. (2002b) 'A viral suppressor of RNA silencing differentially regulates the accumulation of short-interfering RNAs and microRNAs in tobacco', *EMBO Journal* **99**, 15228–15233.

Marathe, R., Anandalakshmi, R., Smith, T.H., Pruss, G.J. and Vance, V.B. (2000) 'RNA viruses as inducers, suppressors and targets of post-transcriptional gene silencing', *Plant Molecular Biology* **43**, 295–306.

Matzke, M., Matzke, A.J. and Kooter, J.M. (2001) 'RNA: guiding gene silencing', *Science* **293**, 1080–1083.

Medina-Escobar, N., Haupt, S., Thow, G., Boevink, P., Chapman, S. and Oparka, K. (2003) 'High-throughput viral expression of cDNA-green fluorescent protein fusions reveals novel subcellular addresses and identifies unique proteins that interact with plasmodesmata', *The Plant Cell* **15**, 1507–1523.

Meins, F., Jr. (2000) 'RNA degradation and models for post-transcriptional gene-silencing', *Plant Molecular Biology* **43**, 261–273.

Mette, M.F., Aufsatz, W., Van Der Winden, J., Matzke, M.A. and Matzke, A.J. (2000) 'Transcriptional silencing and promoter methylation triggered by double-stranded RNA', *EMBO Journal* **19**, 5194–5201.

Mette, M.F., Matzke, A.J. and Matzke, M.A. (2001) 'Resistance of RNA-mediated TGS to HC-Pro, a viral suppressor of PTGS, suggests alternative pathways for dsRNA processing', *Current Biology* **11**, 1119–1123.

Mette, M.F., Van Der Winden, J., Matzke, M. and Matzke, A.J. (2002) 'Short RNAs can identify new candidate transposable element families in *Arabidopsis*', *Plant Physiology* **130**, 6–9.

Mittal, V. (2004) 'Improving the efficiency of RNA interference in mammals', *Nature Reviews Genetics* **5**, 355–365.

Mlotshwa, S., Voinnet, O., Mette, M.F. *et al.* (2002) 'RNA silencing and the mobile silencing signal', *The Plant Cell* (Suppl.), S289–S301.

Morris, K.V., Chan, S.W., Jacobsen, S.E. and Looney, D.J. (2004) 'Small interfering RNA-induced transcriptional gene silencing in human cells', *Science* **305**, 1289–1292.

Mourrain, P., Beclin, C., Elmayan, T. *et al.* (2000) '*Arabidopsis* SGS2 and SGS3 genes are required for posttranscriptional gene silencing and natural virus resistance', *Cell* **101**, 533–542.

Myers, J.W., Jones, J.T., Meyer, T. and Ferrell, J.E., Jr. (2003) 'Recombinant dicer efficiently converts large dsRNAs into siRNAs suitable for gene silencing', *Nature Biotechnology* **21**, 324–328.

Newmark, P.A., Reddien, P.W., Cebria, F. and Sanchez Alvarado, A. (2003) 'Ingestion of bacterially expressed double-stranded RNA inhibits gene expression in planarians', *Proceedings of the National Academy of Sciences USA* **100**(Suppl. 1), 11861–11865.

Nykanen, A., Haley, B. and Zamore, P.D. (2001) 'ATP requirements and small interfering RNA structure in the RNA interference pathway', *Cell* **107**, 309–321.

Olsen, P.H. and Ambros, V. (1999) 'The lin-4 regulatory RNA controls developmental timing in *Caenorhabditis elegans* by blocking LIN-14 protein synthesis after the initiation of translation', *Developmental Biology* **216**, 671–680.

Paddison, P.J., Silva, J.M., Conklin, D.S. *et al.* (2004) 'A resource for large-scale RNA-interference-based screens in mammals', *Nature* **428**, 427–431.

Pal-Bhadra, M., Leibovitch, B.A., Gandhi, S.G. *et al.* (2004) 'Heterochromatic silencing and HP1 localization in *Drosophila* are dependent on the RNAi machinery', *Science* **303**, 669–672.

Palatnik, J.F., Allen, E., Wu, X. *et al.* (2003) 'Control of leaf morphogenesis by microRNAs', *Nature* **425**, 257–263.

Palauqui, J.-C., Elmayan, T., De Borne, F.D., Crété, P., Charles, C. and Vaucheret, H. (1996) 'Frequencies, timing, and spatial patterns of co-suppression of nitrate reductase and nitrite reductase in transgenic tobacco plants', *Plant Physiology* **112**, 1447–1456.

Palauqui, J.-C., Elmayan, T., Pollien, J.-M. and Vaucheret, H. (1997) 'Systemic acquired silencing: transgene-specific post-transcriptional silencing is transmitted by grafting from silenced stocks to non-silenced scions', *EMBO Journal* **16**, 4738–4745.

Palevitz, B.A. and Hepler, P.K. (1985) 'Changes in dye coupling of stomatal cells of *Allium* and *Commelina* demonstrated by microinjection of lucifer yellow', *Planta* **164**, 473–479.

Parizotto, E.A., Dunoyer, P., Rahm, N., Himber, C. and Voinnet, O. (2004) '*In vitro* investigation of the transcription, processing, endonucleolytic activity and functional relevance of the spatial distribution of a plant miRNA', *Genes and Development* **18**, 2237–2242.

Park, W., Li, J., Song, R., Messing, J. and Chen, X. (2002) 'CARPEL FACTORY, a Dicer homolog, and HEN1, a novel protein, act in microRNA metabolism in *Arabidopsis thaliana*', *Current Biology* **12**, 1484–1495.

Parrish, S. and Fire, A. (2001) 'Distinct roles for RDE-1 and RDE-4 during RNA interference in *Caenorhabditis elegans*', *RNA* **7**, 1397–1402.

Parrish, S., Fleenor, J., Xu, S., Mello, C. and Fire, A. (2000) 'Functional anatomy of a dsRNA trigger: differential requirement for the two trigger strands in RNA interference', *Molecular Cell* **6**, 1077–1087.

Pasquinelli, A.E. and Ruvkun, G. (2002) 'Control of developmental timing by microRNAs and their targets', *Annual Review of Cell and Developmental Biology* **18**, 495–513.

Pearce, G., Strydom, D., Johnson, S. and Ryan, C.A. (1991) 'A polypeptide from tomato leaves induces wound-inducible proteinase inhibitor proteins', *Science* **253**, 895–897.

Pfeffer, S., Dunoyer, P., Heim, F., Richards, K.E., Jonard, G. and Ziegler-Graff, V. (2002) 'P0 of Beet western yellows virus is a suppressor of posttranscriptional gene silencing', *Journal of Virology* **76**, 6815–6824.

Pham, J.W., Pellino, J.L., Lee, Y.S., Carthew, R.W. and Sontheimer, E.J. (2004) 'A Dicer-2-dependent 80S complex cleaves targeted mRNAs during RNAi in *Drosophila*', *Cell* **117**, 83–94.

Pickford, A.S. and Cogoni, C. (2003) 'RNA-mediated gene silencing', *Cell and Molecular Life Sciences* **60**, 871–882.

Pickford, A.S., Catalanotto, C., Cogoni, C. and Macino, G. (2002) 'Quelling in *Neurospora crassa*', *Advances in Genetics* **46**, 277–303.

Plasterk, R.H. and Ketting, R.F. (2000) 'The silence of the genes', *Current Opinion in Genetics and Development* **10**, 562–567.

Plasterk, R.H.A. (2002) 'RNA silencing: the genome's immune system', *Science* **296**, 1263–1265.

Provost, P., Dishart, D., Doucet, J., Frendewey, D., Samuelsson, B. and Radmark, O. (2002) 'Ribonuclease activity and RNA binding of recombinant human Dicer', *EMBO Journal* **21**, 5864–5874.

Pruss, G.J., Lawrence, C.B., Bass, T., Li, Q.Q., Bowman, L.H. and Vance, V. (2004) 'The potyviral suppressor of RNA silencing confers enhanced resistance to multiple pathogens', *Virology* **320**, 107–120.

Ratcliff, F., Harrison, B.D. and Baulcombe, D.C. (1997) 'A similarity between viral defense and gene silencing in plants', *Science* **276**, 1558–1560.

Ratcliff, F.G., Macfarlane, S.A. and Baulcombe, D.C. (1999) 'Gene silencing without DNA: RNA-mediated cross-protection between viruses', *The Plant Cell* **11**, 1207–1215.

Reed, J.C., Kasschau, K.D., Prokhnevsky, A.I. *et al.* (2003) 'Suppressor of RNA silencing encoded by Beet yellows virus', *Virology* **306**, 203–209.

Reinhart, B.J. and Bartel, D.P. (2002) 'Small RNAs correspond to centromere heterochromatic repeats', *Science* **297**, 1831.

Reinhart, B.J., Slack, F.J., Basson, M. *et al.* (2000) 'The 21-nucleotide let-7 RNA regulates developmental timing in *Caenorhabditis elegans*', *Nature* **403**, 901–906.

Reinhart, B.J., Weinstein, E.G., Rhoades, M.W., Bartel, B. and Bartel, D.P. (2002) 'MicroRNAs in plants', *Genes and Development* **16**, 1616–1626.

Reynolds, A., Leake, D., Boese, Q., Scaringe, S., Marshall, W.S. and Khvorova, A. (2004) 'Rational siRNA design for RNA interference', *Nature Biotechnology* **22**, 326–330.

Rhoades, M.W., Reinhard, B.J., Lim, L.P., Burge, C.B., Bartel, B. and Bartel, D.P. (2002) 'Prediction of plant microRNA targets', *Cell* **110**, 513–520.

Roberts, A.G., Santa Cruz, S., Roberts, I.M., Prior, D.A.M., Turgeon, R. and Oparka, K.J. (1997) 'Phloem unloading in sink leaves of *Nicotiana benthamiana*: comparison of a fluorescent solute with a fluorescent virus', *The Plant Cell* **9**, 1381–1396.

Roignant, J.Y., Carre, C., Mugat, B., Szymczak, D., Lepesant, J.A. and Antoniewski, C. (2003) 'Absence of transitive and systemic pathways allows cell-specific and isoform-specific RNAi in *Drosophila*', *RNA* **9**, 299–308.

Ruiz, M.T., Voinnet, O. and Baulcombe, D.C. (1998) 'Initiation and maintenance of virus-induced gene silencing', *The Plant Cell* **10**, 937–946.

Ruiz-Medrano, R., Xoconostle-Cázares, B. and Lucas, W.J. (1999) 'Phloem long-distance transport of CmNACP mRNA: implications for supracellular regulation in plants', *Development* **126**, 4405–4409.

Schauer, S.E., Jacobsen, S.E., Meinke, D.W. and Ray, A. (2002) 'DICER-LIKE1: blind men and elephants in *Arabidopsis* development', *Trends in Plant Science* **7**, 487–491.

Schramke, V. and Allshire, R. (2003) 'Hairpin RNAs and retrotransposon LTRs effect RNAi and chromatin-based gene silencing', *Science* **301**, 1069–1074.

Sharp, P.A. (2001) 'RNA interference – 2001', *Genes and Development* **15**, 485–490.

Sijen, T., Fleenor, J., Simmer, F. *et al.* (2001) 'On the role of RNA amplification in dsRNA-triggered gene silencing', *Cell* **107**, 465–476.

Silhavy, D., Molnar, A., Lucioli, A. *et al.* (2002) 'A viral protein suppresses RNA silencing and binds silencing-generated, 21- to 25-nucleotide double-stranded RNAs', *EMBO Journal* **21**, 3070–3080.

Smardon, A., Spoerke, J.M., Stacey, S.C., Klein, M.E., Mackin, N. and Maine, E.M. (2000) 'EGO-1 is related to RNA-directed RNA polymerase and functions in germ-line development and RNA interference in *C. elegans*', *Current Biology* **10**, 169–178.

Song, J.J., Liu, J., Tolia, N.H. *et al.* (2003) 'The crystal structure of the Argonaute2 PAZ domain reveals an RNA binding motif in RNAi effector complexes', *Nature Structural Biology* **10**, 1026–1032.

Song, J.J., Smith, S.K., Hannon, G.J. and Joshua-Tor, L. (2004) 'Crystal structure of Argonaute and its implications for RISC slicer activity', *Science* **305**, 1434–1437.

Szittya, G., Molnar, A., Silhavy, D., Hornyik, C. and Burgyan, J. (2002) 'Short defective interfering RNAs of tombusviruses are not targeted but trigger post-transcriptional gene silencing against their helper virus', *The Plant Cell* **14**, 359–372.

Tabara, H., Yigit, E., Siomi, H. and Mello, C.C. (2002) 'The dsRNA binding protein RDE-4 interacts with RDE-1, DCR-1, and a DExH-box helicase to direct RNAi in *C. elegans*', *Cell* **109**, 861–871.

Takeda, A., Sugiyama, K., Nagano, H. *et al.* (2002) 'Identification of a novel RNA silencing suppressor, NSs protein of Tomato spotted wilt virus', *FEBS Letters* **532**, 75–79.

Taliansky, M.E. and Garcia-Arenal, F. (1995) 'Role of cucumovirus capsid protein in long-distance movement within the infected plant', *Journal of Virology* **69**, 916–922.

Tang, G., Reinhart, B.J., Bartel, D.P. and Zamore, P.D. (2003) 'A biochemical framework for RNA silencing in plants', *Genes and Development* **17**, 49–63.

Thomas, C.L., Leh, V., Lederer, C. and Maule, A.J. (2003) 'Turnip crinkle virus coat protein mediates suppression of RNA silencing in *Nicotiana benthamiana*', *Virology* **306**, 33–41.

Tijsterman, M. and Plasterk, R.H. (2004) 'Dicers at RISC; the mechanism of RNAi', *Cell* **117**, 1–3.

Tijsterman, M., May, R.C., Simmer, F., Okihara, K.L. and Plasterk, R.H. (2004) 'Genes required for systemic RNA interference in *Caenorhabditis elegans*', *Current Biology* **14**, 111–116.

Timmons, L. and Fire, A. (1998) 'Specific interference by ingested dsRNA', *Nature* **395**, 854.

Timmons, L., Tabara, H., Mello, C.C. and Fire, A.Z. (2003) 'Inducible systemic RNA silencing in *Caenorhabditis elegans*', *Molecular Biology of the Cell* **14**, 2972–2983.

Ueki, S. and Citovsky, V. (2001) 'Inhibition of systemic onset of post-transcriptional gene silencing by non-toxic concentrations of cadmium', *Plant Journal* **28**, 283–291.

Ueki, S. and Citovsky, V. (2002) 'The systemic movement of a tobamovirus is inhibited by a cadmium-ion-induced glycine-rich protein', *Nature Cell Biology* **4**, 478–485.

Vaewhongs, A.A. and Lommel, S.A. (1995) 'Virion formation is required for the long-distance movement of red clover necrotic mosaic virus in movement protein transgenic plants', *Virology* **212**, 607–613.

Vaistij, F.E., Jones, L. and Baulcombe, D.C. (2002) 'Spreading of RNA targeting and DNA methylation in RNA silencing requires transcription of the target gene and a putative RNA-dependent RNA polymerase', *The Plant Cell* **14**, 857–867.

Vance, V. and Vaucheret, H. (2001) 'RNA silencing in plants – defense and counterdefense', *Science* **292**, 2277–2280.

Vargason, J.M., Szittya, G., Burgyan, J. and Tanaka Hall, T.M. (2003) 'Size selective recognition of siRNA by an RNA silencing suppressor', *Cell* **115**, 799–811.

Verdel, A., Jia, S., Gerber, S. *et al.* (2004) 'RNAi-mediated targeting of heterochromatin by the RITS complex', *Science* **303**, 672–676.

Verkman, A.S. (2002) 'Solute and macromolecule diffusion in cellular aqueous compartments', *Trends in Biochemical Sciences* **27**, 27–33.

Voinnet, O. (2001) 'RNA silencing as a plant immune system against viruses', *Trends in Genetics* **17**, 449–459.

Voinnet, O. (2002) 'RNA silencing: small RNAs as ubiquitous regulators of gene expression', *Current Opinion in Plant Biology* **5**, 444–451.

Voinnet, O. and Baulcombe, D.C. (1997) 'Systemic signaling in gene silencing', *Nature* **389**, 553.

Voinnet, O., Lederer, C. and Baulcombe, D.C. (2000) 'A viral movement protein prevents spread of the gene silencing signal in *Nicotiana benthamiana*', *Cell* **103**, 157–167.

Voinnet, O., Pinto, Y.M. and Baulcombe, D.C. (1999) 'Suppression of gene silencing: a general strategy used by diverse DNA and RNA viruses of plants', *Proceedings of the National Academy of Sciences USA* **96**, 14147–14152.

Voinnet, O., Vain, P., Angell, S. and Baulcombe, D.C. (1998) 'Systemic spread of sequence-specific transgene RNA degradation in plants is initiated by localized introduction of ectopic promoterless DNA', *Cell* **95**, 177–187.

Volpe, T., Schramke, V., Hamilton, G.L. *et al.* (2003) 'RNA interference is required for normal centromere function in fission yeast', *Chromosome Research* **11**, 137–146.

Volpe, T.A., Kidner, C., Hall, I.M., Teng, G., Grewal, S.I. and Martienssen, R.A. (2002) 'Regulation of heterochromatic silencing and histone H3 lysine-9 methylation by RNAi', *Science* **297**, 1833–1837.

Waigmann, E., Lucas, W., Citovsky, V. and Zambryski, P. (1994) 'Direct functional assay for tobacco mosaic virus cell-to-cell movement protein and identification of a domain involved in increasing plasmodesmal permeability', *EMBO Journal* **91**, 1433–1437.

Waterhouse, P.M. and Helliwell, C.A. (2003) 'Exploring plant genomes by RNA-induced gene silencing', *Nature Reviews Genetics* **4**, 29–38.

Waterhouse, P.M., Wang, M.B. and Lough, T. (2001) 'Gene silencing as an adaptive defence against viruses', *Nature* **411**, 834–842.

Whittaker, G.R. and Helenius, A. (1998) 'Nuclear import and export of viruses and virus genomes', *Virology* **246**, 1–23.

Wille, A. and Lucas, W.J. (1984) 'Ultrastructural and histochemical changes on guard cells', *Planta* **160**, 129–142.

Willmer, C. and Sexton, R. (1979) 'Stomata and plasmodesmata', *Protoplasma* **100**, 113–124.

Winston, W.M., Molodowitch, C. and Hunter, C.P. (2002) 'Systemic RNAi in *C. elegans* requires the putative transmembrane protein SID-1', *Science* **295**, 2456–2459.

Wolf, S., Deom, C.M., Beachy, R.N. and Lucas, W.J. (1989) 'Movement protein of tobacco mosaic virus modifies plasmodesmatal size exclusion limit', *Science* **246**, 377–379.

Xie, Z., Johansen, L.K., Gustafson, A.M. *et al.* (2004) 'Genetic and functional diversification of small RNA pathways in plants', *PLoS Biology* **2**, E104.

Xie, Z., Kasschau, K.D. and Carrington, J.C. (2003) 'Negative feedback regulation of Dicer-Like1 in *Arabidopsis* by microRNA-guided mRNA degradation', *Current Biology* **13**, 784–789.

Yan, K.S., Yan, S., Farooq, A., Han, A., Zeng, L. and Zhou, M.M. (2003) 'Structure and conserved RNA binding of the PAZ domain', *Nature* **426**, 468–474.

Ye, K., Malinina, L. and Patel, D.J. (2003) 'Recognition of small interfering RNA by a viral suppressor of RNA silencing', *Nature* **426**, 874–878.

Yekta, S., Shih, I.H. and Bartel, D.P. (2004) 'MicroRNA-directed cleavage of HOXB8 mRNA', *Science* **304**, 594–596.

Yelina, N.E., Savenkov, E.I., Solovyev, A.G., Morozov, S.Y. and Valkonen, J.P. (2002) 'Long-distance movement, virulence, and RNA silencing suppression controlled by a single protein in hordei- and potyviruses: complementary functions between virus families', *Journal of Virology* **76**, 12981–12991.

Yoo, B.C., Kragler, F., Varkonyi-Gasic, E. *et al.* (2004) 'A systemic small RNA signaling system in plants', *Plant Cell* **16**, 1979–2000.

Yu, D., Fan, B., Macfarlane, S.A. and Chen, Z. (2003) 'Analysis of the involvement of an inducible *Arabidopsis* RNA-dependent RNA polymerase in antiviral defense', *Molecular Plant–Microbe Interactions* **16**, 206–216.

Zambryski, P. and Crawford, K. (2000) 'Plasmodesmata: gatekeepers for cell-to-cell transport of developmental signals in plants', *Annual Review of Cell and Developmental Biology* **16**, 393–421.

Zamore, P.D., Tuschl, T., Sharp, P.A. and Bartel, D.P. (2000) 'Double-stranded RNA directs the ATP-dependent cleavage of mRNA at 21 to 23 nucleotide intervals', *Cell* **101**, 25–33.

Zeng, Y., Yi, R. and Cullen, B.R. (2003) 'MicroRNAs and small interfering RNAs can inhibit mRNA expression by similar mechanisms', *Proceedings of the National Academy of Sciences USA* **100**, 9779–9784.

Zhang, H., Kolb, F.A., Brondani, V., Billy, E. and Filipowicz, W. (2002) 'Human Dicer preferentially cleaves dsRNAs at their termini without a requirement for ATP', *EMBO Journal* **21**, 5875–5885.

Zilberman, D., Cao, X. and Jacobsen, S.E. (2003) 'ARGONAUTE4 control of locus-specific siRNA accumulation and DNA and histone methylation', *Science* **299**, 716–719.

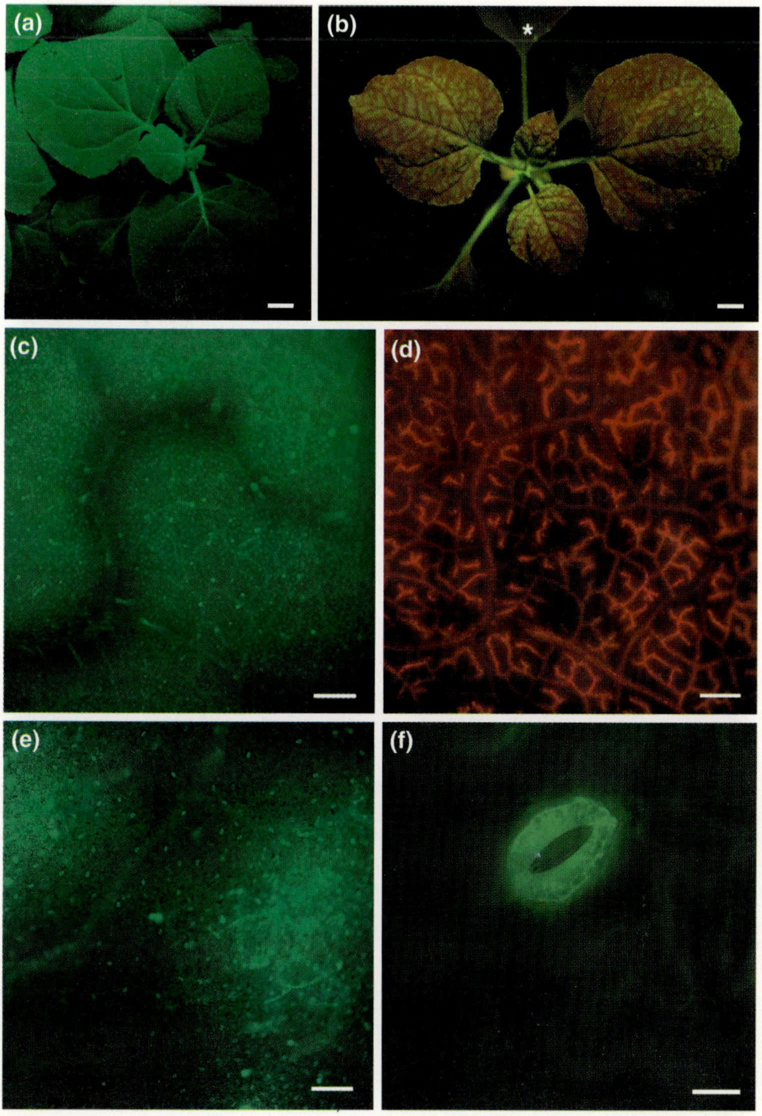


**Plate 10.1** Spread of GFP silencing. (a, b). Upon local infiltration of a leaf of *Nicotiana benthamiana* line 16 c (Ruiz *et al*., 1998) carrying a highly expressed green fluorescent protein (GFP) transgene (a) with a suspension of *Agrobacterium tumefaciens* (agro-infiltration) harboring T-DNA with an inverted repeat GFP construct, silencing spreads systemically from the infiltrated leaf (b, asterisk) into non-infiltrated leaves. Under UV illumination, GFP silencing is revealed by the replacement of green GFP fluorescence by red chlorophyll autofluorescence. The pattern of GFP silencing in non-infiltrated leaves (b) reflects unloading of the silencing signal from the veins and its subsequent cell-to-cell spread into adjacent tissues. (c, d) The silencing signal is unloaded from major (class I–III) veins and not from minor (class IV) veins. The abaxial side of a leaf region containing a class III vein from which GFP silencing spreads into adjacent tissue is imaged for GFP (c) and for texas red dextran that was used for staining the vascular xylem (d). The highly fluorescent spots (in c) are the tips of trichomes. (e, f) The silencing signal does not enter isolated guard cells, which remain highly fluorescent despite GFP silencing in surrounding epidermal cells. Silenced regions surrounding the veins show numerous fluorescent stomata (e). One of the stomata is shown at high magnification (f). Scale bars, 1 cm in (a, b); 100 μm in (c, d, e); 20 μm in (f).

# 11 Techniques for imaging intercellular transport

Karl Oparka and Petra Boevink

## 11.1 Introduction

Plasmodesmata are an integral component of the plant symplasm, and are therefore difficult to study in isolation from the living cells that they interconnect. In addition, plasmodesmata are delicate structures that respond rapidly to wounding by the deposition of callose, which seals off symplasmic connections (Roberts and Oparka, 2003). They are also sensitive to a range of intracellular and extracellular cues that influence the size exclusion limit (SEL) of the pore. In short, many of the approaches used to study plasmodesmata have the potential to influence their functions inadvertently. This sensitivity to manipulation requires that techniques for probing plasmodesmatal functions should be as non-invasive as possible, and that researchers of this structure should be acutely aware of the influence that their experimental treatments might have on plasmodesmatal behaviour.

This chapter describes a number of imaging approaches that are being used to study plasmodesmata, as well as a number of emerging techniques that are likely to offer great potential in imaging intercellular transport in plants. Throughout this chapter, we will stress the need for non-invasive imaging, and where appropriate will point out the relative advantages and pitfalls of the various experimental strategies that are currently in use.

## 11.2 Microinjection

Early studies of intercellular communication in plants made use of techniques that were shown for mammalian gap junctions to yield important information on the nature of molecules that pass from cell to cell. These included the microinjection of fluorescent probes of different molecular mass into individual cells to follow their cell-to-cell movement (Socolar and Loewenstein, 1979). Using this approach, it was established that small fluorescent probes (<1 kDa) moved freely between plant cells, giving an SEL for plasmodesmata that was not dissimilar to that found for gap junctions (Goodwin, 1983). The microinjection of fluorescent probes into plant cells remains a popular means of probing plasmodesmatal functions, and the following sections contain a critique of this important approach.

### *11.2.1 Pressure injection*

Pressure injection operates by applying a positive pressure to the solution held within a microcapillary. Pressure injection may be the appropriate choice when the compounds to be injected are either uncharged, large ($M_r$ 1–10 kDa) or both. Pressure (typically in the range 200–500 kPa) is generated at the back end of the micropipette either through the use of a commercially available pressure–injection system, which delivers a preset pulse of pressure into the injection pipette, or manually through the use of micrometer-based syringes (Oparka and Read, 1994). In studies of plasmodesmata, pressure injection has been used extensively to study intercellular transport of fluorescent probes as large as 20 kDa (Kempers and van Bel, 1997). In addition, pressure injection has been used to study RNA movement between plant cells (Lucas *et al.*, 1995). When considering pressure injection as an option for studying plasmodesmatal permeability, a number of caveats should be considered. First, although skilfully employed in numerous studies, impalement of plant cells with a micropipette is inevitably invasive. In such studies, it is important to monitor the impaled cells for signs of continued viability as assessed, e.g. by continued cytoplasmic streaming. Both impalement alone, which may result in a sudden elevation of cellular $Ca^{2+}$, and the sudden loss of turgor that occurs if the plasma membrane does not seal around the micropipette, may lead to closure of plasmodesmata as a result of cellular wounding (Knoblauch and van Bel, 1998). Recently, significant technical advances have been made in the microinjection of plant cells by reducing the tip diameter of micropipettes as low as 0.1 μm (compared to about 1 μm for conventional tips). Knoblauch (2001) described a gallinstan expansion 'femtosyringe' (GEF) with tip diameters of about 0.1 μm that prevented turgor loss during impalement, minimizing cellular damage. In this unique system, the micropipettes are filled partially with silicone oil, and the remainder of the pipette with gallinstan (a liquid alloy of gallium, indium and tin). Controlled application of warm air to the micropipette causes thermal expansion of the metal, generating high pressure at the micropipette tip and expelling the injection material. With this system, individual chloroplasts have successfully been microinjected (Knoblauch, 2001). The high pressure generated by the GEF system has allowed individual sieve elements to be microinjected (Knoblauch and van Bel, 1998). Significantly, these authors were able to show that in sieve elements, loss of turgor, rather than impalement, was a major determinant in the dispersal of crystalline P-protein aggregates at the sieve-plate pores.

Establishing the precise compartment that has been impaled, although an obvious criterion, is essential in all microinjection studies. However, in numerous published studies, it is not clear within which compartment the micropipette tip resides. Impalements of the vacuole result in non-movement of fluorescent probes, while leakage of dye from the pipette tip to the apoplast during impalement may give the impression of 'false' symplasmic movement. In the case of small ($<1$ kDa) probes, the entry of dye into the nucleus in cells surrounding the impaled cell is usually a good indication that the probe is moving in the symplasm via plasmodesmata.

Potential practitioners of microinjection should not underestimate the difficulty of impaling the exceptionally thin layer of cytoplasm that surrounds the vacuole of mature plant cells, nor the effects of inadvertently releasing turgor pressure during injection.

### *11.2.2 Iontophoretic injection*

Iontophoretic injection involves using an electrical current to introduce charged dye molecules into plant cells. It is best suited for injecting small dyes (<1 kDa) into cells. Several iontophoretic current sources can be obtained commercially but simple devices can also be constructed. The polarity of the current used is the same as the charge on the dye molecule (e.g. negative-going current is used to inject negatively charged dyes such as Lucifer Yellow CH). Either continuous or pulsed currents may be used to introduce dyes into plant cells, but, in general, cells have a better tolerance of low currents for long periods than high currents for short periods of injection. As in the case of pressure injection, the same criteria apply for monitoring cell viability during the injection process.

### *11.2.3 Iontophoretic versus pressure injection*

Because of the growing need to microinject larger molecules into plant cells to study the plasmodesmal SEL, pressure injection is usually the choice in studies of intercellular communication. There has been considerable debate in the literature concerning the efficacy of iontophoretic versus pressure microinjection (for an excellent comparison, see van Bel *et al.*, 1999), and the subject will not be treated in detail here. In principle, both techniques should give similar results, particularly with low-molecular weight probes, and this appears to hold true. However, there have been reports that, as assessed by iontophoretic injection, the SEL of plasmodesmata is decreased, rather than increased (Storms *et al.*, 1998), a finding that contradicts the conventional view that movement proteins (MPs) 'gate' plasmodesmata to a high SEL (reviewed in Roberts and Oparka, 2003). To date, the matter of which injection system is 'best' for intercellular transport studies has not been resolved. However, as shown below, the introduction of a range of alternative approaches for introducing macromolecules into cells makes new choices available to the researcher.

### *11.2.4 Some milestones in microinjection*

Microinjection was one of the first approaches used to probe plasmodesmatal SELs, and the work of Goodwin and colleagues (Goodwin, 1983; Erwee and Goodwin, 1984) was pioneering. These authors microinjected a wide range of FITC-conjugated peptides into plant cells to establish that the basic SEL of plasmodesmata was about 800 Da, and demonstrated that different cells in plant organs were connected by different SELs, establishing that many tissues are separated into symplasmic 'domains' within which cells are connected by a SEL differing from that of surrounding cells. As an extreme example of such a domain, Palevitz and Hepler (1985) were able to show by microinjection of Lucifer Yellow CH into developing guard cells that the guard cell complex becomes completely symplasmically isolated during leaf development. Subsequently, Terry and Robards (1987) showed that the Stokes radius ($R_s$) is the major determinant governing the diffusive movement of small molecules through plasmodesmata. In a controlled pressure injection study, Oparka and Prior (1992) found that total loss of turgor pressure led to the closure of plasmodesmata, while pressure differentials of about 200 kPa (induced between adjoining trichome cells with a pressure probe) was sufficient to reduce cell-to-cell communication of Lucifer Yellow CH.

As interest in plasmodesmatal functions escalated during the late 1980s and 1990s, the microinjection of increasingly larger compounds was attempted to explore the upper limits of plasmodesmal SEL. Much of this work was driven by the increasing evidence that many plant viruses exploit specific MPs to increase plasmodesmal SEL during infection to facilitate passage between cells of the viral genome. Wolf *et al.* (1989) were the first to demonstrate that the leaves of transgenic plants expressing the viral MP of tobacco mosaic virus (TMV) had an increased SEL of about 10 kDA, reflecting the inherent 'gating' capacity of viral MPs. In a subsequent study, Derrick *et al.* (1990) showed that the injection of functional virus particles of tobacco rattle virus into individual leaf trichomes resulted in an increase in the SEL, and were able to examine the kinetics of viral cell-to-cell virus movement. Subsequently, Oparka *et al.* (1997) conducted a detailed study of plasmodesmatal gating in a natural TMV infection site, exploiting a TMV vector that expressed GFP as a fusion to the viral MP to mark recently infected cells. They showed by microinjection of Texas red dextran (10 kDa) that in a natural infection, the gating of plasmodesmata is transitory and restricted to the leading edge of infection. In technically groundbreaking studies, the van Bel group were the first to microinject functional sieve elements (van der Schoot and van Bel, 1989), subsequently demonstrating that the specialized plasmodesmata that interconnect these cells have a large SEL of at least 20 kDa (Kempers and van Bel, 1997). As studies of transcription factors began to reveal that these proteins functioned in a non-cell autonomous manner, the Lucas group (Lucas *et al.*, 1995) demonstrated that KNOTTED 1 (KN1) moved from cell to cell when microinjected into tobacco mesophyll cells, and furthermore trafficked its own mRNA. More recently, a major technical advance has been the microinjection of fluorescent dye into the lumen of the cortical ER, demonstrating that small probes (<500 Da) are able to pass through plasmodesmata via the central desmotubule (Cantrill *et al.*, 1999).

In many studies, microinjection has been the technique of choice when exploring the functions of plasmodesmata in relatively inaccessible organs, such as apical meristems, allowing specific 'target' cells to be injected prior to an examination of intercellular communication. For example, iontophoretic injection has been used successfully to study symplasmic 'fields' in the shoot apical meristem using the low-molecular weight probe, Lucifer Yellow CH (Rinne and van der Schoot, 1998). However, a number of additional approaches are now in common use.

### *11.2.5 Dye loading*

Techniques for loading membrane impermeant dyes into plant cells date back to some of the earliest studies in plant physiology. Schumacher (1933) discovered that fluorescein could be loaded into the phloem and translocated to sink regions of the plant. Fluorescein, in common with a number of fluorescent probes, has unique properties that allow it to pass through cells and become 'trapped' by the alkaline pH of the sieve element (7.5–8.0). Those parameters that confer optimal trapping by the sieve element are based on the molecule's dissociation constant ($pK_a$) and its ability to cross cell membranes, usually assessed by its oil/water partition co-efficient (log $K_{ow}$; Oparka, 1991). For several years, these parameters have formed the basis for the rational design of phloem-mobile xenobiotics such as systemic herbicides (Bromilow and Chamberlain, 1989). Subsequently, it was shown that some derivatives of fluorescein, such as 5(6) carboxyfluorescein have an ideal combination of $pK_a$ and log $K_{ow}$ to permit phloem transport (Grignon *et al.*, 1989). In recent years, dye loading has been facilitated enormously by the generation of ester forms of a wide variety of fluorescent probes. In the ester (uncharged) form, the dye is freely mobile to cross membranes, but once inside the cell, cytosolic esterase activity converts the probe to the anionic form, which is membrane impermeant and unable to leak back out of the cell. Ester loading is now commonplace in cell biology. The ester form of CF, carboxyfluorescein diacetate, is an excellent probe for phloem transport and unloading studies. By applying droplets of CF diacetate to *Arabidopsis* leaves, Oparka *et al.* (1994) were able to follow phloem transport in real time, and were able to observe symplasmic phloem unloading in growing root tips. The unloaded CF moiety was able to move from the terminal protophloem sieve elements to all cells in the developing root meristem. Several ester forms of anionic probes have been shown to be phloem mobile in this way, including the highly fluorescent probe HPTS (Wright *et al.*, 1996). In an elegant study, Gisel *et al.* (1999) used phloem-loaded HPTS to trace symplasmic communication pathways in the shoot apical meristem of *Arabidopsis*, showing that symplasmic communication was reduced in the meristem during the floral transition.

Dye loading is an attractive means of studying plasmodesmatal functions as it is non-invasive and, in the case of phloem-mobile dyes, can be applied a considerable distance away from the cells to be monitored. The use of the phloem as a 'delivery system' for low-molecular weight probes offers several advantages. However, the dyes that can successfully be synthesised in ester form are generally of low molecular weight ($<$500 Da), meaning that only diffusional intercellular transport can be studied. One disadvantage of conventional dye loading, as applied to plasmodesmata, is the difficulty of loading fluorescent dyes into individual cells of interest as only small areas of tissue can be infiltrated.

### *11.2.6 Caged probes*

A major breakthrough in the field of non-invasive fluorescence imaging was the development of 'caged' probes. Caged probes are fluorophores that are rendered nonfluorescent by the addition of a chemical group. The resulting molecular cage is subsequently photolysed, usually by a single burst of short-wavelength (near UV) irradiation, to release the active fluorophore. Caged probes have been used recently to study intercellular communication through animal gap junctions (Leybaert and Sanderson, 2001). Surprisingly, the use of caged probes to study symplasmic transport in plants has only been introduced recently (Martens *et al.*, 2004). However, this approach offers enormous potential as an alternative to microinjection in studying the movement of small solutes between plant cells. Martens *et al.* (2004) infiltrated caged fluorescein by diffusion into onion bulb epidermal cells. The fluorophore (free fluorescein) was subsequently released in the cytosol with a 15-s burst of light (351 nm) from a UV laser, and showed extensive symplasmic transport following its release. Significantly, when the infiltrated cells were plasmolysed prior to photoactivation, the dye did not move from cell to cell when released within a single cell, consistent with plasmolysis disrupting intercellular communication. These data illustrate considerable potential for the field of plasmodesmatal research. Unlike simple dye loading, caged fluorophores can be released within specific cells following UV activation, allowing specific symplasmic pathways to be studied without the need for microinjection. This method will almost certainly find extensive use in probing symplasmic phloem loading pathways in plants. Although caged probes can be released in cells with light from a standard UV lamp, to accurately release the fluorophore in specific cells requires a confocal microscope fitted with a UV laser, and this requirement may ultimately limit the widespread use of the technique.

## 11.3 Fluorescent reporter proteins

The use of fluorescent reporter proteins in cell biology is now widespread, and will not be dealt with in detail here. For a critical evaluation of fluorescent reporter proteins currently in use, the reader is referred to recent reviews on this topic (Brandizzi *et al.*, 2002; Miyawaki *et al.*, 2003). Here, only fluorescent reporters currently used in plasmodesmatal research will be considered.

### *11.3.1 GFP*

Following the introduction of green fluorescent protein (GFP) as a reporter protein (Chalfie *et al.*, 1994), GFP was introduced into plant cells using a viral expression vector, emphasizing its use in tracing viral infections (Baulcombe *et al.*, 1995). The most extensive use of GFP in plant virus studies has been as a fusion protein with viral MPs (MP–GFP). When expressed as GFP fusions, the MPs of several viruses have been shown to retain their capacity to increase plasmodesmal SEL, traffic viral RNA and to target plasmodesmata several cells distant from the original bombarded cell (Huang and Zhang, 1999; Gillespie *et al.*, 2002; Levy and Czosnek, 2003). Thus, for many MPs, fusion to GFP does not appear to interfere with basic MP functions. Perhaps surprisingly, 'free' (i.e. non-fused) GFP (27 kDa) when expressed in the cytoplasm of a variety of cells using different methods shows cell-to-cell movement. This movement of GFP appears to be passive and diffusional, and can probably be explained by the protein's barrel-shaped structure (42 Å long and 24 Å in diameter; Ormö *et al.*, 1996), which gives it a reduced $R_s$ compared to other globular proteins of similar $M_r$. A number of authors have now reported the diffusional movement of free GFP in a range of cell types (Imlau *et al.*, 1999; Oparka *et al.*, 1999; Itaya *et al.*, 2000), and it appears that such non-targeted movement of GFP may be widespread throughout the plant. A consensus has arisen that plant cells may exchange some proteins by non-targeted movement, as well as by a targeted form of movement that involves an interaction with plasmodesmata and an increase in the SEL (for discussion see Chapter 5). The ability of GFP to move passively through plasmodesmata is fortuitous as it means that the protein can be employed as a reporter in the identification of symplasmic domains within developing tissues.

## 11.4 Red fluorescent proteins

### *11.4.1 DsRed*

DsRed was isolated from the coral *Discosoma* and can be excited by green radiation. Unfortunately, DsRed, in common with all *Anthozoan* GFP-like proteins isolated to date, forms obligate tetramers (Zhang *et al.*, 2002). This tetramerization has resulted in fusions of host proteins to DsRed that often display abnormal behaviour (Lauf *et al.*, 2001). The oligomerisation of DsRed does not preclude the use of DsRed in the free (non-fused) form for reporting gene expression or marking cells; however, it does limit its use in fusion-protein applications.

### *11.4.2 mRFP*

Monomeric red fluorescent protein (mRFP) was derived from DsRed2 by directed evolution (Campbell *et al.*, 2002). mRFP matures 10 times faster than its parental protein, and exhibits similar brightness to DsRed in living cells despite its lower quantum yield and photostability. The introduction of mRFP has enabled red-fluorescence labelling protocols that were impossible previously with DsRed (Lauf *et al.*, 2001; Campbell *et al.*, 2002).

In our laboratory, we have used fluorescent reporter proteins extensively, both in the free form and also as fusions to host or viral proteins. In studies of plasmodesmata–virus interactions, the choice of fluorescent reporter is extremely important and may influence the interpretation of results. In one experiment (Roberts *et al.*, unpublished data), we infected plants with TMV vectors expressing GFP, mRFP or DsRed fused to the MP and followed expression patterns on infected leaves with time. The different reporters had a marked effect on the movement of the TMV vectors which showed increasing speed in the order mRFP > GFP > DsRed. It appears that the oligomerisation of the fluorescent protein fused to the MP interferes with its function, perhaps by stearic hindrance. The improved performance of mRFP above GFP can be explained by the tendency of GFP to dimerise (Morise *et al.*, 1974). These data point to a need to select carefully the fluorescent reporter to be used for monitoring virus infection, and emphasise the difficulty in comparing published results using different reporters. The MP–DsRed fusion tended to form substantially larger aggregates on the ER than GFP within infected cells (Gillespie *et al.*, 2002). Furthermore, we have found that MP–DsRed shows a greatly reduced tendency to accumulate within the plasmodesmal pore compared to MP–GFP during infection, suggesting that the accumulation of MP within the plasmodesmal pore may not be an absolute requirement for virus movement. Generally, we have found that the behaviour of mRFP accurately reflects that of GFP when expressed as a fusion to the viral MP, making double-labelling studies with these reporters feasible.

When considering fluorescent reporters as markers for plasmodesmatal SEL, careful consideration needs to be given to the final $R_s$ of the reporter. Although mRFP, GFP and DsRed have the same molecular mass (27 kDa), their oligomerisation state will greatly affect their ability to move through plasmodesmata. For example, we have found that GFP moves freely from cell to cell in many tissues whereas the tetramerisation of DsRed prevents its intercellular spread. Cytosolically expressed (free) GFP and mRFP are useful tools for studying non-targeted protein trafficking between plant cells, particularly in developing tissues.

## 11.5 Biolistic bombardment

The use of ‘gene guns’ or simple hand-held devices for introducing DNA-containing plasmids (biolistic bombardment) became widespread during the 1990s and enabled the genetic transformation of many previously recalcitrant plant species. Biolistic bombardment for transient expression studies was rapidly assimilated into the field of plasmodesmatal research following the introduction of GFP. The technique has been used extensively to study the cell-to-cell trafficking of transcription factors that show directional movement between cell types. The most extensively studies of these has been KNOTTED 1 (KN1), a transcription factor involved in cell fate within meristems, and which traffics in its native context between the L1 and L2 layers of developing meristems (Jackson *et al.*, 1994). In experiments in which KN1 was expressed as a GFP fusion by biolistic bombardment, it retained its capacity to traffic between cells (Kim *et al.*, 2002).

Unlike microinjection, biolistic bombardment is a relatively random process, making it difficult to predict precisely which cells will be ‘hit’ with gold particles. However, it is technically much more facile than microinjection, and inaccessible tissues such as meristems have successfully been bombarded with GFP-fusion constructs (Kim *et al.*, 2002). Furthermore, tissues that are not amenable to other transient expression techniques (e.g. agroinfiltration) may be investigated using bombardment techniques (e.g. sink tissues, stomata and trichomes; P. Boevink, unpublished data). Biolistic bombardment, in common with microinjection, inevitably involves cellular damage as individual gold particles penetrate the cell to deliver the DNA-expressing plasmid. In addition, different results have been obtained with different biolistic approaches (Itaya *et al.*, 2000). Finally, a major limitation of biolistic delivery to study intercellular communication patterns is that the time taken for protein expression from the plasmid, coupled to maturation of the fluorophore (dependent on the reporter protein), can be quite extensive. Accumulation of sufficient fluorophore for detection may lead to a delay of several hours, if not days, before the first traces of fluorescence are observed in bombarded cells, let alone those adjoining the bombarded cell (Figure 11.1a–c). In this sense, biolistic

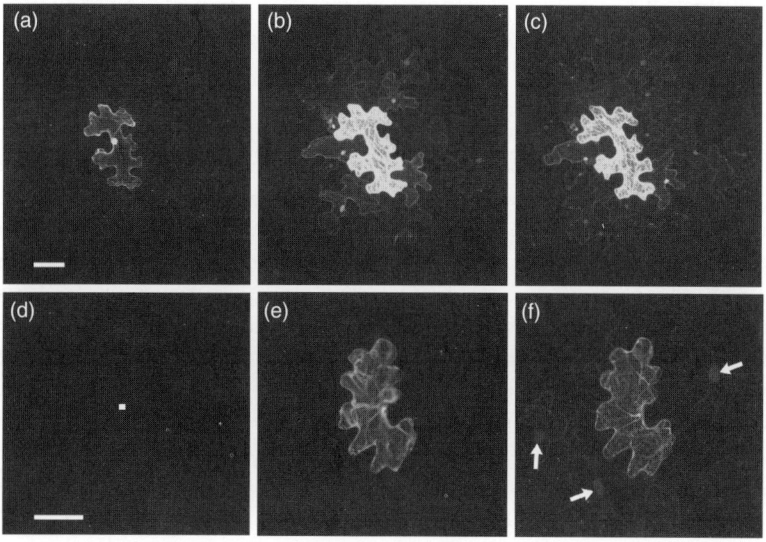


**Figure 11.1** Bombardment and photoactivation for measuring the movement of GFP through plasmodesmata. Cells bombarded with plasmid DNA of pRTL2. GFP take several hours to accumulate detectable fluorescence. (a) Shows a *Nicotiana benthamiana* epidermal cell 6 h post bombardment, and the fluorescence of the bombarded cell increases with time; (b) shows the cell 21 h post bombardment. Movement of the fluorophore into cells surrounding the bombarded cell is detectable within one day (b) and more cells become visible with time as more GFP is synthesised; (c) is 48 h post bombardment. Cells expressing PA–GFP are barely visible before activation (d). After activation of PAGFP in a small patch of cytoplasm and nucleus [small square in (d)] of this *N. tabacum* cell, the fluorescence spreads throughout the cell within seconds (e). Fluorescence may become visible in surrounding cells after 1 or a few hours [arrows in (f) indicate PA–GFP fluorescence in the nuclei of cells surrounding the activated cell] and if the activated PA–GFP is mobile, the fluorescence of the activated cell decreases with time; (f) was taken after 16 h (similar levels of movement were observed after 1 or 2 h in younger leaves). The activated PA–GFP is diluted as it diffuses from the initial cell and is more difficult to detect further from the activated cell; therefore, the observed distance of movement of the probe is partly dependent on the amount initially activated. The degree of movement can be quantified by measuring the loss of fluorescence from the activated cell and the increase of fluorescence of surrounding cells. Scale bars represent 50 μm. Images were taken with a Leica SP2 confocal microscope. Images a–c reproduced with permission from T. Gillespie.

bombardment, unlike microinjection, cannot function as a 'real-time' reporter of intercellular communication. This is especially true if GFP constructs are bombarded into meristematic tissues where rates of cell division may exceed the rate of expression of a fluorescent protein within a single cell. Despite these caveats, biolistic bombardment offers a simple means of examining intercellular protein trafficking in plants, and provides a valuable correlative approach to microinjection in establishing communication domains in plants.

## 11.6 Viral vectors

The high levels of promoter expression and self-replicating nature of plant viruses made them ideally suited to development as expression vectors. The first viral vector to be developed was based on the DNA virus, cauliflower mosaic virus (CaMV; Brisson *et al.*, 1984). Since then, many viruses with different genomic structures and host plants have been adapted to express foreign proteins rapidly and to high levels in plant cells. Viral vectors are introduced into cells by a variety of techniques: manual inoculation using a gentle abrasive, biolistic bombardment, agro-inoculation or agroinfiltration. All of these techniques involve varying degrees of cellular damage at the inoculation site but the viral vector may spread to other parts of the plant and therefore function as a relatively non-invasive system. We have used viral vectors based on tobacco mosaic virus (TMV) extensively in studies of plasmodesmatal functions. When the viral MP gene is deleted, the vector functions as an effective system for the high-level expression of proteins in single cells without interference of the MP functions on plasmodesmatal behaviour (Gillespie *et al.*, 2002). As with bombardment of expression plasmids, there is a delay in the development of detectable fluorescence from viral vectors, although this is usually reduced due to the high levels of expression achieved from viral promoters.

## 11.7 Tissue or cell-specific promoters

Another method used increasingly to study intercellular protein trafficking of GFP and GFP-fusion proteins is the use of tissue-specific promoters. In this approach, transgenic plants are constructed to express free GFP, or a GFP-fusion protein (e.g. GFP-transcription factor) in a given cell or tissue using a promoter that is only expressed in that tissue. For example, the *SUC2* promoter (which normally drives the expression of a companion-cell-specific sucrose transporter) has been used to drive the expression of free GFP in the companion cells (CCs) of source leaves of tobacco and *Arabidopsis* (Imlau *et al.*, 1999; Oparka *et al.*, 1999). The GFP was found to move from the CCs to the sieve elements and subsequently enter the translocation stream. Interestingly, the free GFP was unloaded in sink tissues, including developing roots and leaves, demonstrating that differentiating plant tissues have a high natural SEL (Imlau *et al.*, 1999; Oparka *et al.*, 1999). More recently, the

*SUC2* promoter has been used to express a range of GFP-fusion proteins ranging from free GFP (27 kDa) to a GFP-patatin fusion (70 kDa). It was found that all GFP-fusion proteins moved from the CC to the sieve element, revealing a very high basal SEL for the specialized plasmodesmata that connect the sieve elements with their CCs. However, only free GFP was unloaded from the terminal protophloem sieve elements in roots, and was able to move subsequently into meristematic cells (Stadler *et al.*, submitted). Interestingly, Itaya *et al.* (2002) showed that when the viral MP of cucumber mosaic virus was expressed in CCs using the Commelina yellow mottle virus promoter, the MP–GFP fusion moved out of CCs into several surrounding cells, indicating that viral MPs may expand the SE–CC domain by modifying surrounding plasmodesmata. Cell-specific promoters have found considerable recent utility for studying the movement of transcription factors during plant development. The transcription factor SHORT ROOT (SHR) has been expressed from its native promoter as a GFP fusion in its natural context and ectopically from the promoters *SUC2* (CC-specific), root epidermis and epidermal initials (*GL2* and *WER* promoters; Nakajima *et al.*, 2001; Sena *et al.*, 2004). SHR was found to move one cell layer from stele to endodermis in its native state, and from the epidermis to ground tissue in a *scarecrow* mutant background. However, it did not move out of CCs or the epidermis in a wild-type background. This work has provided a fascinating insight into the complexities of the regulation of movement and function of transcription factors.

As in the case of biolistic bombardment, tissue-specific promoters allow the visualization of symplasmic domains some time after the protein is first expressed. Thus, such patterns reflect the 'net' movement of GFP after several days or weeks of protein expression. Consequently, tissue- or cell-specific promoters cannot be used to monitor real-time communication events between cells.

## 11.8 Photoactivatable GFP

A revolution in non-invasive protein tracking occurred recently with the introduction of photoactivatable GFP (PA–GFP; Patterson and Lippincott-Schwartz, 2002). Although photoactivatable solutes are in widespread use (see earlier), attempts to develop a photoactivatable protein have only recently been fruitful. These authors introduced a mutation of threonine 203 to histidine in wild-type GFP, resulting in a 100-fold increase in fluorescence at 488 nm following activation by a single burst of UV light (413 nm). Before activation, PA–GFP displays a low level of fluorescence. The ability to 'switch on' the fluorescence of a photoactivatable protein makes it an excellent tool for studying protein behaviour in single cells, and importantly in the case of plants, between cells. Significantly, as the fluorescence of the protein comes only after photoactivation, newly synthesized (non-photoactivated) protein pools are invisible and do not complicate the experimental results. PA–GFP can be used to study the kinetics of tagged proteins, their compartmental residency time, and their exchange with non-fluorescent pools. Also, cell lineage can be monitored in a developing organism by imaging the fluorescence dispersion after photoactivation by a single cell or sub-population of cells (Lippincott-Schwartz *et al.*, 2003).

Recently, we have made use of PA–GFP for studies of intercellular protein communication in plants by exploiting the ability of free GFP to move between several cell types. Transgenic *Arabidopsis* plants were generated to express free PA–GFP under control of the 35-s promoter. PA–GFP does not require genetic modification for codon usage in plants as it is derived from an already modified form of GFP (enhanced GFP; eGFP) that has been expressed in plant cells successfully. When examined at 488 nm, levels of PA–GFP fluorescence from individual cells were extremely low, and the protein could only be imaged at high gain settings. Following a 1-s burst of UV light (405 nm) from a blue diode laser, levels of GFP fluorescence at 488 nm increased by several orders of magnitude. In these experiments, the PA–GFP was photoactivated within the nucleus, which provided an easy 'target' for excitation at 405 nm (Figure 11.1d). Within seconds, the PA–GFP had diffused through the nuclear pores into the surrounding cytoplasm (Figure 11.1e). As expected, in fully developed leaf cells, there was little or no movement of PA–GFP between adjoining cells. However, in young leaves PA–GFP moved readily into surrounding cells via plasmodesmata over a period of hours (Figure 11.1f). The ability to release PA–GFP in a wide range of cells within the plant body makes it a powerful tool for studying real-time intercellular protein trafficking in plants. In the future, 2-photon imaging should allow individual cells deep within a plant tissue to be photoactivated without activating fluorescent protein in other cells along the laser path. Furthermore, the ease with which PA–GFP can be fused to proteins of interest (e.g. transcription factors) will permit the expression of the fusion protein in its native (or non-native) context followed by a temporal analysis of its trafficking behaviour. In parallel with the use of low-molecular 'caged' probes, PA–GFP will open up the study of intercellular protein trafficking in plants without the need for microinjection, biolistic bombardment or viral infection.

### *11.8.1 Other photoactivatable proteins*

The Kaede protein has been isolated from *Trachyphyllia geoffroyi* (Ando *et al.*, 2002), and kindling fluorescent protein (KFP) from the sea anemone *Anemonia sulcata* (Chudakov *et al.*, 2003). Like PA–GFP, these proteins share the common characteristic of displaying low levels of fluorescence before photoactivation and higher levels after photoactivation. In the case of Kaede, photoactivation results in a 2000-fold increase in its red-to-green fluorescence, making it useful in dual wavelength studies of protein dynamics. In the case of KFP, the protein undergoes a reversible photoconversion from the nonfluorescent state to a bright red fluorescent form ('kindling'). Interestingly, this reversible response can be controlled by both the light intensity level and total light dose. KFP kindled by a short burst of light relaxes back to the nonfluorescent state with a half-life of about 50 s. KFP is kindled by green light (which for plant tissues may be less damaging than UV irradiation) and quenched in blue light. Upon long exposure to green light, KFP becomes irreversibly kindled and does not fade for up to a year after kindling (Chudakov *et al.*, 2003). Furthermore, irreversibly kindled KFP is not quenched by irradiation with blue light. The above proteins have yet to find utility in plant studies, and offer attractive alternatives to PA–GFP in studies of intercellular protein trafficking. However, both Kaede and KFP associate into tetrameric forms and appear to have low quantum yields, limiting their use as protein fusion tags. Engineering of more fluorescent, monomeric forms of these proteins would greatly enhance their utility.

## 11.9 Fluorescence recovery after photobleaching

Fluorescence recovery after photobleaching (FRAP) has been used extensively to study the diffusive properties of molecules in living cells. In this technique, a region of interest is selectively photobleached with a high-intensity laser pulse and the fluorescence recovery into the bleached region is monitored over time with low-intensity laser light. Depending on the nature of the protein being studied, fluorescence recovery can result from protein diffusion, binding/dissociation or transport requiring molecular motors (Lippincott-Schwartz *et al.*, 2003). One way to photobleach using FRAP studies is to define a region of interest at the highest possible zoom setting on a confocal laser scanning microscope, with the laser power set to maximum, thereby greatly increasing the radiation over a small area of cell. Many modern confocal microscopes have an acousto-optical tunable filter (AOTF) that rapidly attenuates the laser as it scans a field, and allows the user to photobleach virtually any pattern or shape within the cell (Lippincott-Schwartz *et al.*, 2003).

FRAP studies offer great potential in studies of plasmodesmatal function. Schulz (1999) demonstrated the potential of this approach by loading fluorescein diacetate into young pea roots to release free fluorescein. Cells containing fluorescein were selectively photobleached with a laser and symplasmic dye transport was monitored into the photobleached region. Similarly, FRAP has been used to selectively photobleach the unloading zone in *Arabidopsis* roots to examine the kinetics of phloem unloading (Oparka *et al.*, 1994). FRAP studies also offer potential in studying fluorescent MP kinetics following photobleaching. Several variants of native GFP (e.g. eGFP) are particularly sensitive to photobleaching and make valuable tools in FRAP studies (Lippincott-Schwartz *et al.*, 2003). Recently, we fused the MP of TMV to eGFP and expressed the fusion from a full-length TMV vector. As expected, the MP–GFP fusion was targeted to plasmodesmata at the leading edge of infection. Single plasmodesmata, or small groups of plasmodesmata, were then selectively photobleached with a laser. Fluorescence recovery occurred within the bleached plasmodesmata, indicating a continuous trafficking of MP into the plasmodesmal pore. This recovery was severely inhibited in the presence of the metabolic inhibitor, sodium azide, suggesting that MP trafficking to plasmodesmata is energy dependent (Wright *et al.*, unpublished data). In additional experiments involving specific inhibitors, applied prior to photobleaching, it was shown that MP trafficking requires a functional ER/actin network to target plasmodesmata (Wright *et al.*, unpublished data). FRAP studies such as these are likely to be useful in monitoring continuous protein movement into plasmodesmata, and for examining protein turnover within the plasmodesmal pore. As novel plasmodesmatal proteins continue to be identified (Medina-Escobar *et al.*, 2003), FRAP will provide an important tool for examining the kinetics of targeting of different proteins to the plasmodesmal pore. FRAP approaches are likely to be of particular use in viral MP studies where the earliest stages of viral infection are difficult to image due to either slow maturation of the fused fluorophore, or the presence of extremely low levels of MP during early infection.

## 11.10 FlAsH and ReAsH

Many modern imaging techniques are based on the fusion of fluorescent proteins to host proteins of interest (e.g. viral MPs). As outlined above, the 'palette' of fluorescent proteins now available is extensive; however, nearly all are about the same size as GFP (27 kDa). There is increasing interest in the development of smaller fluorescent tags but attempts to reduce the sizes of existing fluorescent proteins through mutagenesis have been unsuccessful (Miyawaki *et al.*, 2003). Interestingly, many viral MPs and transcription factors are able to pass through plasmodesmata when fused to GFP, suggesting that the trafficking capacity of the native transport protein is not affected by fusion to GFP (reviewed in Roberts and Oparka, 2003). However, for many proteins, it seems inevitable that fusion to GFP will affect either their plasmodesmal targeting or their subsequent passage through the pore. As the field of plasmodesmatal research expands towards the identification and characterization of novel plasmodesmatal proteins, there is a growing need to find alternative tags for tracking proteins that move through plasmodesmata.

Recently, a number of recombinant proteins have successfully been labelled with small organic fluorophores within living cells. Genetic modification of proteins by addition of tetra-cysteine motifs has allowed the localization of these proteins by labelling them with bi-arsenic containing fluorophores (Griffin *et al.*, 1998). Tsien and colleagues (Griffin *et al.*, 1998) introduced two arsenoxide groups into fluorescein to form FlAsH, which binds with high affinity to tetra-cysteines containing the rare sequence CCXXCC. A host (plasmodesmal) protein is fused to this short peptide sequence and expressed within cells using an appropriate delivery vector (Figure 11.2). The FlAsH label, which is membrane permeant and virtually nonfluorescent, is then introduced into cells and becomes highly fluorescent on binding covalently to the CCXXCC motif (Griffin *et al.*, 1998). The potential toxicity of the arsenic groups is alleviated by mixing the FlAsH with 1,2-ethanedithiol (EDT) before introducing it into cells. For studies on plasmodesmata, FlAsH labelling of proteins would greatly reduce the $S_r$ of a fusion protein and should not impair the protein's natural functions. Recently, a red analogue of FlAsH has been synthesized using the red fluorophore resorufin (ReAsH; Adams *et al.*, 2002). In an ingenious

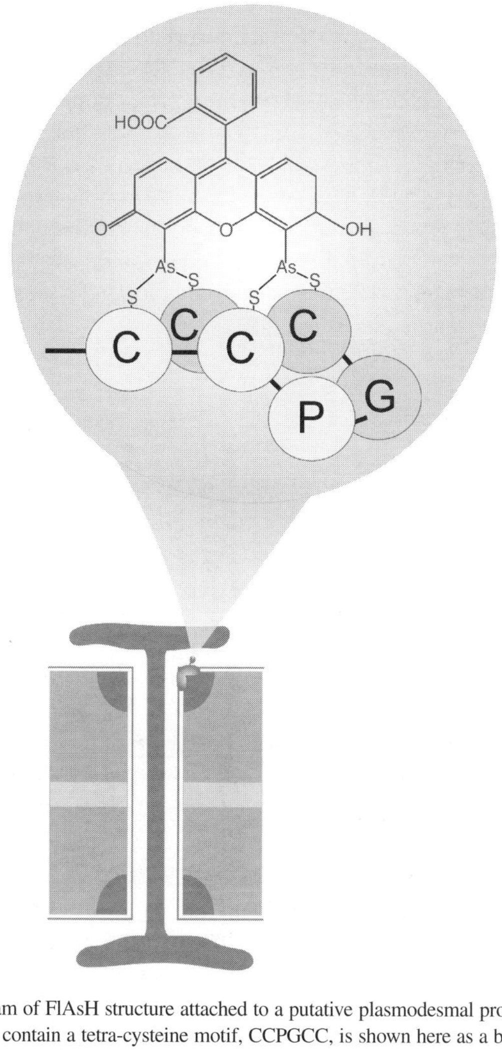


**Figure 11.2** Diagram of FlAsH structure attached to a putative plasmodesmal protein. A plasmodesmal protein engineered to contain a tetra-cysteine motif, CCPGCC, is shown here as a bent oval located at the neck of a plasmodesmata with the tetra-cysteine motif as a small protruding circle expanded in the shaded circle above. The binding of the bi-arsenical fluorescein derivative, FlAsH, to the four cysteine residues is illustrated. The two arsenics are thought to be clamped between the four cysteines located on either side of an α-helix.

combined use of FlAsH and ReAsH, Gaietta *et al.* (2002) determined the mechanism by which connexin 43 – a subunit of gap junction channels – is added to and removed from gap junction plaques. By engineering the CCXXCC motif into connexin 43, and then alternately labelling cells with FlAsH and ReAsH, different pools of the protein could be followed over time. They showed that by monitoring the pools of old (green) and young (red) connexin 43, the gap junction protein moves from the periphery to the centre of the gap junction plaque. ReAsH also has potential uses in correlative fluorescence and electron microscopy. Following intense illumination of fixed cells, ReAsH catalyses the photo-oxidation of diaminobenzidine, resulting in the formation of electron-opaque precipitates that can be visualized by electron microscopy (Gaietta *et al.*, 2002). The strategy of multiple labelling with FlAsH, ReAsH and the recently developed blue fluorophore, ChoXaSH (Adams *et al.*, 2002), has great potential in studies of plasmodesmatal development. For example, transgenic roots expressing a plasmodesmatal protein, or viral MP, fused to the CCXXCC motif could be incubated alternately with the above fluorophores to reveal dynamic patterns in plasmodesmatal development. In the above approach, early-formed plasmodesmata would appear green, and later-formed plasmodesmata red, while the most recently formed plasmodesmata would be blue (Figure 11.3). Because FlAsH is a membrane permeant derivative of fluorescein,

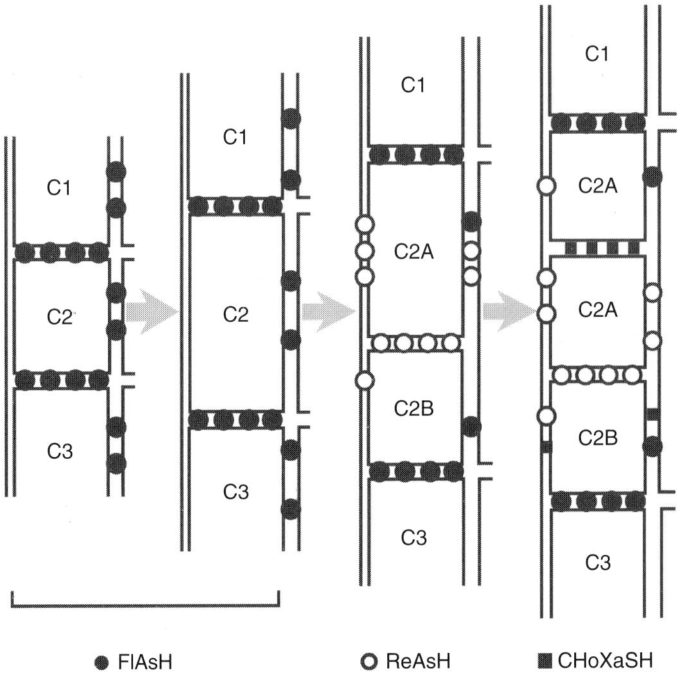


**Figure 11.3** Using FlAsH, ReAsH and CHoXaSH to study plasmodesmal development. We suggest that files of root cells in transgenic plants expressing a plasmodesmal protein tagged with a tetra-cysteine motif could be sequentially labelled with FlAsH (filled circles), ReAsH (open circles) and ChoXaSH (filled squares). The newly developed plasmodesmata at each stage would be labelled with a different coloured fluorescein derivative, revealing the patterns of plasmodesmal development in growing roots. During the first labelling procedure, all of the plasmodesmata containing the tagged protein would become labelled with FlAsH (filled circles). After washing off the FlAsH substrate, the roots would be allowed to grow for a time and then be exposed to ReAsH. New plasmodesmata formed after the FlAsH labelling would thus be labelled with ReAsH (open circles). Again, the substrate would be washed off and the roots allowed to grow once more. Finally, the roots would be treated with ChoXaSH and any new plasmodesmata formed since the ReAsH labelling would be revealed (filled squares).

it may also be possible to load FlAsH into the phloem and use the translocation system as an *in vivo* delivery system to cells containing peptide-tagged proteins.

## 11.11 Ablation techniques

### *11.11.1 Laser ablation*

A number of studies have successfully used ablation techniques to study cell-to-cell communication in plants. Tirlapur and König (1999) adapted the laser ablation technique to create a sensitive means of delivering fluorescent dye to cells deep in developing roots. They used near-infrared (NIR) femtosecond laser pulses as a means of dye loading the membrane-impermeant probe propidium iodide (PI) into meristematic cells in *Arabidopsis* roots without causing any perceptible damage to the cells. PI (668 Da) was introduced into the apoplast of the root and loaded into single cells with a NIR laser pulse. The highly focused NIR laser beam was able to penetrate and permeabilise single cells deep in the tissue (e.g. quiescent centre cells) without perturbing cells along the path of the laser beam. Following laser permeabilisation, PI leaked into the treated cell and subsequently moved via plasmodesmata to neighbouring cells. Interestingly, when PI was loaded into central cells, it moved into initials of the columella, cortex and pericycle. Conversely, the central cells remained isolated from dye import when the above cells were first loaded with PI. These results suggested that quiescent centre cells are differentially coupled to neighbouring initials, and that unidirectional transport of small solutes may occur between these cell types.

### *11.11.2 Diphtheria toxin*

An interesting genetic approach to studying intercellular communication during development is the use of cell- or tissue-specific promoters to express the diphtheria toxin A chain. This protein is highly toxic (Yamaizumi *et al.*, 1978) and the encoding gene has been used successfully to ablate root cap cells (Tsugeki and Fedoroff, 1999) and petal and stamen primordia (Nilsson *et al.*, 1998). Using this approach in the *Arabidopsis* embryo, Weijers *et al.* (2003) were able to disable a strictly defined set of cells within the embryo by expression of the diphtheria toxin. Subsequent non-cell autonomous effects on development were then monitored to reveal direct functional interactions. Ablation of the protoderm revealed apical control of basal embryo development, while ablation of the endosperm revealed that it has a role in regulating seed size and embryo patterning. Although in the above study, intercellular communication and signalling may have been interrupted by means other than altered symplasmic continuity, ablation of specific cells by production of toxins provides a further interesting approach to studying the development and interactions between specific tissue domains. A refinement of the use of diphtheria toxin ablation would be to link inducible promoter elements, such as tetracycline or ethanol-regulated elements (Love *et al.*, 2000; Deveaux *et al.*, 2003), to the tissue-specific elements. Such an approach would allow normal development to occur up to the point of induction (or removal of suppression) of expression of the toxin.

## 11.12 Quantum dots

Quantum dots are semiconductor nanocrystals that are exceptionally photostable over a wide range of excitation and emission wavelengths, and which may be used to allow long-term, multicolour imaging of live cells (Chan *et al.*, 2002). To date, their use in biological systems has been limited by their lack of biocompatibility, although new advances in surface coating chemistry have helped to alleviate this problem (Dubertret *et al.*, 2002). Quantum dots provide several advantages over organic fluorophores and protein tags, including narrow and tuneable emission spectra that can be varied according to the size and material composition of the particles. These properties allow close spacing of the quantum dots without spectral overlap (Miyawaki *et al.*, 2003). Recent advances in the development of nanomaterials have allowed quantum dots to be conjugated to biorecognition molecules, such as antibodies and DNA (Dubertret *et al.*, 2002; Wu *et al.*, 2003). To date, both cell surface proteins and the endocytic compartments of live animal cells have been labelled with quantum dot bioconjugates (Jaiswal *et al.*, 2003; Wu *et al.*, 2003). Quantum dots have recently been encapsulated into phospholipid micelles and injected into *Xenopus laevis* embryos, and their fluorescence followed until the tadpole stage in a cell-autonomous manner, indicating that quantum dots are stable and non-toxic inside cellular compartments (Dubertret *et al.*, 2002).

It is premature to assess the utility of quantum dots in studies of intercellular communication in plants, and their potential is mentioned here as a future prospect for live-cell imaging. The extremely small dimensions of quantum dots (2–10 nm diameter) are about the size of a 'typical' protein. Conjugation of quantum dots to small proteins with known intercellular trafficking capacity may provide a stable, long-term means of assessing intercellular communication patterns. Biotinylated host proteins can be linked to quantum dots that have been surface coated in streptavidin (Jaiswal *et al.*, 2003; Wu *et al.*, 2003). The applications of quantum dots in plant cells are likely to be limited by the means of delivery, and the introduction of useful quantities of quantum dots into living plant cells will probably require microinjection.

## 11.13 Summary

As the field of plasmodesmatal research expands in the post-genomics era, a range of new techniques will be required to keep pace with the need for functional assessment of novel plasmodesmatal proteins, and for the analysis of the intercellular trafficking of macromolecules. Conventional microinjection and biolistic approaches, although still in widespread use, are likely to be superseded by a wide range of non-invasive approaches for studying plasmodesmatal functions. The current revolution in fluorescence imaging, and the development of increasingly smaller and brighter fluorophores, holds much promise in future studies of plant intercellular communication.

## References


Adams, S.R., Campbell, R.E., Gross, L.A. *et al.* (2002) 'New biarsenical ligands and tetracysteine motifs for protein labeling *in vitro* and *in vivo*: synthesis and biological applications', *Journal of the American Chemical Society* **124**(21), 6063–6076.

Ando, R., Hama, H., Yamamoto-Hino, M., Mizuno, H. and Miyawaki, A. (2002) 'An optical marker based on the UV-induced green-to-red photoconversion of a fluorescent protein', *Proceedings of the National Academy of Sciences USA* **99**(20), 12651–12656.

Baulcombe, D.C., Chapman, S. and Santa Cruz, S. (1995) 'Jellyfish green fluorescent protein as a reporter for virus infections', *Plant Journal* **7**(6), 1045–1053.

Brandizzi, F., Fricker, M. and Hawes, C. (2002) 'A greener world: the revolution in plant bioimaging', *Nature* **3**, 520–530.

Brisson, N., Paszkowski, J., Penswick, J.R., Gronenborn, B., Potrykus, I. and Hohn, T. (1984) 'Expression of a bacterial gene in plants by using a viral vector', *Nature* **310**, 511–514.

Bromilow, R.H. and Chamberlain, K. (1989) Designing molecules for systemicity, in *Mechanisms and Regulation of Transport Processes* (eds R.K. Atkin and D.R. Clifford), British Plant Growth Regulator Group, Monograph 18, pp. 113–128.

Campbell, R.E., Tour, O., Palmer, A.E. *et al.* (2002) 'A monomeric red fluorescent protein', *Proceedings of the National Academy of Sciences USA* **99**, 7877–7882.

Cantrill, L.C., Overall, R.L. and Goodwin, P.B. (1999) 'Cell-to-cell communication via plant endomembranes', *Cell Biology International* **23**(10), 653–661.

Chalfie, M., Tu, Y., Euskirchen, G., Ward, W.W. and Prasher, D.C. (1994) 'Green fluorescent protein as a marker for gene expression', *Science* **263**, 802–805.

Chan, W.C.W., Maxwell, D.J., Gao, X., Bailey, R.E., Han, M. and Nie, S. (2002) 'Luminescent quantum dots for multiplexed biological detection and imaging', *Biotechnology* **13**, 40–46.

Chudakov, D.M., Belousov, V.V., Zaraisky, A.G. *et al.* (2003) 'Kindling fluorescent proteins for precise *in vivo* photolabeling', *Nature Biotechnology* **21**, 191–194.

Derrick, P.M., Barker, H. and Oparka, K.J. (1990) 'Effect of virus infection on symplastic transport of fluorescent tracers in *Nicotiana clevelandii* leaf epidermis', *Planta* **181**, 555–559.

Deveaux, Y., Peaucelle, A., Roberts, G.R. *et al.* (2003) 'The ethanol switch: a tool for tissue-specific gene induction during plant development', *Plant Journal* **36**, 918–930.

Dubertret, B., Skourides, P., Norris, D.J., Noireaux, V., Brivanlou, A.H. and Libchaber, A. (2002) '*In vivo* imaging of quantum dots encapsulated in phospholipid micelles', *Science* **298**, 1759–1762.

Erwee, M.G. and Goodwin, P.B. (1984) 'Characterization of the *Egeria densa* leaf symplast: response to plasmolysis, deplasmolysis and to aromatic amino acids', *Protoplasma* **122**, 162–168.

Gaietta, G., Deerinck, T.J., Adams, S.R. *et al.* (2002) 'Multicolor and electron microscopic imaging of connexin trafficking', *Science* **296**, 503–507.

Gillespie, T., Boevink, P., Haupt, S. *et al.* (2002) 'Functional analysis of a DNA-shuffled movement protein reveals that microtubules are dispensible for cell-to-cell movement of tobacco mosaic virus', *The Plant Cell* **14**, 1207–1222.

Gisel, A., Barella, S., Hempel, F.D. and Zambryski, P.C. (1999) 'Temporal and spatial regulation of symplasmic trafficking during development in *Arabidopsis thaliana* apices', *Development* **126**, 1879–1889.

Goodwin, P.B. (1983) 'Molecular size limit for movement in the symplast of the *Elodea* leaf', *Planta* **157**, 124–130.

Griffin, B.A., Adams, S.R. and Tsien, R.Y. (1998) 'Specific covalent labeling of recombinant protein molecules inside live cells', *Science* **281**, 269–272.

Grignon, N., Touraine, B. and Durand, M. (1989) '6(5) Carboxyfluorescein as a tracer of phloem sap translocation', *American Journal of Botany* **76**, 871–877.

Huang, M. and Zhang, L. (1999) 'Association of the movement protein of alfalfa mosaic virus with the endoplasmic reticulum and its trafficking in epidermal cells of onion bulb scales', *Molecular Plant–Microbe Interactions* **12**(8), 680–690.

Imlau, A., Truernit, E. and Sauer, N. (1999) 'Cell-to-cell and long-distance trafficking of the green fluorescent protein in the phloem and symplastic unloading of the protein into sink tissues', *The Plant Cell* **11**, 309–322.

Itaya, A., Liang, G., Woo, Y.-M., Nelson, R.S. and Ding, B. (2000) 'Nonspecific intercellular protein trafficking probed by green-fluorescent protein in plants', *Protoplasma* **213**, 165–175.

Itaya, A., Ma, F., Qi, Y. *et al.* (2002) 'Plasmodesmata-mediated selective protein traffic between "symplasmically isolated" cells probed by a viral movement protein', *The Plant Cell* **14**, 2071–2083.

Jackson, D., Veit, B. and Hake, S. (1994) 'Expression of maize *KNOTTED1* related homeobox genes in the shoot apical meristem predicts patterns of morphogenesis in the vegetative shoot', *Development* **120**, 405–413.

Jaiswal, J.K., Mattoussi, H., Mauro, J.M. and Simon, S.M. (2003) 'Long-term multiple color imaging of live cells using quantum dot bioconjugates', *Nature Biotechnology* **21**, 47–51.

Kempers, R. and van Bel, A.J.E. (1997) 'Symplasmic connections between sieve element and companion cell in the stem of *Vivia faba* L. have a molecular exclusion limit of at least 10 kDa', *Planta* **201**, 195–201.

Kim, J.Y., Yuan, Z., Cilia, M., Khalfan-Jagani, Z. and Jackson, D. (2002) 'Intercellular trafficking of a *KNOTTED1* green fluorescent protein fusion in the leaf and shoot meristem of *Arabidopsis*', *Proceedings of the National Academy of Sciences USA* **99**(6), 4103–4108.

Knoblauch, M. (2001) 'Microinjection', in *Plant Cell Biology* (ed. C. Hawes), Oxford University Press, Oxford, pp. 143–157.

Knoblauch, M. and van Bel, A.J.E. (1998) 'Sieve tubes in action', *The Plant Cell* **10**, 35–50.

Lauf, U., Lopez, P. and Falk, M.M. (2001) 'Expression of fluorescently tagged connexins: a novel approach to rescue function of oligomeric DsRed-tagged proteins', *FEBS Letters* **498**, 11–15.

Levy, A. and Czosnek, H. (2003) 'The DNA-B of the non-phloem-limited bean dwarf mosaic virus (BDMV) is able to move the phloem-limited Abutilon mosaic virus (AbMV) out of the phloem, but DNA-B of AbMV is unable to confine BDMV to the phloem', *Plant Molecular Biology* **53**(6), 789–800.

Leybaert, L. and Sanderson, M.J. (2001) 'Intercellular calcium signaling and flash photolysis of caged compounds: a sensitive method to evaluate gap junctional coupling', *Methods in Molecular Biology* **154**, 407–430.

Lippincott-Schwartz, J., Altan-Bonnet, N. and Patterson, G.H. (2003) 'Photobleaching and photoactivation: following protein dynamics in living cells', *Nature Cell Biology* **5** (Suppl.), S7–S14.

Love, J., Scott, A.C. and Thompson, W.F. (2000) 'Stringent control of transgene expression in *Arabidopsis thaliana* using the Top10 promoter system', *Plant Journal* **21**, 579–588.

Lucas, W.J., Bouché-Pillon, S., Jackson, D.P. *et al.* (1995) 'Selective trafficking of KNOTTED1 homeodomain protein and its mRNA through plasmodesmata', *Science* **270**, 1980–1983.

Martens, H.J., Hansen, M. and Schulz, A. (2004) 'Caged probes: a novel tool in studying symplasmic transport in plant tissues', *Protoplasma* **223**, 63–66.

Medina-Escobar, N., Haupt, S., Thow, G., Boevink, P., Chapman, S. and Oparka, K.J. (2003) 'High-throughput viral expression of cDNA-green fluorescent protein fusions reveals novel subcellular addresses and identifies unique proteins that interact with plasmodesmata', *The Plant Cell* **15**, 1507–1523.

Miyawaki, A., Sawano, A. and Kogure, T. (2003) 'Lighting up cells: labelling proteins with fluorophores', *Nature Cell Biology* **5** (Suppl.), S1–S7.

Morise, H., Shimomura, O., Johnson, F.H. and Winant, J. (1974) 'Intermolecular energy transfer in the bioluminescent system of *Aequorea*', *Biochemistry* **13**(12), 2656–2662.

Nakajima, K., Sena, G., Nawy, T. and Benfey, P.N. (2001) 'Intercellular movement of the putative transcription factor SHR in root patterning', *Nature* **413**, 307–311.

Nilsson, O., Wu, E., Wolfe, D.S. and Weigel, D. (1998) 'Genetic ablation of flowers in transgenic *Arabidopsis*', *Plant Journal* **15**, 799–804.

Oparka, K.J. (1991) 'Uptake and compartmentation of fluorescent probes by plant cells', *Journal of Experimental Botany* **42**, 565–579.

Oparka, K.J. and Prior, D.A.M. (1992) 'Direct evidence for pressure-generated closure of plasmodesmata', *Plant Journal* **2**(5), 741–750.

Oparka, K.J. and Read, N.D. (1994) The use of fluorescent probes for studies of living plant cells, in *Plant Cell Biology: A Practical Approach*, IRL Press, Oxford, pp. 27–50.

Oparka, K.J., Duckett, C.M., Prior, D.A.M. and Fisher, D.B. (1994) 'Real-time imaging of phloem unloading in the root tip of *Arabidopsis*', *Plant Journal* **6**(5), 759–766.

Oparka, K.J., Prior, D.A.M., Santa Cruz, S., Padgett, H.S. and Beachy, R.N. (1997) 'Gating of epidermal plasmodesmata is restricted to the leading edge of expanding infection sites of tobacco mosaic virus (TMV)', *Plant Journal* **12**(4), 781–789.

Oparka, K.J., Roberts, A.G., Boevink, P. *et al.* (1999) 'Simple, but not branched, plasmodesmata allow the nonspecific trafficking of proteins in developing tobacco leaves', *Cell* **97**, 743–754.

Ormö, M., Cubitt, A.B., Kallio, K., Gross, L.A., Tsien, R.Y. and Remington, S.J. (1996) 'Crystal structure of the *Aequorea victoria* green fluorescent protein', *Science* **273**, 1392–1395.

Palevitz, B.A. and Hepler, P.K. (1985) 'Changes in dye coupling of stomatal cells of *Allium* and *Commelina* demonstrated by microinjection of Lucifer yellow', *Planta* **164**, 473–479.

Patterson, G.H. and Lippincott-Schwartz, J. (2002) 'A photoactivatable GFP for selective photolabeling of proteins and cells', *Science* **297**, 1873–1877.

Rinne, P.L.H. and van der Schoot, C. (1998) 'Symplasmic fields in the tunica of the shoot apical meristem coordinate morphogenetic events', *Development* **125**, 1477–1485.

Roberts, A.G. and Oparka, K.J. (2003) 'Plasmodesmata and the control of symplastic transport', *Plant Cell and Environment* **26**, 103–124.

Schulz, A. (1999) Physiological control of plasmodesmal gating, in *Plasmodesmata. Structure, Function, Role in Cell Communication* (eds A.J.E. van Bel and W.J.P. van Kesteren), Springer, Berlin, pp. 173–204.

Schumacher, W. (1933) 'Untersuchungen uber die Wanderung des fluoresceins in den Siebrohren', *Jahrbuch für Wissenschaftliche Botanik* **77**, 685–732.

Sena, G., Jung, J.W. and Benfey, P.N. (2004) 'A broad competence to respond to SHORT ROOT revealed by tissue-specific ectopic expression', *Development* **131**, 2817–2826.

Socolar, S.J. and Loewenstein, W.R. (1979) Cell-to-cell junctional transmission, in *Methods in Membrane Biology 10* (ed. E.D. Korn), Plenum Press, New York, pp. 123–179.

Stadler, R., Wright, K.M., Lauterbach, C. *et al.* (2005) 'Transgenic expression of GFP fusions in *Arabidopsis* reveals extensive non-specific protein trafficking between companion cells and sieve elements, and identifies a novel post-phloem domain in roots', *Plant Physiology* **41**, 319–331.

Storms, M.M.H., van der Schoot, C., Prins, M., Kormelink, R., van Lent, J.W.M. and Goldbach, R.W. (1998) 'A comparison of two methods of microinjection for assessing altered plasmodesmal gating in tissues expressing viral movement proteins', *Plant Journal* **13**(1), 131–140.

Terry, B.R. and Robards, A.W. (1987) 'Hydrodynamic radius alone governs the mobility of molecules through plasmodesmata', *Planta* **171**, 145–157.

Tirlapur, U. and König, K. (1999) 'Near-infrared femtosecond laser pulses as a novel non-invasive means for dye-permeation and 3D imaging of localized dye-coupling in the *Arabidopsis* root meristem', *Plant Journal* **20**, 363–370.

Tsugeki, R. and Fedoroff, N.V. (1999) 'Genetic ablation of root cap cells in *Arabidopsis*', *Proceedings of the National Academy of Sciences USA* **96**(22), 12941–12946.

van Bel, A.J.E., Günther, S. and van Kesteren, W.J.P. (1999) Plasmodesmata, a maze of questions, in *Plasmodesmata. Structure, Function, Role in Cell Communication* (eds A.J.E. van Bel and W.J.P. van Kesteren), Springer, Berlin, pp. 1–26.

van der Schoot, C. and van Bel, A.J.E. (1989) 'Glass microelectrode measurements of sieve tube membrane potentials in internode discs and petiole strips of tomato (*Solanum lycopersicum* L.)', *Protoplasma* **149**, 144–154.

Weijers, D., van Hamburg, J.-P., van Rijn, E., Hooykaas, P.J.J. and Offringa, R. (2003) 'Diptheria toxin-mediated cell ablation reveals interregional communication during *Arabidopsis* seed development', *Plant Physiology* **133**, 1882–1892.

Wolf, S., Deom, C.M., Beachy, R.N. and Lucas, W.J. (1989) 'Movement protein of tobacco mosaic virus modifies plasmodesmatal size exclusion limit', *Science* **246**, 377–379.

Wright, K.M., Horobin, R.W. and Oparka, K.J. (1996) 'Phloem mobility of fluorescent xenobiotics in *Arabidopsis* in relation to their physicochemical properties', *Journal of Experimental Botany* **47**, 1179–1187.

Wu, C., Li, X., Yuan, W. *et al.* (2003) 'Development of enhancer trap lines for functional analysis of the rice genome', *Plant Journal* **35**, 418–427.

Yamaizumi, M., Mekada, E., Uchida, T. and Okada, Y. (1978) 'One molecule of diphtheria toxin fragment A introduced into a cell can kill the cell', *Cell* **15**, 245–250.

Zhang, J., Campbell, R.E., Ting, A.Y. and Tsien, R.Y. (2002) 'Creating new fluorescent probes for cell biology', *Nature Reviews Molecular Cell Biology* **3**, 906–918.

# 12 Electrical signalling via plasmodesmata

Aart J.E. van Bel and Katrin Ehlers

## 12.1 Introduction

Having discovered intercellular corridors, Eduard Tangl (1879) established the concept of intercellular trafficking, but it was a long time before his ideas were probed seriously (see Chapter 1). The suggestion that plasmodesmata were open channels for solute transfer encountered a great deal of scepticism. Therefore proponents of plasmodesmal functionality (e.g. Arisz, 1960) faced the task of convincing their adversaries and sceptics. Most of the initial experimental approaches departed from the assumption that plasmodesmata were the 'weak' spots in the high-resistant lipid membrane layers, and should be low-resistance pathways for electric currents between cells. Differences in electrical conductance between plasma membranes with and without plasmodesmata were taken as an argument for the open-state of plasmodesmata. Experiments on electrical conductivity have shown that plasmodesmata are functional, but have not solved the question as to whether electrical conductivity itself has a function in plants.

## 12.2 Electrical coupling as an initial probe for symplasmic cell coupling

As a general approach to demonstrate intercellular coupling, a current pulse was injected into one cell via an injector microelectrode, and its strength was recorded in a neighbouring cell by a receiver electrode. Spanswick and Costerton (1967) injected a current into one of the linearly arranged cells of *Nitella translucens*, which was picked up by a microelectrode several cells away from the injected cell. They calculated that electrical resistance over the cell array was 350 times lower than the resistance over the plasma membrane of the injected cell. It was concluded that the high electrical conductance of the intervening cell membranes was due to the presence of plasmodesmata.

In a later experiment with variegated *Oenothera* leaves, transient plasma-membrane depolarisations induced by light were immediately propagated to other green mesophyll cells kept in darkness, or to cells free of chloroplasts (Brinckmann and Lüttge, 1974). The profile and size of the potential wave in the treated and the distant cells was similar. Apparently, plasmodesmata allowed the propagation of electrical signals that could be perceived in distant cells.

In a third classic approach, Overall and Gunning (1982) showed that intercellular electrical conductance was linearly correlated with the number of plasmodesmata in *Azolla* roots. These roots grow by successive divisions of a tip cell that continuously produces spirally positioned merophyte daughter cells until a determinate length has been reached (Gunning, 1978). Electrical coupling between the tip cell and adjoining daughter cells declined with the growth of the root (Overall and Gunning, 1982). As an explanation for this phenomenon, the number of plasmodesmata in the division walls, that turned out to be directly related to the degree of electrical coupling, decreased with every cell division (Gunning, 1978).

## 12.3 Physical essentials of electrical coupling measurements

Electrical coupling can be measured in the most simple and reliable manner in filamentous structures as the currents can only move one-way. However, staminal hairs (Tucker and Tucker, 1993) and multicellular leaf hairs (Oparka and Prior, 1992) have been used for dye coupling studies, but not for the investigation of electrical coupling. In such linear structures, the coupling ratio ($V_{\text{receptor}}/V_{\text{injector}}$) is calculated from the depolarization ($V_{\text{injector}}$) induced by current injection via an injector electrode and the depolarization ($V_{\text{receptor}}$) recorded by a receiver electrode. As was amply discussed by van Rijen *et al.* (1999), quantitative information on electrical coupling between cells can be readily obtained only when all intercellular pathways have the same resistance, and all membranes possess identical resistance values. Hence, calculations on cell pairs are relatively easy (Holdaway-Clarke *et al.*, 1996), but become increasingly difficult with an increasing length of the path. However, the problems regarding quantification of the measurements in filamentous structures are minor compared to those with two-dimensional or even three-dimensional cell systems.

As described for animal systems, the only accurate method to quantify electrical conductance between neighbouring plant cells is the use of a dual voltage clamp system (van Rijen *et al.*, 1999). Even under these 'ideal' conditions, data must be corrected for errors imposed by series resistance and membrane resistance (Goldsmith and Goldsmith, 1978; van Rijen *et al.*, 1999). When using two single-barrelled electrodes, membrane resistances turned out to vary between 16 MΩ and 4 GΩ (van Rijen *et al.*, 1999). Therefore correction for current leakage through the plasma membrane is necessary for cell–cell contacts in low-resistance membranes. In contrast to the high conductance of the large plasmodesmata in *Chara*, which is in the order of 5–10 μS (Côté *et al.*, 1987; Ding and Tazawa, 1989; Holdaway-Clarke *et al.*, 1996), plasmodesmal conductance of higher plant cells was a factor 100 (equivalent to 10–20 MΩ) lower (Spanswick, 1972; Overall and Gunning, 1982; Holdaway-Clarke *et al.*, 1996). As the electrical resistance of the microelectrodes is often as low as 10–20 MΩ (e.g. van der Schoot and van Bel, 1989, 1990), a correction for series resistance is compulsory under these conditions.

Apart from the theoretical limitations, the nature of the plant cell wall and the presence of a large vacuole present practical barriers for a flawless measurement of electrical coupling. Voltage-clamping can only be done by the technically difficult insertion of double-barrelled microelectrodes, since the use of patch pipettes is impossible given the presence of the cell wall (van Rijen *et al.*, 1999). Further, resistance and elasticity of the plant cell wall demands such a vigorous impalement that the microelectrode tip is mostly pushed into the vacuole. To ensure that both electrodes are inserted into the cytosolic compartment, intracellular co-injection of fluorochromes is required. A special device has been designed for simultaneous measurement of electrical potentials and intracellular injection of dyes (Kempers *et al.*, 1999).

## 12.4 Membrane potential mapping as a tool for recognition of symplasmic domain borders

Apart from the above examples, electrical coupling has been demonstrated in a variety of tissues such as *Elodea* leaves and *Avena* coleoptiles (Spanswick, 1972), *Chara* nodal/internodal cells (Côté *et al.*, 1987; Ding and Tazawa, 1989; Reid and Overall, 1992), clumps of corn suspension cells (Holdaway-Clarke *et al.*, 1996), *Arabidopsis* root hairs (Lew, 1994, 1996) and *Lupinus* stem phloem (van Bel and van Rijen, 1994).

It is to be expected, however, that the plant does not provide an infinite continuum for electrical conductance. Intracellular injection of fluorescent dyes shows that symplasmic discontinuity not only exists at cell wall interfaces with truncated plasmodesmata (Wille and Lucas, 1984; Erwee *et al.*, 1985; Palevitz and Hepler, 1985), but also at interfaces with plasmodesmata such as at the cell wall between epidermal cells and basal hair cells (Palevitz and Hepler, 1985; Terry and Robards, 1987) which contains numerous plasmodesmata (Gunning and Hughes, 1976). Thus, transient or permanent electrical non-conductance may occur at several interfaces despite the presence of plasmodesmata.

Symplasmic demarcation between cell groups has been demonstrated by means of conventional electrophysiology. Successive microelectrode impalement into all cells of a tissue led to the production of membrane-potential maps (van der Schoot and van Bel, 1990). Membrane-potential differences of more than 10 mV were supposed to indicate electrical discontinuity between cell groups. Iontophoretic injection of fluorochromes confirmed the correctness of this concept and demonstrated that stem tissues are subdivided in a number of radial, longitudinal and concentric domains (van der Schoot and van Bel, 1990).

Electrical discontinuity infers the existence of symplasmic cell domains which may have some degree of autonomy (Ehlers and van Bel, 1999). Localisation of electrical discontinuity is a useful tool to identify symplasmic domains. However, the method has a limited validity, as adjoining symplasmic domains that would possess similar membrane potentials cannot be distinguished. Moreover, it is not possible to monitor transient (un)coupling events. With all evidence in favour of controlled electrical (un)coupling of cells, the question remains if electrical cell coupling is a side-effect of the open-state of plasmodesmata, or has a function of its own.

## 12.5 Interrelations between current injection and plasmodesmatal permeability

Before making assessments on the function of electrical coupling, it may be useful to consider potential flaws in the existing concept of electrical coupling:

(1) What may have been overlooked thus far is that current injection may create transient opening/closure of plasmodesmata giving incorrect information on *the vivo* status of the pore. Thus, current injection, in itself, may affect plasmodesmal gating.

(2) That current injection influences the plasmodesmal status may also hold for coupling experiments in which fluorochromes have been injected by iontophoresis, e.g. for the identification of symplasmic domains (van der Schoot and van Bel, 1990). Furthermore, absence of dye coupling does not necessarily exclude electrical coupling, since electrical transport requires a much narrower pathway than molecular transport. Thus, dye coupling and electrical coupling experiments can provide different information. This may impact on the interpretation of dye experiments showing symplasmic segregation between cells; electrical coupling may still be possible at these borders.

(3) The previous considerations may explain inconsistencies with regard to plasmodesmatal coupling between sieve elements/companion cells (SE/CCs) and phloem parenchyma cells (PPCs) in transport phloem. Dye coupling experiments (van der Schoot and van Bel, 1990; Oparka *et al.*, 1994; van Bel and van Rijen, 1994) and membrane potential measurements (van der Schoot and van Bel, 1990; van Bel and van Rijen, 1994) suggest an almost perfect symplasmic segregation between these cells. However, electrical uncoupling between SE/CCs and PPCs is not perfect and amounts to 10–15% of a maximal electrical coupling (van Bel and van Rijen, 1994). In addition, lateral distribution of electrical potential waves propagating along the sieve tubes has been observed (Rhodes *et al.*, 1996). In both cases, the few plasmodesmata between SE/CCs and PPCs (Kempers *et al.*, 1998), challenged by a sudden change in electropotential, may partly open up giving way for an electrical current to move laterally through the plasmodesmata (Rhodes *et al.*, 1996). The link between electrical disturbance and plasmodesmatal opening/closure is a potential key event in electrical signalling that deserves further research.

## 12.6 Potential functions of electrical cell–cell (un)coupling

The function of electrical discontinuity may be more obvious than that of electrical coupling. Symplasmic segregation enables the existence of two adjacent domains, each with an autonomous electrical processing. Hence, these domains can build up different membrane potentials with inherently different driving forces for proton-driven substrate uptake and different conditions for ion-channel gating. This leads to a situation in which two competing symplasmic domains at either side of an apoplasmic barrier can specifically accumulate materials from the apoplasmic space.

For instance, symplasmic segregation imposed by truncation of existing plasmodesmata (Wille and Lucas, 1984) is functional in the stomatal apparatus. The isolation of stomatal cells from the subsidiary cells enables autonomous regulation of ion channels giving rise to opening/closure of stomata (e.g. Majore *et al.*, 2002). As another example, the SE plasmodesmata towards other cells are truncated (Kempers *et al.*, 1998) with exception of those to the CC, during the ontogeny of the SE/CC complexes in stem phloem of tomato (K. Ehlers, unpublished results). Asymmetric truncation of SE plasmodesmata probably induces the exclusive symplasmic exchange of materials and messages between SE and CC. In transport phloem, plasmodesmata at the interface between CCs and PPCs are generally rare (Kempers *et al.*, 1998), and are often impermeable to chemical compounds, as indicated by absence of Lucifer Yellow (van der Schoot and van Bel, 1990) and CFDA (Oparka *et al.*, 1994; Knoblauch and van Bel, 1998) transport from the SE/CCs to the PPCs. Supposedly, isolation of SE/CCs allows a strict regulation of the sieve-tube content along the phloem pathway (see list item 2, e.g.; van Bel, 2003).

It is difficult to assess the function of electrical coupling, as well-documented and complete experimental evidence is scarce, but one can make the following rational guesses:

(1) Electrical coupling will eliminate membrane potential differences, so that cells belonging to the same symplasmic domain can act as a co-ordinated unit. Such a compensatory co-operation may be a general feature of symplasmic domains.

(2) Electrical coupling may also supplement insufficient electrical achievements of cells. An example may be electrical coupling between SEs and CCs in transport phloem (Figure 12.1). The rate of electrical coupling between the two is high (van Bel and van Rijen, 1994) with the membrane potential of CCs slightly more negative than that of SEs (van der Schoot and van Bel, 1989; J.B. Hafke, unpublished results). As the number of proton pumps in the SE plasma membrane is low compared to that in the CC plasma membrane (Bouché-Pillon *et al.*, 1994; Langhans *et al.*, 2001), the low proton-motive force generated by the SE itself may be supplemented by electrical charges (produced by the CC) that flow via the plasma membrane that lines the unilaterally branched plasmodesmata to the SE (van Bel, 1996).

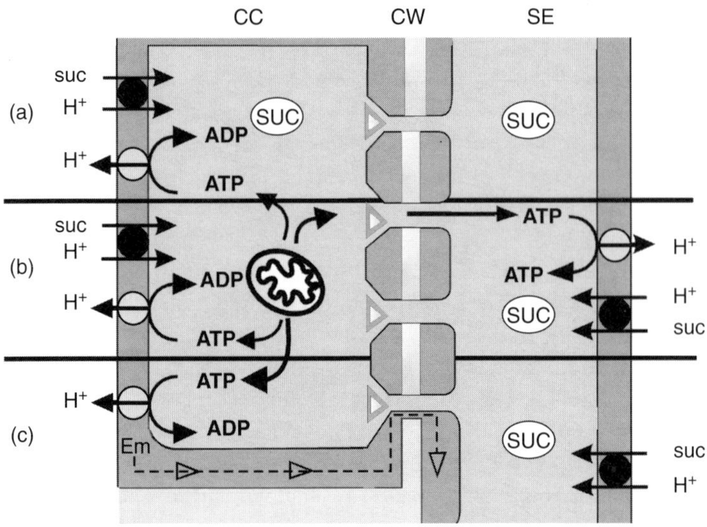


**Figure 12.1** Potential roles of branched plasmodesmata (PPUs) between sieve element (SE) and companion cell (CC) in establishing the accumulation capacity of SEs. Sugar uptake (e.g. sucrose, SUC) by SEs, the transport components of SE/CC complexes (SE/CCs), may result from various modes of plasmodesma-mediated interplay between SEs and CCs. (a) $H^+$-ATPases (proton pumps, open circles) and sucrose/proton symporters (black circles) are mainly localised on the CC-plasma membrane. Sucrose accumulation in CCs is driven by the proton-motive force (PMF) resulting from proton extrusion. Accumulated SUC is transferred via PPUs to the SEs. (b) Proton pumps are both localised on the CC- and SE-membranes. ATP molecules produced by numerous active mitochondria in CCs move through PPUs to SEs and energise proton pumps in the plasma membrane of SEs and CCs. The resulting PMF drives sucrose/proton symporters localised in the plasma membrane of SEs. (c) Proton pumps are mainly localised on the CC plasma membrane, sucrose/proton symporters mainly on the plasma membrane of SEs. The electrogenic potential (Em) generated by proton pumps located in the plasma membrane of CCs propagates via the plasma membrane lining of PPUs to the plasma membrane lining of SEs to make up for the presumptive insufficiency in PMF. The supplementary electrogenic potential allows SEs to retrieve sucrose according to model 1(b) (after van Bel, 1996).

It is relevant for a scenario in which the CCs supplement the electrogenic potential of SEs as to whether the plasmodesmata between CCs and PPCs are open or closed. In transport phloem, the difference between the membrane potentials of SE/CCs and PPCs is considerable and varies between approximately 25 mV in *Lupinus* (van Bel and van Rijen, 1994) and *Ricinus* (van Bel and Kempers, 1991), approximately 75 mV in *Salix* (van Bel and Kempers, 1991) and *Lycopersicon* (van der Schoot and van Bel, 1990; Rhodes *et al.*, 1996) up to 100 mV in *Mimosa* (Sikaoka, 1962). Should the PPCs become symplasmically coupled to the CCs, as is the case under sink-limiting conditions (Patrick and Offler, 1996), the charge supplementation of the SEs should become largely reduced as much of the electricity generated by the CCs would leak to the PPC symplasm.

(3) Major disturbances (by changes in light, temperature, pH, touch or wounding) induced at any site of a symplasmic domain are reported as an electrical wave to all other cells of the domain. One-to-one copies of waveforms induced by light propagated to neighbouring cells suggest instantaneous electrical cell-to-cell information (Brinckmann and Lüttge, 1974).

(4) Electrical signalling along the sieve tubes probably represents a special case of electrical signalling via plasmodesmata, because sieve tubes can be regarded as longitudinal arrays of cells between which enlarged plasmodesmata are operative (Esau and Thorsch, 1985; van Bel, 2003). Thus far, long-distance electrical signalling provides the best example of functional electrical cell coupling.

## 12.7 Relationships between electropotential waves, plasmodesmatal conductance and physiological activities

Several examples seem to illustrate the far-reaching consequences of electrical coupling via plasmodesmata. It has been reported that the respiration of the ovary increases following the pollen-triggered action potential that is transmitted down the style of *Lilium* and *Incarvillia* (Sinyukin and Britikov, 1967). Similarly, electrical potential waves have been observed by stimulation of *Hibiscus* stigmas using heat or cold shocks which propagate to the ovary and exert effects on the metabolism of the ovary (Fromm *et al.*, 1995). It appears that after the pollen lands on the stigma, the ovary metabolism becomes prepared for the arrival of the pollen tube by means of an electrical signal, which is speculated to follow the plasmodesmatal pathway. Touching the trigger hairs of the transformed trap leaf of *Dionaea* leads to a potential wave that brings about rapid closure of the leaf halves (Hodick and Sievers, 1989). The process starts with the bending of large sensory cells in the joint of the three trigger hairs positioned in the centre of the leaf half (Jacobson, 1965). The change in mechanical pressure induces a release of calcium into the cytosol of the sensor cells (Hodick and Sievers, 1988). The rise of cytosolic calcium, possibly released from whorls of ER in the sensor cells (Buchen *et al.*, 1983), brings about a membrane depolarisation from −160 to −50 mV (Hodick and Sievers, 1988). Subsequently, a depolarisation wave spreads with a velocity of approximately 20 cm $s^{-1}$ which results in the snapping of the trap (Sibaoka, 1966). The exact pathway along which the potential wave progresses is unknown. Most likely, the potential wave is propagated via the numerous plasmodesmata in the basal walls of the sensory cells (Buchen *et al.*, 1983) towards the cells responsible for the snapping. The site of action is most likely not in the main vein region, but rather in the epidermal cells (Hodick and Sievers, 1989).

When the glandular heads of the large outer *Drosera* leaf tentacles are touched by insects, the plasma membrane of the epidermal cells is depolarized (Williams and Pickard, 1972a). If the depolarisation exceeds a certain threshold, a train of action potentials is initiated that propagate with a speed of about 5 mm $s^{-1}$ along the tentacle stalk (Williams and Pickard, 1972a). Abundant plasmodesmatal connectivity in the transverse walls of the stalk cells allows rapid propagation of an action potential towards the stalk bases (Williams and Pickard, 1974). As soon as the action potential reaches the stalk base, a bending of the tentacles is triggered (Williams and Pickard, 1972b). Concerted bending of the stalks conveys the entangled insect to the shorter tentacles in the centre of the leaf, where it is digested by fluid excreted by the tentacle heads.

Intriguing questions remain concerning the existence of trains of depolarization spikes and their function. The capacity for producing spikes has been reported for several species (Pickard, 1974). The spike trains have been observed to occur spontaneously or to be triggered by external stimuli (Williams and Pickard, 1972a). For instance, the frictional resistance encountered in the pea epicotyl is related to the appearance of spike trains. Rubbing with a tiny rod induces similar spike trains which do not seem to propagate to other cells (Pickard, 1971). Symplasmic isolation – perhaps by spike-induced plasmodesmatal closure – may contain the effects of the membrane deflections to specific cells. It is not excluded that plasmodesmata may open up in response to electrical stimulation, making the way free for signal propagation in other signalling events.

To gain thorough knowledge of the relationship between induction of cell-to-cell electrical signals, plasmodesmatal transport and functional effects, it may be rewarding to repeat a number of earlier experiments using precisely located intracellular electrodes, along with the recording of plasmodesmatal gating in tissues with a proven excitability. In the context of propagation of potential waves through tissues, a long-forgotten observation may be important. In *Azolla* roots, electrical coupling was stronger in the acropetal than in the basipetal direction which suggest a polarity of plasmodesmatal electrical conductance (Overall and Gunning, 1982). Polarity of plasmodesmal conductance was also observed in the linear *Vallisneria* leaves, where basipetal transport exceeded acropetal transport of amino acids (Arisz, 1969). This polar behaviour of plasmodesmata is one of the many items that should also be investigated with regard to electrical coupling.

## 12.8 Role of plasmodesmata in the propagation of electropotential waves and long-distance signalling

Electrical potential waves are thought to confer information between plant parts. Best known are the long-distance signals triggered by touch that give rise to the movement of distant tissues or organs (e.g. Simons, 1992). Touching *Mimosa* leaves triggers a potential wave along the vascular system that causes shifts in water between flexor and extensor cells in the pulvinus. As a result, the petiole bends downwards and the leaflets fold down. It has been postulated that the electrical signals in *Mimosa* are propagated along the sieve tubes (Sibaoka, 1962; Samejima and Sibaoka, 1982; Fromm, 1991). Similar electrical waves along the plant axis, designated action potentials or wound potentials, have been described in non-excitable plants (e.g. Paszewski and Zawadski, 1973, 1974, 1976a,b; Zawadski, 1980; Jones and Wilson, 1982). These electrical signals, which are induced by electroshock, cold shock, changes in mechanical pressure, wounding or damage by leaf-biting insects are preferably conducted via sieve tubes (e.g. Eschrich *et al.*, 1988; Fromm and Bauer, 1994; Rhodes *et al.*, 1996).

The events in the *Mimosa* pulvinus provide convincing evidence for an association between electropotential propagation and physiological reaction. In the petiole, the vascular bundle is surrounded by a sclerenchyma sheath which confers symplasmic isolation that restricts electrical propagation to the vascular channels (Fleurat-Lessard and Roblin, 1982). Upon arrival of the signal in the pulvinus, the electropotential wave may spread via the numerous, mostly branched, plasmodesmata in the collenchyma sheath (Fleurat-Lessard and Bonnemain, 1978) to the thin-walled extensor and flexor parenchyma cells. There, voltage-dependent ion channels facilitate massive efflux and influx, respectively, of ions and associated water movements giving rise to leaflet bending (Satter and Galston, 1981). Interestingly, the movement of the flexor and extensor cells may be accelerated by contraction of actin microfibrils (Fleurat-Lessard *et al.*, 1988).

In analogy to electropotential waves in *Chara* cells (Kourie, 1994), potassium, chloride and calcium channels may be involved in the presumptive self-amplifying electrical signal propagation in higher plants (Fromm and Spanswick, 1993; Fromm and Bauer, 1994). The depolarization/repolarization potential waves have been designated 'action potentials', since they show the characteristic waveform of action potentials in animals. Moreover, the potential waves in plants and action potentials in animals both follow an all-or-none principle (Shiina and Tazawa, 1986; Fromm and Spanswick, 1993) and possess a refractory period (Fromm and Spanswick, 1993). Yet, as far as we are aware, the nature of the potential waves in plants is dubious, since self-amplification of the electrical signal (the main prerequisite for an action potential) has not been checked.

Given the loss of electrical current during each plasmodesmal passage (e.g. Overall and Gunning, 1982), one must conclude that an electrotonic propagation should be rapidly extinguished. It seems almost imperative therefore that a potential wave is permanently amplified, or seeks low-resistance pathways for propagation over longer distances. No wonder then that electropotential waves make their way preferentially through the low-resistance sieve-tube conduits (Eschrich *et al.*, 1988; Fromm and Bauer, 1994; Rhodes *et al.*, 1996). However, the waves also spread in a lateral direction, most likely via the plasmodesmata.

Accompanying depolarisation waves, which become weaker with the distance from the SEs, have been recorded in phloem parenchyma, cortex and epidermal cells (Rhodes *et al.*, 1996). Provided that the lateral distribution of electrical signals impacts on cells close to the sieve tubes, it may have far-reaching consequences for systemic signalling (see next section).

## 12.9 Are integrated electrical and chemical long-distance signals mediated by plasmodesmata during systemic induced resistance?

In solanaceous plants, cold shocks and electroshocks both result in the production of the defensive proteinase inhibitor (PI) which interferes with digestion processes in insects (Wildon *et al.*, 1992) and in a 10–15-times enhancement of *pin* mRNA (the *pin*-genes are responsible for the production of PI) (Stankovic and Davies, 1997). Production of PIs is under the indirect control of systemin – an 18 amino acid peptide – detected in the phloem sap of tomato (Narváez-Vásquez *et al.*, 1995). After systemin has interacted with a membrane receptor of the target cells, a signalling cascade that results in the production of jasmonic acid is elicited (Ryan and Moura, 2002). Jasmonic acid activates the expression of several genes (Li *et al.*, 2002; Ryan and Moura, 2002) finally leading to the expression of *pin*-genes i.e. the synthesis of PIs (Pearce *et al.*, 1991; Wildon *et al.*, 1992).

Given the eventual PI-production induced by either shocks (Wildon *et al.*, 1992; Stankovic and Davies, 1997) or application of systemin (Pearce *et al.*, 1991), the question about whether electrical and chemical signalling are integrated arises. Systemin is proteolytically processed from prosystemin, a protein synthesised in PPCs (Narváez-Vásquez and Ryan, 2004; see Chapter 8). It is imaginable that lateral distribution of an electropotential wave propagated through sieve tubes (cf. Rhodes *et al.*, 1996) triggers the transformation of prosystemin into systemin in PPCs (Narvaez-Vazquez and Ryan, 2004). Systemin may then be trafficked via the plasmodesmata between PPCs and CCs, and subsequently via the pore-plasmodesma units (PPUs) to the SEs, and be transported with the phloem stream to the target cells (Figure 12.2). Specific mechanisms for macromolecular trafficking have been found for the plasmodesmata between PPCs and CCs (Itaya *et al.*, 2002). Furthermore, the diameter of PPUs may be large enough to allow passage of systemin (Kempers and van Bel, 1997; Oparka, 2004).

It is important to underline the role of plasmodesmata in these complex signalling events. According to the present model, the wound-induced signalling cascade depends on the high degree of electrical coupling via the sieve pores (van Bel and van Rijen, 1994), the electrical conductance of plasmodesmata in lateral directions and the ability for macromolecular trafficking (Lucas *et al.*, 2001) in the symplasmic pathway between PPC and SE (Figure 12.2b). In view of long-distance wound signalling events (which are apparently quicker than mass-flow phloem transport), a parallel mode of signal transfer has been postulated (Wildon *et al.*, 1992). It is imaginable that propagation of an electropotential wave along the sieve tubes results in a calcium influx in the target cells. Systemin activity has been related to an increase in intracellular $Ca^{2+}$ which seems to be an early step in triggering PI production (Moyen *et al.*, 1998). Similarly, the propagation of the electropotential wave pushes a calcium wave through the sieve tubes which would almost immediately elicit signal cascades in remote tissues (Figure 12.2a).

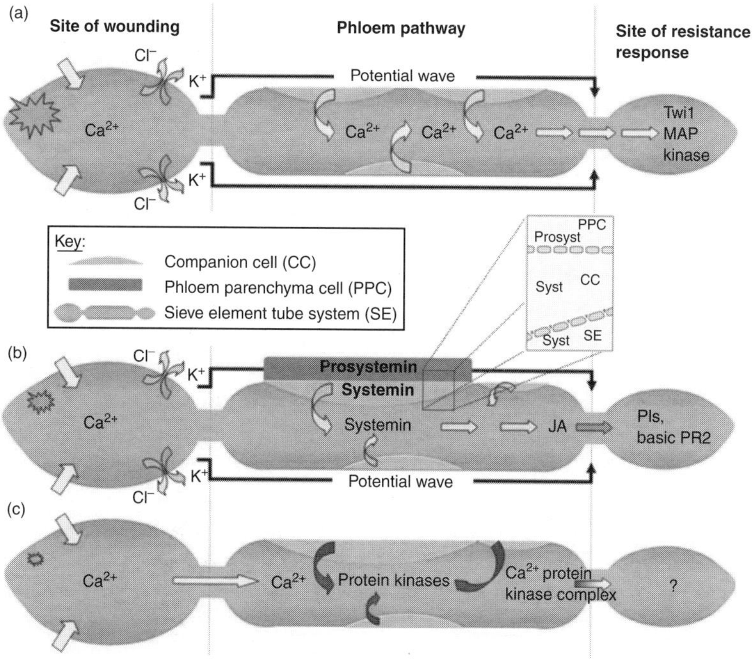


**Figure 12.2** Speculative multiple roles of plasmodesmata in wound-induced, long-distance signalling cascades. (a) Electrical signalling along the phloem pathway has been invoked to explain wound-induced reactions which are too rapid for mass-flow phloem transport (Wildon *et al.*, 1992). Electropotential waves along the phloem have been found in several non-excitable plants (Zawadski, 1980; Jones and Wilson, 1982; Eschrich *et al.*, 1988; Fromm and Bauer, 1994; Rhodes *et al.*, 1996). Propagation of an electropotential wave induces calcium release into the target cells or is accompanied by a wave of calcium released into the sieve tubes which arrives at the target cells. Electrical signals as well as $Ca^{2+}$-waves are candidates for induction of MAP-kinase (Seo *et al.*, 1995) or *Twi1* expression (O'Donell *et al.*, 1998) in adjacent non-wounded leaves several minutes after wounding. *Twi1* is a gene encoding for a protein with homology to glucosyltransferases. Sieve pores being transformed plasmodesmata present low-resistance corridors for a quick propagation of the electropotential wave (Esau and Thorsch, 1985). (b) More durable is a chemical cascade by which prosystemin in the phloem parenchyma cells (PPCs) is transformed into systemin (Narvaez-Vasquez and Ryan, 2004), which is transported to the companion cells (CCs) and finally released into sieve elements (SEs). Provided that systemin transport is symplasmic, this process renders obvious functions to the specific plasmodesmata between CC and phloem parenchyma (Kempers *et al.*, 1998; Chen *et al.*, 2000; Itaya *et al.*, 2002) and the large-diameter pore-plasmodesma units between CCs and SEs (Kempers and van Bel, 1997). It appears that systemin is involved in the amplification of the chemical signal rather than that it functions as the signal itself. At present, jasmonic acid seems the most likely candidate for triggering resistance expression (proteinase inhibitors, PIs, basic pathogen-related proteins, PR2) in target cells (Li *et al.*, 2002). Electrical and chemical signalling cascades may be interlinked; the electropotential wave may initiate the production of systemin and induce the opening of plasmodesmata to allow transport of systemin to the CCs. (c) In case of slight wounding, the threshold for an electropotential wave may not be exceeded, but calcium may be locally released into SEs at the site of damage and interact with calcium-dependent protein kinases, which are permanently delivered into the SEs by macromolecular trafficking (Yoo *et al.*, 2002). In the target tissues, calcium–protein-kinase complexes must make their way through plasmodesmal networks that are able to traffic these macromolecules (Lucas *et al.*, 2001; Foster *et al.*, 2002).

## 12.10 Concluding remarks

Electrical signalling via plasmodesmata is a phenomenon that has hardly been investigated. It appears that plasmodesmata are relays in a communication network that is modulated and integrated by electropotential waves (cf. Davies, 1985). What may have been overlooked thus far is that electrical pulses are able to modulate plasmodesmal gating. Integrative activities may therefore include electricity-induced transient opening or closure of plasmodesmata, which would temporarily prevent or allow intercellular traffic.

As a prime function, electrical conductivity of plasmodesmata seems to be a means for a rapid dispersal of information on relevant changes between cells, tissues and plant organs. What is perceived in one part of the symplasmic domain is quickly communicated to the other parts. The electrical events in reaction to pollen deposition onto the stigma (Sinyukin and Britikov, 1967; Fromm *et al.*, 1995) may be an example of what happens when abrupt changes are experienced in one part of a tissue or an organ.

Where the significance of the information is beyond the local level, electrical conduction via phloem tissues is used to confer the information over long distances. Sieve tubes can be regarded as an extremely long symplasmic domain, with symplasmic connections having extremely large functional diameters which can reach many other symplasmic domains. In view of the structural and functional homology of the sieve pores and plasmodesmata, electrical events along sieve tubes may bear some similarity to what happens on the local scale in electrically coupled tissues.

With electrical coupling, we probably just see a glimpse of a multifunctional and sophisticated signalling system, the complexity of which we have not yet fully understood. The integrative function of electrical signalling in combination with other messaging systems, enables the plant to take the appropriate measures for survival. By a variety of signalling systems, the plant is able to orchestrate information on continuous changes in light, temperature and humidity in various plant parts under the permanent attack of pests within the natural habitat.

## References

Arisz, W.H. (1960) 'Symplasmatischer Salztransport in *Vallisneria*-Blättern', *Protoplasma* **52**, 309–343.

Arisz, W.H. (1969) 'Intercellular polar transport and the role of the plasmodesmata in coleoptiles and *Vallisneria* leaves', *Acta Botanica Neerlandica* **18**, 14–38.

Bouché-Pillon, S., Fleurat-Lessard, P., Fromont, J.-C., Serrano, R. and Bonnemain, J.-L. (1994) 'Immunolocalization of the plasma membrane $H^+$-ATPase in minor veins of *Vicia faba* in relation to phloem loading', *Plant Physiology* **105**, 691–697.

Brinckmann, E. and Lüttge, U. (1974) 'Lichtabhängige Membranpotentialschwankungen und deren interzelluläre Weiterleitung bei panaschierten Photosynthesemutanten von *Oenothera*', *Planta* **119**, 45–57.

Buchen, B., Hensel, D. and Sievers, A. (1983) 'Polarity in mechanoreceptor cells of trigger hairs of *Dionaea muscipula* Ellis', *Planta* **158**, 458–468.

Chen, N.-H., Su, C.-L., Carter, S.A. and Nelson, R.S. (2000) 'Vascular invasion routes and systemic accumulation patterns of tobacco mosaic virus in *Nicotiana benthamiana*', *Plant Journal* **23**, 349–362.

Côté, R., Thain, J. and Fensom, D.S. (1987) 'Increase in electrical resistance of plasmodesmata of *Chara* induced by an applied pressure gradient across the node', *Canadian Journal of Botany* **65**, 509–511.

Davies, E. (1985) 'Action potentials as multifunctional signals in plants: a unifying hypothesis to explain apparently disparate wound responses', *Plant Cell and Environment* **10**, 623–631.

Ding, D.Q. and Tazawa, M. (1989) 'Influence of cytoplasmic streaming and turgor pressure gradient on the transnodal transport of rubidium and electrical conductance of *Chara corallina*', *Plant and Cell Physiology* **30**, 739–748.

Ehlers, K. and van Bel, A.J.E. (1999) The physiological and developmental consequences of plasmodesmal connectivity, in *Plasmodesmata. Structure, Function, Role in Cell Communication* (eds A.J.E. van Bel and W.J.P. van Kesteren), Springer, Berlin, Heidelberg, New York, pp. 243–260.

Erwee, M.G., Goodwin, P.B. and van Bel, A.J.E. (1985) 'Cell–cell communication in the leaves of *Commelina cyanea* and other plants', *Plant, Cell and Environment* **8**, 173–178.

Esau, K. and Thorsch, J. (1985) 'Sieve plate pores and plasmodesmata, the communication channels of the symplast: ultrastructural aspects and developmental relations', *American Journal of Botany* **72**, 1641–1653.

Eschrich, W., Fromm, J. and Evert, R.F. (1988) 'Transmission of electrical signals in sieve tubes of zucchini plants', *Botanica Acta* **101**, 327–331.

Fleurat-Lessard, P. and Roblin, G. (1982) 'Comparative histocytology of the petiole and the main pulvinus in *Mimosa pudica* L.', *Annals of Botany* **50,** 83–92.

Fleurat-Lessard, P., Roblin, G., Bonmort, J. and Besse, C. (1988) 'Effects of colchicine, vinblastine, cytochalasin B and phalloidin on the seismonastic movement of *Mimosa pudica* leaf and on motor cell structure', *Journal of Experimental Botany* **39**, 209–221.

Fleurat-Lessard, P.M. and Bonnemain, J.-L. (1978) 'Structural and ultrastructural characteristics of the vascular apparatus of the sensitive plant *Mimosa pudica* L.', *Protoplasma* **94**, 127–143.

Foster, T.M., Lough, T.J., Emerson, S.J. *et al.* (2002) 'A surveillance system regulates selective entry of RNA into the shoot apex', *The Plant Cell* **14**, 1497–1508.

Fromm, J. (1991) 'Control of phloem unloading by action potentials in *Mimosa*', *Physiologia Plantarum* **83**, 529–533.

Fromm, J. and Bauer, T. (1994) 'Action potentials in maize sieve tubes change phloem translocation', *Journal of Experimental Botany* **45**, 463–469.

Fromm, J. and Spanswick, R. (1993) 'Characteristics of action potentials in willow (*Salix viminalis* L.)', *Journal of Experimental Botany* **44**, 1119–1125.

Fromm, J., Hajirezaei, M. and Wilke, I. (1995) 'The biochemical response of electrical signaling in the reproductive system of *Hibiscus* plants', *Plant Physiology* **109**, 375–384.

Goldsmith, T.H. and Goldsmith, M.H.M. (1978) 'The integration of intracellular measurements of membrane potential, resistance, and coupling in cells of higher plants', *Planta* **143**, 267–274.

Gunning, B.E.S. (1978) 'Age-related and origin-related control of the numbers of plasmodesmata in cell walls of developing *Azolla* roots', *Planta* **143**, 181–190.

Gunning, B.E.S. and Hughes, J.E. (1976) 'Quantitative assessment of symplastic transport of pre-nectar into trichomes of *Abutilon* nectarines', *Australian Journal of Plant Physiology* **3**, 619–637.

Hodick, D. and Sievers, A. (1988) 'The action potential of *Dionaea muscipula* Ellis', *Planta* **174**, 8–18.

Hodick, D. and Sievers, A. (1989) 'On the mechanism of trap closure of Venus flytrap (*Dionaea muscipula* Ellis)', *Planta* **179**, 32–42.

Holdaway-Clarke, T.L., Walker, N.A. and Overall, R.L. (1996) 'Measurement of the electrical resistance of plasmodesmata and membranes of corn suspension cells', *Planta* **199**, 537–544.

Itaya, A., Ma, F., Qi, Y. *et al.* (2002) 'Plasmodesma-mediated selective protein traffic between "symplasmically isolated" cells probed by a viral movement protein', *The Plant Cell* **14**, 2071–2083.

Jacobson, S.L. (1965) 'Receptor response in the Venus' flytrap', *Journal of General Physiology* **49**, 117–129.

Jones, C. and Wilson, J.M. (1982) 'The effects of temperature on action potentials in the chill sensitive seismonastic plants *Biophytum sensitivum*', *Journal of Experimental Botany* **33**, 313–320.

Kempers, R. and van Bel, A.J.E. (1997) 'Symplasmic connections between sieve element and companion cell in the stem phloem of *Vicia faba* L. have a size exclusion limit of at least 10 kDa', *Planta* **201**, 195–201.

Kempers, R., Ammerlaan, A. and van Bel A.J.E. (1998) 'Symplasmic constriction and ultrastructural features of the sieve element/companion cell complex in the transport phloem of apoplasmically and symplasmically phloem-loading species', *Plant Physiology* **116**, 271–278.

Kempers, R., Prior, D.A.M., Oparka, K.J., Knoblauch, M. and van Bel, A.J.E. (1999) 'Integration of controlled intracellular pressure microinjection, iontophoresis, and membrane potential measurement', *Plant Biology* **1**, 61–67.

Knoblauch, M. and van Bel, A.J.E. (1998) 'Sieve tubes in action', *The Plant Cell* **10**, 53–50.

Kourie, J.I. (1994) 'Transient $Cl^-$ and $K^+$ currents during the action potential in *Chara inflata*. Effects of external sorbitol, cations and ion channel blockers', *Plant Physiology* **106**, 651–660.

Langhans, M., Ratajczak, R., Lützelschwab, M. *et al.* (2001) 'Immunolocalization of plasma membrane $H^+$-ATPase and tonoplast-type pyrophosphatase of the sieve element-companion cell complex in the stem of *Ricinus communis* L.', *Planta* **213**, 11–19.

Lew, R.R. (1994) 'Regulation of electrical coupling between *Arabidopsis* root hairs', *Planta* **193**, 67–73.

Lew, R.R. (1996) 'Pressure regulation of the electrical properties of growing *Arabidopsis thaliana* L. root hairs', *Plant Physiology* **112**, 1089–1100.

Li, L., Li, C., Lee, G.I. and Howe, G.A. (2002) 'Distinct roles for jasmonate synthesis and action in the systemic wounding response of tomato', *Proceedings of the National Academy of Sciences USA* **99**, 6416–6421.

Lucas, W.J., Yoo. B.-C. and Kragler, F. (2001) 'RNA as a long-distance information macromolecule in plants', *Nature Reviews Molecular Cell Biology* **2**, 849–857.

Majore, I., Wilhelm, B. and Marten, I. (2002) 'Identification of $K^+$ channels in the plasma membrane of maize subsidiary cells', *Plant and Cell Physiology* **43**, 844–852.

Moyen, C., Hammond-Kosack, K.E., Jones, J., Knight, M.R. and Johannes, E. (1998) 'Systemin triggers an increase of cytoplasmic calcium in tomato mesophyll cells: $Ca^{2+}$ mobilization from intra- and extracellular compartments', *Plant Cell and Environment* **21**, 1101–1111.

Narváez-Vásquez, J. and Ryan, C.A. (2004) 'The cellular localization of prosystemin: a functional role for phloem parenchyma in systemic wound signalling', *Planta* **218**, 360–369.

Narváez-Vásquez, J., Pearce, G., Orozco-Cardenas, M.L., Franceschi, V.R. and Ryan, C.A. (1995) 'Autoradiographic and biochemical evidence for the systemic translocation of systemin in tomato plants', *Planta* **195**, 593–600.

O'Donnell, P.J., Truesdale, M.R., Calvert, C.M., Dorans, A., Roberts, M.R. and Bowles, D.J. (1998) 'A novel tomato gene that rapidly responds to wound- and pathogen-related signals', *Plant Journal* **14**, 137–142.

Oparka, K.J. (2004) 'Getting the message across: how do plant cells exchange macromolecular complexes?', *Trends in Plant Science* **9**, 33–41.

Oparka, K.J. and Prior, D.A.M. (1992) 'Direct evidence for pressure generated closure of plasmodesmata', *Plant Journal* **2**, 741–746.

Oparka, K.J., Duckett, C.M., Prior, D.A.M. and Fisher, D.B. (1994) 'Real-time imaging of phloem unloading in the root tip of *Arabidopsis*', *Plant Journal* **6**, 759–766.

Overall, R.L. and Gunning, B.E.S. (1982) 'Intercellular communication in *Azolla* roots. I. Electrical coupling', *Protoplasma* **111**, 151–160.

Palevitz, B.A. and Hepler, P.K. (1985) 'Changes in dye coupling of stomatal cells of *Allium* and *Commelina* demonstrated by the injection of Lucifer yellow', *Planta* **164**, 473–479.

Paszewski, A. and Zawadski, T. (1973) 'Action potentials in *Lupinus angustifolius* L. shoots', *Journal of Experimental Botany* **24**, 804–809.

Paszewski, A. and Zawadski, T. (1974) 'Action potentials in *Lupinus angustifolius* L. shoots. II Determination of the strength-duration relation and the all-or-nothing law', *Journal of Experimental Botany* **25**, 1097–1103.

Paszewski, A. and Zawadski, T. (1976a) 'Action potentials in *Lupinus angustifolius* L. shoots. III Determination of the refractory periods', *Journal of Experimental Botany* **27**, 369–374.

Paszewski, A. and Zawadski, T. (1976b) 'Action potentials in *Lupinus angustifolius* L. shoots. IV Application of thermal stimuli and investigations on the conduction pathways of the excitation', *Journal of Experimental Botany* **27**, 859–863.

Patrick, J.W. and Offler, C.E. (1996) 'Post-sieve element transport of photoassimilates in sink regions', *Journal of Experimental Botany* **47**, 1165–1177.

Pearce, G., Strydom, D., Johnson, S. and Ryan, C.A. (1991) 'A polypeptide from tomato leaves induces wound-inducible proteins', *Science* **213**, 895–898.

Pickard, B.G. (1971) 'Action potentials resulting from mechanical stimulation of pea epicotyls', *Planta* **97**, 106–115.

Pickard, B.G. (1974) 'Electrical signals in higher plants', *Naturwissenschaften* **61**, 60–64.

Reid, R.J. and Overall, R.L. (1992) 'Intercellular communication in *Chara:* factors affecting transnodal electrical resistance and solute fluxes', *Plant Cell and Environment* **15**, 507–517.

Rhodes, J.D., Thain, J.F. and Wildon, D.C. (1996) 'The pathway for systemic electrical signal conduction in the wounded tomato plant', *Planta* **200**, 50–57.

Ryan, C.A. and Moura, D.S. (2002) 'Systemic wound signalling in plants: a new perception', *Proceedings of the National Academy of Sciences USA* **99**, 6519–6520.

Samejima, M. and Sibaoka, T. (1982) 'Membrane potentials and resistances of excitable cells in the petiole and the main pulvinus of *Mimosa pudica*', *Plant and Cell Physiology* **23**, 459–465.

Satter, R.L. and Galston, A.W. (1981) 'Mechanism of control of leaf movements', *Annual Review of Plant Physiology* **32**, 83–110.

Seo, S., Okamoto, M., Seto, H., Ishikuza, K., Sano, H. and Ohashi, Y. (1995) 'Tobacco MAP kinase: a possible mediator in wound signal transduction pathways', *Science* **270**, 1988–1992.

Shiina, T. and Tazawa, M. (1986) 'Action potential in *Luffa cylindrica* and its effects on elongation growth', *Plant and Cell Physiology* **27**, 1981–1989.

Sibaoka, T. (1962) 'Excitable cells in *Mimosa*', *Science* **137**, 226–227.

Sibaoka, T. (1966) 'Action potentials in plant organs', *Symposia of the Society of Experimental Biology* **20**, 49–74.

Simons, P. (1992) *The Action Plant. Movement and Nervous Behaviour in Plants*, Blackwell Publishing, Oxford.

Sinyukin, A.M. and Britikov, E.A. (1967) 'Action potentials in the reproductive system of plants', *Nature* **215**, 1278–1280.

Spanswick, R.M. (1972) 'Electrical coupling between cells of higher plants: a direct demonstration of intercellular communication', *Planta* **102**, 215–227.

Spanswick, R.M. and Costerton, J.W.F. (1967) 'Plasmodesmata in *Nitella translucens*: structure and electrical resistance', *Journal of Cell Science* **2**, 451–464.

Stankovic, B. and Davies, E. (1997) 'Intercellular communication in plants: electrical stimulation of proteinase inhibitor gene expression in tomato', *Planta* **202**, 402–406.

Tangl, E. (1879) 'Über offene Kommunikationen zwischen den Zellen des Endosperms einiger Samen', *Jahrbuch der Wissenschaften Botanik* **12**, 170–190.

Terry, B.R. and Robards, A.W. (1987) 'Hydrodynamic radius alone governs the mobility of molecules through plasmodesmata', *Planta* **171**, 145–157.

Tucker, E.B. and Tucker, J.E. (1993) 'Cell-to-cell diffusion selectivity in staminal hairs of *Setcreasea purpurea*', *Protoplasma* **174**, 36–44.

van Bel, A.J.E. (1990) 'Symplastic isolation of the sieve element-companion cell complex in the phloem of *Ricinus communis* and *Salix alba* stems', *Planta* **183**, 69–76.

van Bel, A.J.E. (1996) 'Interaction between sieve element and companion cell and the consequences for photoassimilate distribution. Two structural hardware frames with associated physiological software packages', *Journal of Experimental Botany* **49**, 1129–1140.

van Bel, A.J.E. (2003) 'The phloem, a miracle of ingenuity', *Plant, Cell and Environment* **26**, 125–149.

van Bel, A.J.E. and van Rijen, H.V.M. (1994) 'Microelectrode-recorded development of symplasmic autonomy of the sieve element/companion cell complex in the stem phloem of *Lupinus luteus*', *Planta* **192**, 165–175.

van der Schoot, C. and van Bel, A.J.E. (1989) 'Glass microelectrode measurements of sieve tube membrane potentials in internode discs and petiole strips of tomato (*Solanum lycopersicum* L.)', *Protoplasma* **149**, 144–154.

van der Schoot, C. and van Bel, A.J.E. (1990) 'Mapping membrane potential differences and dye coupling in internodal tissues of tomato (*Solanum lycopersicum* L.)', *Planta* **182**, 9–21.

van Rijen, H.V.M., Wilders, R. and Jongsma, H.J. (1999) Electrical coupling, in *Plasmodesmata. Structure, Function, Role in Cell Communication* (eds A.J.E. van Bel and W.J.P. van Kesteren), Springer, Berlin, Heidelberg, New York, pp. 51–65.

Wildon, D.C., Thain, J.F., Minchin, P.E.H. *et al.* (1992) 'Electrical signalling and systemic proteinase inhibitor induction in the wounded plant', *Nature* **360**, 62–65.

Wille, A.C. and Lucas, W.J. (1984) 'Ultrastructural and histochemical studies on guard cells', *Planta* **160**, 129–142.

Williams, S.E. and Pickard, B.G. (1972a) 'Properties of action potentials in *Drosera* tentacles', *Planta* **103**, 222–240.

Williams, S.E. and Pickard, B.G. (1972b) 'Receptor potentials and action potentials in *Drosera* tentacles', *Planta* **103**, 193–221.

Williams, S.E. and Pickard, B.G. (1974) 'Connections and barriers between cells of *Drosera* tentacles in relation to their electrophysiology', *Planta* **116**, 1–16.

Yoo, B.-C., Lee, J.-Y. and Lucas, W.J. (2002) 'Analysis of the complexity of protein kinases within the phloem sieve tube system', *Journal of Biological Chemistry* **277**, 15325–15332.

Zawadski, T. (1980) 'Action potentials in *Lupinus angustifolius* L. shoots. V Spread of excitation in the stem, leaves and root', *Journal of Experimental Botany* **31**, 1371–1377.

# 13 Regulation of plasmodesmal conductance

Terena L. Holdaway-Clarke

## 13.1 Introduction

In the preceding chapters, much has been written about the targeted movement of macromolecules from cell to cell by plasmodesmata. This chapter deals with the regulation of movement of ions and small molecules through these intercellular channels, as measured by the intercellular electrical conductance (or resistance) and the diffusion of small dyes from cell to cell. A recent review by Roberts and Oparka (2003) covered the regulation of both diffusional and facilitated transport through plasmodesmata. However, electrophysiological techniques for measuring plasmodesmal conductance have not been adequately reviewed in recent times. Thus, this chapter will present a more expansive treatment of electrophysiological techniques for studying plasmodesmata, with briefer notes on other approaches, and an emphasis on more recent findings. When teamed with proteomic approaches, the combination of electrophysiology, dye studies and electron microscopy may prove crucial in elucidating more precisely the functional role of specific proteins in controlling the plasmodesmal size exclusion limit (SEL).

## 13.2 Monitoring plasmodesmal conductance

### *13.2.1 Terminology*

'Conductance' is primarily an electrical property of a substance – how well it conducts electricity – but it can also be used in a wider sense to mean the act of being a conduit, or passageway through which substances may flow or diffuse. Both of these definitions can apply to plasmodesmata, for they are indeed intercellular conduits for diffusion of a wide variety of uncharged molecules, but they are also capable of conducting ions from cell to cell. The electrical resistance (or conductance, the mathematical inverse of resistance) of any material is dependent on the cross-sectional area through which current is passed; doubling the cross-sectional area of a conductor doubles its electrical conductance and halves its electrical resistance. As a result, the dimensions of the cytoplasmic sleeve, as measured by electron microscopy or indicated by dye movement studies, should relate directly to the ease with which ions may move from cell to cell via the cytoplasmic sleeves of plasmodesmata, and vice versa.

The electrical conductance of plasmodesmata is a product of their cross-sectional area and the number of ions available to carry charge. Since the ionic strength of the cytoplasm and the cytoplasmic annulus (likely to be dominated by the 100 mM [$K^+$] of the cytoplasm) are unlikely to change markedly, the electrical resistance of plasmodesmata is an indication of how easily ions move in response to the 'push' of an applied electric current, but will also reflect the ease with which molecules smaller than the SEL of the plasmodesmata will diffuse from cell to cell. Since there may be many plasmodesmata between two neighbouring cells, the electrical resistance of the cell-to-cell junction will be dependent on the total cross-sectional area of the cytoplasmic sleeves of all plasmodesmata, i.e. upon the number of plasmodesmata present and the cross-sectional area of the cytoplasmic annulus of each.

### *13.2.2 Monitoring plasmodesmata as diffusion conduits*

Crawford and Zambryski (2000) have suggested that plasmodesmata fluctuate between three possible states: closed, open and dilated (an extended-open conformation). Different techniques are most appropriate for investigating plasmodesmata undergoing transitions between various states. Tracing of the intercellular movement of labelled macromolecules, larger than the commonly accepted 'basal' SEL of 1 kDa (Tucker, 1982; Erwee and Goodwin, 1983, 1984; Goodwin, 1983), including GFP-fusion proteins, can only indicate whether plasmodesmata are in the dilated state or not. Passage of smaller dye molecules indicate open plasmodesmata. In order to determine the state of plasmodesmata in a cell using dye techniques, it is necessary to use a 'multi-probe' approach and observe the movement of both small (<1 kDa) and larger molecules simultaneously (e.g. Crawford and Zambryski, 2000). Electrophysiological studies can monitor changes in intercellular conductance over time – from open or dilated to closed – but notably this technique is not ideal for revealing shifts in plasmodesmatal state from open to dilated or vice versa (see Section 13.2.6 for reasons). Thus, monitoring of dye movement and electrophysiological investigations give different insights into the function of plasmodesmata as intercellular conduits.

### *13.2.3 Transport through the desmotubule*

Although the cytoplasmic sleeve has been most investigated as the path by which the bulk of cell-to-cell movement occurs, there are numerous reports of intercellular transport via the desmotubule. Grabski *et al.* (1993) observed intercellular movement of labeled lipids in the endoplasmic reticulum (ER), but not in the plasma membrane (PM), indicating that at least some molecules can diffuse along the lipids of the desmotubule. Small dye molecules injected or taken up into the endomembrane system, possibly ER, move to neighbouring cells via the desmotubule (Lazzaro and Thomson, 1996; Cantrill *et al.*, 1999). Sugar transport has also been reported to occur through the lumen of the desmotubule (Gamalei *et al.*, 1994). Notably, Crawford and Zambryski (2000) found that GFP localized to the ER did not diffuse to neighbouring cells, suggesting that the desmotubule may not be capable of dilating sufficiently to accommodate macromolecules.

### *13.2.4 Dye and GFP-fusion studies*

Initially, diffusion and iontophoresis were used to get small dyes (<1 kDa) inside plant cells to monitor intercellular communication (Tucker, 1982, 1988, 1990a, 1990b, 1993; Tucker and Spanswick, 1985; Erwee *et al.*, 1985; Erwee and Goodwin, 1983, 1984, 1985). Later, pressure injection techniques were used to cytoplasmically load macromolecules (e.g. Ding, 1997). However, Storms *et al.* (1998) have reported different results depending on whether iontophoresis or pressure injection was used to load the dyes, and suggested that the force of pressure injection could 'blow' plasmodesmata open so that larger molecules move intercellularly. Non-invasive transfection systems employing microprojectile bombardment (Oparka *et al.*, 1999; Crawford and Zambryski, 2000) or leaf-dip (Wu *et al.*, 2003) for intercellular transport studies have also been used. Yet another new method of uncaging fluorescent probes *in situ* (Martens *et al.*, 2004) seems to be an attractive tool for future studies on cell-to-cell movement of macromolecules. With plasmodesmata capable of changing their SEL in response to various stimuli, clearly the less invasive a system is for measuring plasmodesmata, the better. Techniques for imaging intercellular transport are dealt with in Chapter 11.

It now seems that macromolecular trafficking and diffusion cannot be entirely separated, since dilating of plasmodesmata by viral movement proteins (MPs) also increases the SEL for diffusion, and some macromolecules, including a transcription factor, LEAFY, appear to move intercellularly by diffusion alone (Crawford and Zambryski, 2000; Wymer *et al.*, 2001; Wu *et al.*, 2003). In the *Arabidopsis* apex, the SEL of the secondary plasmodesmata connecting tissue layers is greater than 74 kDa, which is the size of GFP fused to the floral identity transcription factor, LEAFY (Wu *et al.*, 2003). This protein seems to move through the secondary plasmodesmata between non-clonally related cells by diffusion, rather than by targeted trafficking. It would appear that LEAFY does not have any specific movement domains or, if it does, that there are redundant movement signals, as various truncations of LEAFY fused to GPF all show intercellular movement.

### *13.2.5 Inferences from electron microscopy*

Electron micrographs can give some clues as to the conductivity of plasmodesmata. For example, Overall and Gunning (1982) used measurements of the dimensions of the cytoplasmic sleeve to calculate the theoretical resistance of a single plasmodesma, which was two orders of magnitude greater than the actual measured electrical resistance (see Table 13.1). Some powerful ultrastructural studies have linked altered permeability of plasmodesmata with ultrastructural changes. For example, increased SEL of plasmodesmata treated with an actin-depolymerizing drug corresponded to

**Table 13.1** Measurements and estimates of plasmodesmal resistance and junctional conductance

| Tissue type | Electrical resistance of a single plasmodesma (G Ω) | Specific resistance (Ω · m$^2$) or conductance (Sm$^{-2}$) of shared wall | References |
|---|---|---|---|
| *Zea mays* suspension culture sister-cell pairs | 53 ± 11 | 0.038 ± 0.011 Ω · m$^2$<br>26.32 Sm$^{-2}$ | Holdaway-Clarke *et al.*, 1996 |
| *Azolla* root: | | | Overall and Gunning, 1982 |
| measured | 440–840 | | |
| theoretical | 2.05–2.85 | | |
| *Elodea* leaf | | 0.91 × 10$^{-6}$ Ω · m$^2$<br>1 × 10$^6$ Sm$^{-2}$ | Spanswick, 1972 |

wider funnels at the base of plasmodesmata (White *et al.*, 1994; Ding *et al.*, 1996). In addition, callose inhibition results in differently shaped plasmodesmata that are more open at the neck (Radford *et al.*, 1998), while over-expression of the callose-degrading enzyme β-1-3-glucanase results in increased tobacco mosaic virus (TMV) spread in tobacco plants, indicating a role for callose in the regulation of plasmodesmata (Bucher *et al.*, 2001).

### *13.2.6 Measuring plasmodesmal conductance/electrical resistance*

As shown in the preceding chapter (Chapter 12), electrical signals can, and do move via plasmodesmata, so the capacity of plasmodesmata to conduct electric current will determine the extent to which electrical signals may propagate from cell to cell. Electrophysiological measurements of intercellular resistance may also provide information on how fast changes in plasmodesmal conductance occur, as measurements have been made as often as every 5 s, and over a period of up to 30 min (Holdaway-Clarke *et al.*, 1996). Measurements such as these have shown that plasmodesmal resistance may increase significantly in a matter of seconds in response to externally applied treatments (Holdaway-Clarke *et al.*, 2000). These short-term changes are not easily measured using other methods.

The resistance of intercellular junctions is determined by putting electrodes on both sides of the junction, and measuring changes in voltage across the PM due to the injected current. The amount of current that passes through the junction is determined by the number, and individual conductance, of the plasmodesmata therein. Since membranes are non-polar, insulating structures, ions may move most freely through the non-membranous parts of plasmodesmata, i.e. the cytoplasmic annulus and possibly the lumen of the desmotubule. In the only instance in which PM resistivity and that of a cellular junction has been measured, the resistivity of plasmodesmata was less than one-tenth that of the PM (Holdaway-Clarke *et al.*, 1996).

While electrophysiological approaches are most effective at measuring changes in relatively high-resistance (low-conductance) situations, they become less reliable the lower the measured resistance becomes. Electrical resistances are high when plasmodesmata are closed and low when they are open; hence, measurements of low-junctional resistance become dominated by noise as the measured resistance approaches zero (i.e. as electrical conductance increases).

### *13.2.7 Time resolution*

The greatest advantage of electrophysiological approaches in the study of plasmodesmata is that they are able to monitor fast changes in junctional resistance between cells over tens of minutes in a clearly quantitative way. Although the kinetics of dye movement have, in some instances, been quantified (Tucker and Spanswick, 1985; Tucker, 1993; Tucker and Tucker, 1993), different techniques such as fluorescence recovery after photobleaching (FRAP) are required to determine a change in the SEL of plasmodesmata from one point in time to another.

### *13.2.8 Difficulties and some solutions*

In order for plasmodesmal resistance to be properly calculated, all of the possible paths the injected current may take to get to ground must be known, and the magnitude of all other resistances in the system, such as those of the PM and tonoplast, must be measured. The complex cellular geometries of most plants have thus made measurements of the electrical properties of plasmodesmata difficult. Apart from the *Azolla* root tip system used by Overall and Gunning (1982), there are only two simple cell geometries in which it is possible to solve for junctional resistance by impaling only two cells: a two-cell system and a single-ended file (see Figures 13.1 and 13.2). The mathematical solutions for these cellular geometries are given in the Appendix. In these cell systems, it is necessary to determine the cellular compartment into which the current-injecting electrode has been inserted. This may be done by the iontophoretic injection of dye by the current-injecting electrode (Holdaway-Clarke *et al.*, 1996).

### *13.2.9 Coupling ratio*

For experimental systems in which it is not possible to determine all the paths the injected current may take, the coupling ratio has been used as a measure of intercellular communication (see Table 13.2). This is the ratio of the voltage response to an injected current in one cell to the voltage response in a neighbouring cell distal to the point of injection. Because the coupling ratio is dependent not only on plasmodesmal conductance, but also on the resistivity of the PM, it is really only a 'second-best' measure of intercellular communication. For example, early work by Drake (1979) indicated that metabolic inhibition by azide or cyanide decreased electrical coupling of oat coleoptiles, but this could have been a result of changes in membrane resistance which would alter the proportion of current moving through plasmodesmata, rather than changes in plasmodesmal resistance itself.

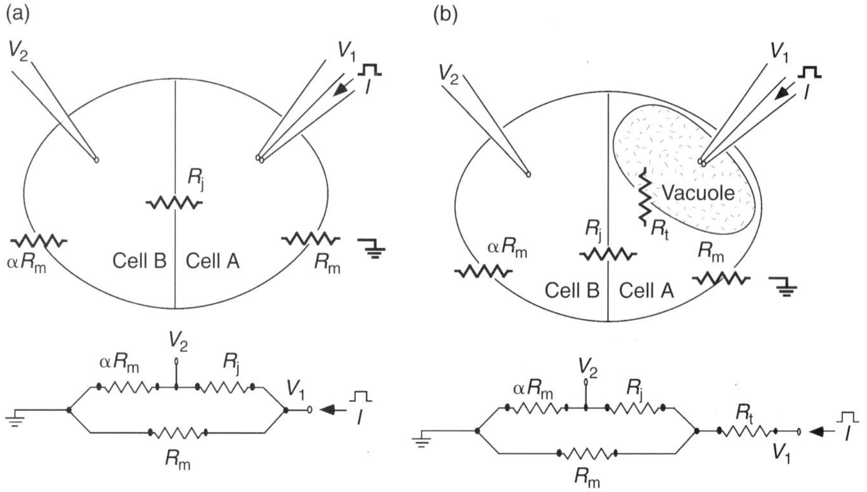


**Figure 13.1** The two circuit models for two plant sister-cells. The electrode tip may be on either side of the vacuolar membrane. (a) Cytoplasmic impalement: the injected current may take two different paths to ground; either via the plasma membrane (PM) of cell A or through the plasmodesmata at the cell junction and then the PM of cell B. (b) When the electrode is in the vacuole, all current must traverse the tonoplast before entering the cytoplasm and taking a path to ground.

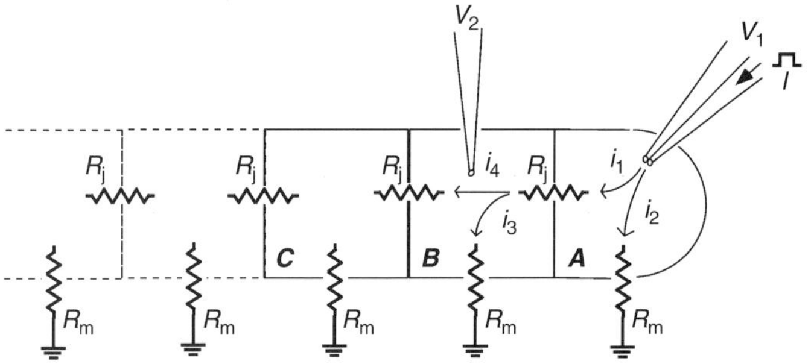


**Figure 13.2** Circuit model of a single-ended 'infinite' file of cells in which the current injecting electrode is in the cytoplasm of the terminal cell. $I$, $i_1$, $i_2$, $i_3$, $i_4$ are the magnitude of current pulses that pass, respectively, down the electrode, from cell A to cell B through plasmodesmata, from cell A to ground via the plasma membrane (PM), from cell B to ground via the PM and from cell B to cell C via plasmodesmata.

### *13.2.10 Patch-clamping?*

van Rijen and Wilders (1999) suggested that cell-attached patch-clamping of sister-cell pairs may be the most elegant way to measure the electrical properties of the plasmodesmata between them, but there are major physical obstacles in the

**Table 13.2** Measurements of coupling ratios

| Tissue type | Coupling ratio | References |
|---|---|---|
| *Elodea* leaf | $0.2 \pm 0.04$ | Spanswick, 1972 |
| *Tiranea bogotensis* roots | 0.2–0.8 | Vorob'ev *et al.*, 1981 |
| *Azolla* roots | 0.1–0.33 | Gunning and Overall, 1983 |
| Wheat roots, cortical cells | $0.059 \pm 0.06$ (SD) | Zhang and Tyerman, 1991 |
| *Arabidopsis* root hairs | 0–0.9, average = 0.32<br>$\leq 0.05$ | Lew, 1994;<br>Meharg *et al.*, 1994 |
| *Zea mays* suspension culture sister cells | $0.84 \pm 0.04$ | Holdaway-Clarke *et al.*, 1996 |

application of this technique. The major problem is that of accessing the PM. Most plant patch-clamping studies are done on cell patches removed from protoplasts, although there are a few cases in which the PM has been accessed by microsurgery on the cell wall (Laver, 1991; Hendriksen and Assmann, 1997; Miedema *et al.*, 1999). Such microsurgery on the cell wall, however, requires plasmolysis, which may interfere with the structure and function of plasmodesmata, at least in the short term (Drake *et al.*, 1978; Oparka *et al.*, 1994). So far, it appears that this method has not been successfully employed to measure the electrical conductivity of plasmodesmata.

## 13.3 Regulation of plasmodesmal conductance

Dye studies have yielded the bulk of information on the factors regulating diffusion of small molecules through plasmodesmata, but more recent electrophysiological studies have shown that plasmodesmata can open and close very quickly indeed in response to fast changes in cytoplasmic calcium ($[Ca^{2+}]_{cyt}$) (Holdaway-Clarke *et al.*, 2000). Results from electrophysiological and dye-transport studies indicate that not all plasmodesmata have the same SEL, or mechanisms of regulation. Structurally and functionally, algal plasmodesmata differ from their higher plant counterparts, and primary and secondary plasmodesmata also have different transport properties, as do the various specialized plasmodesmata discussed in Chapter 4.

### *13.3.1 Algal plasmodesmata*

The size of the giant Characean cells makes them ideal for transport studies using radio-labelled ions and electrophysiological investigations, but Characean plasmodesmata are structurally different from those of higher plants. In general, they consist of PM-lined pores (Robards, 1971; Bostrom and Walker, 1975), with desmotubules (Spanswick and Costerton, 1967) or ER being seen occasionally (Cook and Graham, 1993). As a result of these structural differences, charophyte plasmodesmata may be regulated differently from those of higher plants. A summary of factors influencing cell-to-cell communication in algal cells is presented in Table 13.3. The role of plasmodesmatal coupling in algal cell differentiation has been reviewed by Kwiatkowska (1999).

### *13.3.2 Plasmodesmata in higher plants*

Plasmodesmata are regulated in plant development and morphogenesis (see Chapter 5). Some regulation consists of insertion of plasmodesmata in walls where there were none, such as in gynoecium development (van der Schoot *et al.*, 1995) or reduction in the numbers of plasmodesmata, as occurs between tissue layers in early floral development of *Iris* (Bergmans *et al.*, 1997) or in stomatal development (Palevitz and Hepler, 1985). Plasmodesmata are also dynamic, and change their conductivity over time as required by their developmental program. For example, cells in different stages of mitosis are transiently uncoupled (Ehlers and Kollmann, 1996), while cells in the shoot apical meristem that are well connected to each other are all similar metabolically and together constitute a morphogenic field (Rinne and van der Schoot, 1998). In regenerating tobacco thin cell layers, the extent of cell-to-cell communication increases in the initial stages of culture and redifferentiation, and cycles up and down within the first week of growth in newly formed shoots (Cantrill *et al.*, 2001). At present, it is not known what factors are involved in uncoupling cells at various points in development. However, several cellular

**Table 13.3** Factors regulating plasmodesmata in algal cells

| Treatment | Effect on plasmodesmata | References |
|---|---|---|
| Turgor pressure gradient | Nodal resistance increase (Charophyte) | Coté *et al.*, 1987; Ding and Tazawa, 1989; Reid and Overall, 1992 |
| Increasing $[Ca^{2+}]_{cyt}$ | No effect (*Chara* whorl cells) | Reid and Overall, 1992 |
| | Decreased communication (*Chara corallina*) | Shepherd and Goodwin, 1992 |
| Increasing $pH_{cyt}$ | No effect (*Chara* whorl cells) | Reid and Overall, 1992 |
| Decreasing $pH_{cyt}$ | Reduced transport of ions (Charophyte) | Trebacz *et al.*, 1988; Reid and Overall, 1992 |
| Reduced ATP | Reduced transnodal transport (Charophyte) | Williams and Fensom, 1975; Reid and Overall, 1992 |
| Transnodal potential difference | No effect (Charophyte) | Reid and Overall, 1992 |
| Synchronously dividing cells | Plasmodesmata 'unplugged' (*Chara*) | Kwiatkowska and Maszewski, 1986 |
| Asynchronously dividing cells or cells of different types | Plasmodesmata 'plugged' (*Chara*) | Kwiatkowska and Maszewski, 1986 |

parameters are known to be able to reduce intercellular communication, including $[Ca^{2+}]_{cyt}$, plasmolysis and pressure differentials across the cell-to-cell junction, while metabolic inhibition appears to increase symplasmic coupling. The regulation of plasmodesmata involved in phloem loading, which appears to be regulated by source and sink strengths, probably by mechanisms involving pressure differentials between cells, has been reviewed by Schulz (1999).

### *13.3.3 Factors decreasing conductance*

#### *13.3.3.1 Divalent cations/$[Ca^{2+}]_{cyt}$*

For over 20 years, elevated $[Ca^{2+}]_{cyt}$ has been known as a factor that can reduce plasmodesmal permeability (see Table 13.4, and references therein). Many different methods have been used both to raise the intracellular concentrations of the ion including direct, iontophoretic injection of $Ca^{2+}$ (Erwee and Goodwin, 1983, 1984; Lew, 1994; Holdaway-Clarke *et al.*, 2000), use of calcium iontophore (Tucker, 1990a) and injection of $Ca^{2+}$-loaded BAPTA buffer (Tucker, 1990a; Lew, 1994). Indirect methods of increasing $[Ca^{2+}]_{cyt}$ have also been used such as treatment with mastoporan (Tucker and Boss, 1996), injection of inositol polyphosphates (Tucker, 1988; Lew, 1994) and cold (Holdaway-Clarke *et al.*, 2000). Whether monitored by dye passage or electrophysiological methods, the results are fairly consistent for higher plant plasmodesmata – that elevated $[Ca^{2+}]_{cyt}$ reduces cell-to-cell coupling and SEL. One exceptions is found in the study by Lew (1994) which concluded that iontophoretic injection of $InsP_3$ or $Ca^{2+}$-BAPTA had no effect on the electrical coupling of *Arabidopsis* root hairs. Another exception was found in experiments with butyric acid, which both acidifies the cytoplasm and raises $[Ca^{2+}]_{cyt}$ and yet has no observable effect on junctional resistance (Holdaway-Clarke *et al.*, 2001). This result indicates that elevations in $[Ca^{2+}]_{cyt}$ and cytoplasmic acidification may have antagonistic effects on plasmodesmata (for further discussion, see Section 13.4.2 on centrin and $Ca^{2+}/pH_{cyt}$ regulation of plasmodesmata).

Unfortunately, very few investigations have quantified the changes in $[Ca^{2+}]_{cyt}$ that occur with treatments which alter plasmodesmal permeability. Our work with corn suspension cells showed that even physiological increase in $[Ca^{2+}]_{cyt}$, such as those induced by exposure to cold, which raises average $[Ca^{2+}]_{cyt}$ from $107 \pm 16$ to $202 \pm 32$ nM, has a significant effect on the calculated resistance of a single plasmodesma, increasing it from $47 \pm 21$ to $193 \pm 80$ GΩ (Holdaway-Clarke *et al.*, 2000). This increase in plasmodesmal resistance by a factor of four indicates that the cross-sectional area of the cytoplasmic annulus decreases to one-fourth of its pre-treatment area with a doubling in $[Ca^{2+}]_{cyt}$.

Plasmodesmata appear to respond very quickly to elevated $[Ca^{2+}]_{cyt}$, as plasmodesmata that are open (low junctional resistance) prior to iontophoresis of $Ca^{2+}$ have been observed to close instantly (showing high junctional resistance) during iontophoresis of $Ca^{2+}$, and open again once iontophoresis has ceased (Holdaway-Clarke *et al.*, 2000). Notably, iontophoresis of $K^+$ did not produce similar increases in junctional resistance. This increase in resistance does not appear

**Table 13.4** Factors influencing plasmodesmal conductance in higher plants

| Factor | Effect | Technique | References |
|---|---|---|---|
| *Elevated calcium* | | | |
| Iontophoresis of $Ca^{2+}$ | Decreased SEL | Dye iontophoresis | Erwee and Goodwin, 1983, 1984 |
| Iontophoresis of $Ca^{2+}$ | Fast increase in $R_j$ during iontophoresis | Electrophysiology | Holdaway-Clarke *et al.*, 2000 |
| Diffusion of $Ca^{2+}$ from electrode | Small increase in $R_j$ | Electrophysiology | Holdaway-Clarke *et al.*, 2000 |
| $Ca^{2+}$ iontophoresis to non-physiological levels | Decreased coupling | Electrophysiology | Lew, 1994 |
| Cold (increases $[Ca^{2+}]_{cyt}$) | Plasmodesmal resistance doubles | Electrophysiology | Holdaway-Clarke *et al.*, 2000 |
| $Ca^{2+}$ -loaded BAPTA | Reduced transport | Dye iontophoresis | Tucker, 1990a |
| Ionophore A23187 | Reduced transport | Dye iontophoresis | Baron-Epel *et al.*, 1998; Tucker, 1990a |
| $IP_2$, $IP_3$ | Reduced transport | Dye iontophoresis | Tucker, 1988 |
| Mastoporan (increases $[Ca^{2+}]_{cyt}$) | Reduced transport | Dye iontophoresis | Tucker and Boss, 1996 |
| *Decreasing $pH_{cyt}$* | | | |
| Iontophoresis of $H^+$ | Decreased coupling | Electrophysiology | Lew, 1994 |
| Butyric acid | No effect* | Electrophysiology | Holdaway-Clarke *et al.*, 2001 |
| *Metabolic inhibition* | | | |
| Anaerobiosis, azide | Increased SEL | Dye studies | Cleland *et al.*, 1994 |
| Azide | Increased diffusion and SEL | Dye studies | Tucker, 1993 |
| Hypoxia | No effect on coupling | Electrophysiology | Zhang and Tyerman, 1997 |
| *Plasmolysis* | | | |
| Mannitol | Uncoupling | Dye iontophoresis | Drake *et al.*, 1978; Erwee and Goodwin, 1984 |
| Deplasmolysis | Increased SEL | Dye iontophoresis | Erwee and Goodwin, 1984 |
| *Pressure differentials* | Reduction in transport dependent on magnitude of pressure differential | Dye and micropressure probe | Oparka and Prior, 1992 |
| *Developmental stage* | | | |
| Asynchronous division of neighbours | Uncoupling | Dye studies | Ehlers and Kollman, 1996 |

* Note that $[Ca^{2+}]_{cyt}$ increased concomitantly with acidification of the cytoplasm in these experiments.

to stop $Ca^{2+}$ from moving through plasmodesmata as we have observed waves of elevated $Ca^{2+}$ moving from cell-to-cell by loading neighbouring *Tradescantia* stamen hair cells with FURA-2 dextran and injecting $Ca^{2+}$ by iontophoresis into one of the cells (Holdaway-Clarke and Hepler, unpublished results). This is an interesting result as Tucker and Tucker (1993) concluded from their experiments that cations should not move easily through plasmodesmata.

#### *13.3.3.2 Plasmolysis and pressure differentials*

Plasmolysis has been shown to disrupt plasmodesmata so that cell-to-cell transport of injected dyes no longer occurs (Erwee and Goodwin, 1984), but recovery occurs within 10 min of return to normal bathing medium. During plasmolysis, the cytoplasmic annulus and desmotubule become constricted such that dye transport is no longer possible (Oparka *et al.*, 1994).

Work by Tilney *et al.* (1990) demonstrated that strong plasmolysis may break plasmodesmatal connections between cells, and Drake *et al.* (1978) showed that electrical coupling between cells that had been plasmolyzed then deplasmolyzed did not return for at least 1 h. Further, Storms *et al.* (1998) found that pressure injection of dyes can increase the SEL of plasmodesmata in the presence of viral MP and suggest that this may be a result of the pressure injection itself. Plasmodesmata not exposed to MP did not exhibit any alterations in SEL as a result of pressure injection, but this does indicate that pressure can act upon the very structure of plasmodesmata.

### *13.3.4 Factors increasing plasmodesmal conductance*

#### *13.3.4.1 Deplasmolysis*

Erwee and Goodwin (1984) found that in *Egeria*, the SEL of plasmodesmata markedly increases following plasmolysis and deplasmolysis of the tissue: from 665 to 1678 Da, allowing transport of larger dye conjugates that were blocked from intercellular transport under control conditions. They also report that following plasmolysis and deplasmolysis, plasmodesmata are relatively insensitive to elevations in $[Ca^{2+}]_{cyt}$, with iontophoretic injection of $Ca^{2+}$ merely restricting intercellular dye movement rather than blocking it completely (Erwee and Goodwin, 1984).

#### *13.3.4.2 Metabolic inhibition*

Metabolic inhibition has been shown to increase plasmodesmal permeability in a number of cases. For example, azide increased both the SEL and cell-to-cell diffusion coefficient of variously sized small fluorescent probes in staminal hairs of *Stetcreasea purpurea* (Tucker, 1993). Cleland and co-workers (1994) described an increase in the SEL of plasmodesmata of wheat root cortical or epidermal cells from less than 1 kDa to between 5 and 10 kDa after 5–30 min of metabolic inhibition induced by either azide or $N_2$. Interestingly, short-term oxygen deficiency had no effect on the coupling ratio of wheat root cortical cells or their membrane resistance (Zhang and Tyerman, 1997) indicating that, in this case, plasmodesmal conductivity was not affected, but this may have been due to the short-term nature of the oxygen deficiency which may not have allowed the ATP level to fall below that critical for control of plasmodesmata. These authors had earlier suggested that anaerobiosis might result in dilation of plasmodesmata as they found that the hydraulic conductivity of wheat root cortical cells decreased under conditions of low $O_2$ (Zhang and Tyerman, 1991).

Metabolic inhibition has other cytoplasmic effects, such as increasing $[Ca^{2+}]_{cyt}$ (Gilroy *et al.*, 1989). Although elevated $[Ca^{2+}]_{cyt}$ is associated with restriction of plasmodesmal aperture, it appears that the effect of reduced ATP overrides any effect of raised $[Ca^{2+}]_{cyt}$. Alternatively, since $pH_{cyt}$ is also lowered in response to inhibition of metabolism (Reid *et al.*, 1985; Felle, 1987), it could be that any possible antagonistic effects of raised $[Ca^{2+}]_{cyt}$ and lowered $pH_{cyt}$ would also come into play in the situation of lowered ATP, resulting in no restriction of the SEL, but rather a dilation.

### *13.3.5 Other possibilities*

#### *13.3.5.1 Cytoplasmic pH*

It is unclear whether cytoplasmic pH ($pH_{cyt}$) is a regulator of plasmodesmata. Electrophysiological investigations by Lew (1994) showed that changing $pH_{cyt}$ by iontophoretic injection of $H^+$ or $OH^-$ causes a small decrease in the electrical coupling between root hairs of *Arabidopsis*, but the magnitude of the changes in $pH_{cyt}$ were not measured. However, Lew observed obvious cellular damage as a result of iontophoresis of $H^+$ and this may be the reason for decrease in coupling ratio rather than lowered cytoplasmic pH per se. Acidification of the cytoplasm by 0.8 units with butyric acid had no effect on the electrical resistance of plasmodesmata despite a concomitant increase in $[Ca^{2+}]_{cyt}$ to levels that had previously been shown to effect an increase in plasmodesmal resistance (Holdaway-Clarke *et al.*, 2001). In many circumstances, $pH_{cyt}$ and $[Ca^{2+}]_{cyt}$ appear to be interregulated (Felle, 1988a,b; Suzaki *et al.*, 1996) but it seems that $pH_{cyt}$ and $[Ca^{2+}]_{cyt}$ may act antagonistically on plasmodesmata.

#### *13.3.5.2 Voltage gating*

Voltage differences imposed across cell-to-cell junctions in animal cells may close gap junction channels in a fraction of a second (Spray *et al.*, 1979). Small voltage differences, such as those imposed by iontophoresis and electrophysiological measurements of electrical coupling, do not appear to diminish the conductivity of plasmodesmata. Full investigation of this cellular parameter has yet to be undertaken, and will be crucially relevant to long-distance electrical signalling in plants (Chapter 12).

## 13.4 Mechanisms

While proteomic, immunolocalization and electron microscopy studies are elucidating the structural components of plasmodesmata, functional studies provide insights into the mechanisms by which plasmodesmal conductance may be controlled, and the parameters within which they function. It is helpful to consider current models of plasmodesmal structure in the light of what is known about the regulation of diffusion through plasmodesmata, the factors that influence plasmodesmal conductance and the time taken for regulating factors to take effect.

### *13.4.1 Myosin*

Models by Blackman *et al.* (1999) and Oparka (2004) both have actin spiraling around the desmotubule and myosin attached to both the actin filament and the PM, at least at some stages in transport. As a motor protein, myosin VIII could certainly respond quickly to an appropriate stimulus, and with its calmodulin-binding sites has the potential to respond quickly to changes in $[Ca^{2+}]_{cyt}$, as has been observed (Holdaway-Clarke *et al.*, 2000) and could potentially control diffusion through plasmodesmata. The combination of motor and cytoskeletal components in plasmodesmata that extend out into the cell body make it seem likely that these elements are more closely involved in macromolecular trafficking from cell-to-cell than in the control of intercellular diffusional transport.

### *13.4.2 Centrin and $Ca^{2+}/pH_{cyt}$ regulation of plasmodesmata*

A centrin-like protein has been localized at the necks of higher plant plasmodesmata (Blackman *et al.*, 1999). Centrin is a contractile protein which contracts upon exposure to elevated $[Ca^{2+}]_{cyt}$ (Sanders and Salisbury, 1994). Blackman *et al.* (1999) have proposed two models in which centrin is involved in the regulation of cell-to-cell communication by plasmodesmata. In their first model, centrin nanofilaments connect the ER to the PM, and upon contraction, draw the ER and PM together, restricting the space between the membranes and stopping transport through the cytoplasmic annulus. Their other model has centrin traversing the cytoplasmic neck from PM to PM in a stellate shape, and contraction of the protein drawing the PM inwards against turgor pressure and sealing the cytoplasmic annulus. Models that propose a $Ca^{2+}$-sensitive element at the neck of plasmodesmata are consistent with the findings that plasmodesmata are disrupted at the neck region following plasmolysis (Tilney *et al.*, 1990) and are less responsive to elevated $[Ca^{2+}]_{cyt}$ after plasmolysis and subsequent deplasmolysis than in control situations (Erwee and Goodwin, 1984). One would envisage that any centrin cross-bridges would be broken or dispersed following plasmolysis and may take some time to reform upon deplasmolysis. The involvement of a centrin-like protein in plasmodesmata may make sense of our somewhat confusing finding that acidification of the cytoplasm by external application of butyric acid does not close plasmodesmata despite concomitant increases in $[Ca^{2+}]_{cyt}$ (Holdaway-Clarke *et al.*, 2001). While $Ca^{2+}$ binding is essential for centrin contraction, the role of phosphorylation is less clear. Elevated $[Ca^{2+}]_{cyt}$ induces dephosphorylation of centrin and contraction of flagellar roots in *Tetraselmis striata*, but acid shock, consisting of exposure to an external medium of pH less than 4.0, induces increased centrin phosphorylation, and flagellar roots remain extended (Martindale and Salisbury, 1990). Thus, there is the possibility of elevated $[Ca^{2+}]_{cyt}$ and acidified $pH_{cyt}$ acting antagonistically on centrin-like proteins in plasmodesmata to produce no observable net effect on intercellular resistance.

A kinase associated with cell walls has been shown to be developmentally regulated and its activity synchronized with the maturation of plasmodesmata (Citovsky *et al.*, 1993). This kinase phosphorylates the 30-kDa MP of tobacco mosaic virus, which interacts with plasmodesmata and increases the SEL. Yahalom *et al.* (1998) have also found a $Ca^{2+}$-dependent protein kinase associated with plasmodesmata. Evidence thus points to phosphorylation being involved in the regulation of both macromolecular and diffusional transport through plasmodesmata.

### *13.4.3 How do plasmodesmata dilate?*

Both myosin and centrin are clear possible candidates for reversible mechanisms by which elevated $[Ca^{2+}]_{cyt}$ effects restriction of plasmodesmal aperture from the ‘open’ to the ‘closed’ state. How plasmodesmata dilate to permit intercellular diffusion of macromolecules up to 74 kDa (Wu *et al.*, 2003) has not been definitively established.

Schulz (1999) suggested that the extension of spokes may be responsible for controlling plasmodesmal dilation, but the mechanism seems unclear if the spokes are myosin VIII as has been recently modelled (Oparka, 2004). Myosin spokes are modelled to connect the actin spiraling around the desmotubule to the PM along the length of plasmodesmata (Blackman *et al.*, 1999), and may let go of the membrane to transport macromolecular cargo from cell to cell (Oparka, 2004). The release of myosin from the PM (rather than extension of the myosin spokes per se) could result in the widening of plasmodesmata and increased SEL during facilitated transport as the PM is no longer held tight to the actin filaments spiraling through from cell to cell (see Chapter 1 for more detail of plasmodesmal structure as it relates to the regulation of the SEL).

Metabolic inhibition dilates plasmodesmata but the mechanism by which this is achieved is far from clear. As motor proteins, myosins require ATP to move, and shortage of cellular ATP would see myosin within plasmodesmata become stuck in a rigor-like state. As an ATP-independent protein modification, dephosphorylation of a plasmodesmal protein is likely to be the mechanism by which ATP shortage dilates plasmodesmata and allows ATP to enter from neighbouring cells (Schulz, 1999). If dephosphorylation of myosin VIII caused it to be released from the PM, then this would allow the PM to move further from the desmotubule, thereby increasing the width of the cytoplasmic annulus and thus the SEL.

## 13.5 Conclusion

Both dye studies and electrophysiological investigations have been crucial in elucidating factors that control the diffusion of ions and molecules through plasmodesmata, showing how wide and how quickly plasmodesmal permeability may change. These type of studies, and those using non-invasive transgenic techniques to monitor the movement of molecules from cell to cell, will continue to provide crucial information on the ways in which plasmodesmata function. Thus far, elevated $[Ca^{2+}]_{cyt}$ appears to be the main factor causing plasmodesmata to reversibly close, with callose deposition being the mechanism for more long-term closure, probably initiated by sustained elevated $[Ca^{2+}]_{cyt}$. For dilation of plasmodesmata from the open (SEL $\leq$ 1 kDa) state to a wider configuration permitting diffusion of very large macromolecules, protein dephosphorylation appears to be the likely mechanism.

As more structural components of plasmodesmata come to light, their discovery prompts the asking of further questions as to how they are involved in plasmodesmal function. At present, there are still many gaps in our knowledge of the interaction of facilitated macromolecular trafficking and simple diffusion through plasmodesmata. For example: can plasmodesmata dilated by viral MP still close in response to elevated $[Ca^{2+}]_{cyt}$? Do low-level increases in $[Ca^{2+}]_{cyt}$ have any effect on macromolecular trafficking? Although much data point to the involvement of protein kinases and phosphatases in the regulation of both macromolecular and diffusional transport through plasmodesmata, specific target structures within plasmodesmata have not been conclusively identified. The major challenge in this ever-widening field of plasmodesmal biology, is to bring together data from all types of studies: proteomics, dye studies, EM studies, electrophysiology, and fit them together into a comprehensive and cohesive model of plasmodesmal structure and function.

## References


Baron-Epel, O., Hernandez, D., Jiang, L.W., Meiners, S. and Schindler, M. (1988) 'Dynamic continuity of cytoplasmic and membrane compartments between plant cells', *Journal of Cell Biology* **106**, 715–721.

Bergmans, A.C.J., de Boer, A.D., Derksen, J.W.M. and van der Schoot, C. (1997) 'The symplasmic coupling of $L_2$-cells diminishes in early floral development of *Iris*', *Planta* **203**, 245–252.

Blackman, L.M., Harper, J.D.I. and Overall, R.L. (1999) 'Localisation of a centrin-like protein to higher plant plasmodesmata', *European Journal of Cell Biology* **78**, 297–304.

Bostrom, T.E. and Walker, N.A. (1975) 'Intercellular transport in plants I. The rate of transport of chloride and the electric resistance', *Journal of Experimental Botany* **26**, 767–782.

Bucher, G.L., Tarina, C., Heinlein, M., Di Serio, F., Meins, F. and Iglesias, V.A. (2001) 'Local expression of enzymatically active class I beta-1,3-glucanase enhances symptoms of TMV infection in tobacco', *Plant Journal* **28**, 361–369.

Cantrill, L.C., Overall, R.L. and Goodwin, P.B. (1999) 'Cell-to-cell communication via plant endomembranes', *Cell Biology International* **23**, 653–661.

Cantrill, L.C., Overall, R.L. and Goodwin, P.B. (2001) 'Changes in symplastic permeability during adventitious shoot regeneration in tobacco thin cell layers', *Planta* **214**, 206–214.

Citovsky, V., McLean, B.G., Zupan, J.R. and Zambryski, P. (1993) 'Phosphorylation of tobacco mosaic virus cell-to-cell movement protein by a developmentally regulated plant cell wall associated protein kinase', *Genes and Development* **7**, 904–910.

Cleland, R.E., Fujiwara, T. and Lucas, W.J. (1994) 'Plasmodesmal-mediated cell-to-cell transport in wheat roots is modulated by anaerobic stress', *Protoplasma* **178**, 81–85.

Cook, M.E. and Graham, L.E. (1993) 'Ultrastructure of intercellular connections in *Chara zylanica* as revealed by high pressure freezing', *Journal of Phycology* **29** (Suppl.), 8.

Coté, R., Thain, J.F. and Fensom, D.S. (1987) 'Increase in electrical resistance of plasmodesmata of *Chara* induced by an applied pressure gradient across nodes', *Canadian Journal of Botany* **65**, 509–511.

Crawford, K.M. and Zambryski, P.C. (2000) 'Subcellular localization determines the availability of non-targeted proteins to plasmodesmatal transport', *Current Biology* **10**, 1032–1040.

Ding, B. (1997) 'Cell-to-cell transport of macromolecules through plasmodesmata: a novel signalling pathway in plants', *Trends in Cell Biology* **7**, 5–9.

Ding, B., Kwon, M.-O. and Warnberg, L. (1996) 'Evidence that actin filaments are involved in controlling the permeability of plasmodesmata in tobacco mesophyll', *Plant Journal* **10**, 157–164.

Ding, D.-Q. and Tazawa, M. (1989) 'Influence of cytoplasmic streaming and turgor pressure gradient on the transnodal transport of rubidium and electrical conductance in *Chara corallina*', *Plant Cell Physiology* **30**, 739–748.

Drake, G. (1979) 'Electrical coupling, potentials, and resistances in oat coleoptiles: effects of azide and cyanide', *Journal of Experimental Botany* **30**, 719–725.

Drake, G., Carr, D.J. and Anderson, W.P. (1978) 'Plasmolysis, plasmodesmata, and the electrical coupling of oat coleoptile cells', *Journal of Experimental Botany* **29**, 1205–1214.

Ehlers, K. and Kollmann, R. (1996) 'Regulation of the symplasmic contact between physiologically different cells', *III International Workshop on Basic and Applied Research in Plasmodesmal Biology*, Israel, March 10–16, 1996, 77–81.

Erwee, M.G. and Goodwin, P.B. (1983) 'Characterisation of *Egeria densa* Planch. leaf symplast: inhibition of the intercellular movement of fluorescent probes by group II ions', *Planta* **158**, 320–328.

Erwee, M.G. and Goodwin, P.B. (1984) 'Characterization of the *Egeria densa* leaf symplast: response to plasmolysis, deplasmolysis and to aromatic amino acids', *Protoplasma* **122**, 162–168.

Erwee, M.G. and Goodwin, P.B. (1985) 'Symplast domains in extrastelar tissues of *Egeria densa* Planch.', *Planta* **163**, 9–19.

Erwee, M.G., Goodwin, P.B. and van Bel, A.E.J. (1985) 'Cell–cell communication in the leaves of *Commelina cyanea* and other plants', *Plant, Cell and Environment* **8**, 173–178.

Felle, H. (1987) 'Proton transport and pH control in *Sinapis alba* root hairs: a study carried out with double-barrelled pH micro-electrodes', *Journal of Experimental Botany* **38**, 340–354.

Felle, H. (1988a) 'Auxin causes oscillations of cytosolic free calcium and pH in *Zea mays* coleoptiles', *Planta* **174**, 495–499.

Felle, H. (1988b) 'Cytoplasic free calcium in *Riccia fluitans* L. and *Zea mays* L. : interaction of $Ca^{2+}$ and pH', *Planta* **176**, 248–255.

Gamalei, Y.V., van Bel, A.J.E., Pakhomova, M.V. and Sjutkina, A.V. (1994) 'Effects of temperature on the conformation of the endoplasmic reticulum and on starch accumulation in leaves with the symplasmic minor-vein configuration', *Planta* **194**, 443–453.

Gilroy, S., Hughes, W.A. and Trewavas, A.J. (1989) 'A comparison between quin-2 and aequorin as indicators of cytoplasmic calcium levels in higher plant cell protoplasts', *Plant Physiology* **90**, 482–491.

Goodwin, P.B. (1983) 'Molecular size limit for movement in the symplast of the *Elodea* leaf', *Planta* **157**, 124–130.

Grabski, S., de Feijter, A.W. and Schindler, M. (1993) 'Endoplasmic reticulum forms a dynamic continuum for lipid diffusion between contiguous soybean root cells', *The Plant Cell* **5**, 25–38.

Gunning, B.E.S. and Overall, R.L. (1983) 'Plasmodesmata and cell-to-cell transport in plants', *BioScience* **33**, 260–265.

Hendriksen, G.H. and Assmann, S.M. (1997) 'Laser-assisted patch clamping: a methodology', *Pflügers Archive – European Journal of Physiology* **433**, 832–841.

Holdaway-Clarke, T.L., Walker, N.A., Hepler, P.K. and Overall, R.L. (2000) 'Physiological elevations in cytoplasmic free calcium by cold or ion injection result in transient closure of higher plant plasmodesmata', *Planta* **210**, 329–335.

Holdaway-Clarke, T.L., Walker, N.A. and Overall, R.L. (1996) 'Measurement of the electrical resistance of plasmodesmata and membranes of corn suspension-culture cells', *Planta* **199**, 537–544.

Holdaway-Clarke, T.L., Walker, N.A., Reid, R.J., Hepler, P.K. and Overall, R.L. (2001) 'Cytoplasmic acidification with butyric acid does not alter the ionic conductivity of plasmodesmata', *Protoplasma* **215**, 184–190.

Kwiatkowska, M. (1999) Plasmodesmal coupling and cell differentiation in algae, in *Plasmodesmata. Structure, Function, Role in Cell Communication* (eds A.E.J. van Bel and W.J.P. van Kesteren), Springer-Verlag, Heidelberg, pp. 205–224.

Kwiatkowska, M. and Maszewski, J. (1986) 'Changes in the occurrence and ultrastructure of plasmodesmata in antheridia of *Chara vulgaris* L. during different stages of spermatogenesis', *Protoplasma* **132**, 179–188.

Laver, D. (1991) 'A surgical method for accessing the plasma membrane of *Chara australis*', *Protoplasma* **161**, 79–84.

Lazzaro, M.D. and Thomson, W.W. (1996) 'The vacuolar-tubular continuum in living trichomes of chickpea (*Cicer arietinum*) provides a rapid means of solute delivery from base to tip', *Protoplasma* **193**, 181–190.

Lew, R. (1994) 'Regulation of electrical coupling between *Arabidopsis* root hairs', *Planta* **193**, 67–73.

Martens, H.J., Hansen, M. and Schulz, A. (2004) 'Caged probes: a novel tool in studying symplasmic transport in plant tissues', *Protoplasma* **223**, 63–66.

Martindale, V.E. and Salisbury, J.L. (1990) 'Phosphorylation of algal centrin is rapidly responsive to changes in the external milieu', *Journal of Cell Science* **96**, 395–402.

Meharg, A.A., Maruousset, L. and Blatt, M.R. (1994) 'Cable correction of membrane currents recorded from root hairs of *Arabidopsis thaliana* L.', *Journal of Experimental Botany* **45**, 1–6.

Miedema, H., Henriksen, G.H. and Assmann, S.M. (1999) 'A laser microsurgical method of cell wall removal allows detection of large-conductance ion channels in the guard cell plasma membrane', *Protoplasma* **209**, 58–67.

Oparka, K.J. (2004) 'Getting the message across: how do plant cells exchange macromolecular complexes?', *Trends in Plant Science* **9**, 33–41.

Oparka, K.J. and Prior, D.A.M. (1992) 'Direct evidence for pressure-generated closure of plasmodesmata', *Plant Journal* **2**, 741–750.

Oparka, K.J., Prior, D.A.M. and Crawford, J.W. (1994) 'Behaviour of plasma membrane, cortical ER and plasmodesmata during plasmolysis of onion epidermal cells', *Plant Cell and Environment* **17**, 163–171.

Oparka, K.J., Roberts, A.G., Boevink, P. *et al.* (1999) 'Simple, but not branched, plasmodesmata allow the nonspecific trafficking of proteins in developing tobacco leaves', *Cell* **97**, 743–754.

Overall, R.L. and Gunning, B.E.S. (1982) 'Intercellular communication in *Azolla* roots: II Electrical coupling', *Protoplasma* **111**, 151–160.

Palevitz, B.A. and Hepler, P.K. (1985) 'Changes in dye coupling of stomatal cells of *Allium* and *Commelina* demonstrated by microinjection of lucifer yellow', *Planta*, **164**, 473–479.

Radford, J.E., Vesk, M. and Overall, R.L. (1998) 'Callose deposition at plasmodesmata', *Protoplasma* **201**, 30–37.

Reid, R.J. and Overall, R.L. (1992) 'Intercellular communication in *Chara*: factors affecting transnodal electrical resistance and solute fluxes', *Plant Cell and Environment* **15**, 507–517.

Reid, R.J., Loughman, B.C. and Ratcliffe, R.G. (1985) '$^{31}$P-NMR measurement of inhibitor-induced changes in cytoplasmic pH in maize root tips', *Journal of Experimental Botany* **36**, 889–897.

Rinne, P.L.H. and van der Schoot, C. (1998) 'Symplasmic fields in the tunica of the shoot apical meristem coordinate morphogenetic events', *Development* **125**, 1477–1485.

Robards, A.W. (1971) 'The ultrastructure of plasmodesmata', *Protoplasma* **72**, 315–323.

Roberts, A.G. and Oparka, K.J. (2003) 'Plasmodesmata and the control of symplastic transport', *Plant, Cell and Environment* **26**, 103–124.

Sanders, M.A. and Salisbury, J.L. (1994) 'Centrin plays an essential role in microtubule severing during flagellar excision in *Chlamydomonas reinhardtii*', *Journal of Cell Biology* **124**, 795–805.

Schulz, A. (1999) Physiological control of plasmodesmal gating, in *Plasmodesmata. Structure, Function, Role in Cell Communication* (eds A.E.J. van Bel and W.J.P. van Kesteren), Springer-Verlag, Heidelberg, pp. 173–204.

Shepherd, V.A. and Goodwin, P.B. (1992) 'Seasonal patterns of cell-to-cell communication in *Chara corallina* Klein ex Wild: I Cell-to-cell communication in vegetative lateral branches during winter and spring', *Plant Cell and Environment* **15**, 137–150.

Spanswick, R.M. (1972) 'Electrical coupling between higher plant cells: a direct demonstration of intercellular communication', *Planta* **102**, 215–227.

Spanswick, R.M. and Costerton, J. (1967) 'Plasmodesmata in *Nitella translucens*: structure and electrical resistance', *Journal of Cell Science* **2**, 451–464.

Spray, D.C., Harris, A.L. and Bennet, M.V.L. (1979) 'Voltage dependence of junctional conductance in early amphibian embryos', *Science* **204**, 432–434.

Storms, M.M.H., van der Schoot, C., Prins, M., Kormelink, R., Van Lent, J.W.M. and Goldbach, R.W. (1998) 'A comparison of two methods of microinjection for assessing altered plasmodesmal gating in tissues expressing viral movement proteins', *Plant Journal* **13**, 131–140.

Suzaki, E., Suzaki, T., Jackson, S.L. and Hardham, A.R. (1996) 'Changes in intracellular pH during zoosporogenesis in *Phytophthora cinnamomi', Protoplasma* **191**, 79–83.

Tilney, L.G., Cooke, T.J., Connelly, P.S. and Tilney, M.S. (1990) 'The distribution of plasmodesmata and its relationship to morphogenesis in fern gametophytes', *Development* **110**, 1209–1221.

Trebacz, K., Fensom, D.S., Harris, A.L., Harris, A. and Zawadzki, T. (1988) 'Transnodal transport of $^{14}C$ in *Nitella flexilis*. III Further studies on dissolved inorganic carbon movements in tandem cells', *Journal of Experimental Botany* **39**, 1561–1573.

Tucker, E.B. (1982) 'Translocation in the staminal hairs of *Setcreasea purpurea*. I. A study of cell ultrastructure and cell-to-cell passage of molecular probes', *Protoplasma* **113**, 193–201.

Tucker, E.B. (1988) 'Inositol bisphosphate and inositol trisphosphate inhibit cell-to-cell passage of carboxyfluorescein in staminal hairs of *Setcreasea purpurea*', *Planta* **174**, 358–363.

Tucker, E.B. (1990a) 'Calcium-loaded 1,2-bis (2-aminophenoxy) ethane -N,N,N′,N′-tetraacetic acid blocks cell-to-cell diffusion of carboxyfluorescein in staminal hairs of *Setcreasea purpurea*', *Planta* **182**, 34–38.

Tucker, E.B. (1990b) Analytical studies of dye-coupling between plant cells, in *Parallels in Cell to Cell Junctions in Plants and Animals* (eds A.W. Robards, W.J. Lucas, J.D. Pitts, H.J. Jongsma and D. Spray), Springer-Verlag, Berlin, pp. 239–248.

Tucker, E.B. (1993) 'Azide treatment enhances cell-to-cell diffusion in staminal hairs of *Setcreasea purpurea*', *Protoplasma* **174**, 45–49.

Tucker, E.B. and Boss, W.F. (1996) 'Mastoparan-induced intracellular $Ca^{2+}$ fluxes may regulate cell-to-cell communication in plants', *Plant Physiology* **111**, 459–467.

Tucker, E.B. and Spanswick, R.M. (1985) 'Translocation in the staminal hairs of *Setcreasea purpurea* II. Kinetics of intercellular transport', *Protoplasma* **128**, 167–172.

Tucker, E.B. and Tucker, J.E. (1993) 'Cell-to-cell diffusion selectivity in staminal hairs of *Setcreasea purpurea*', *Protoplasma* **174**, 36–44.

van der Schoot, C., Dietrich, M.A., Storms, M., Verbeke, J.A. and Lucas, W.J. (1995) 'Establishment of a cell-to-cell communication pathway between separate carpels during gynoecium development', *Planta* **195**, 450–455.

van Rijen, H.V.M. and Wilders, R. (1999) Electrical coupling, in *Plasmodesmata. Structure, Function, Role in Cell Communication* (eds A.E.J. van Bel and W.J.P. van Kesteren), Springer-Verlag, Heidelberg, pp. 51–65.

Vorob'ev, L.N., Tarkhanov, K.A. and Vakhmistrov, D.B. (1981) 'Use of the electrical coupling factor for quantitative estimation of symplastic communications', *Soviet Plant Physiology* **28**, 495–502.

White, R.G., Badelt, K., Overall, R.L. and Vesk, M. (1994) 'Actin associated with plasmodesmata', *Protoplasma* **180**, 169–184.

Williams, E.J. and Fensom, D.S. (1975) 'Axial and transnodal movement of $^{14}C$, $^{22}Na$ and $^{36}Cl$ in *Nitella translucens*', *Journal of Experimental Botany* **26**, 783–807.

Wu, X.L., Dinneny, J.R., Crawford, K.M. *et al.* (2003) 'Modes of intercellular transcription factor movement in the *Arabidopsis* apex', *Development* **130**, 3735–3745.

Wymer, C.L., Fernandez-Abalos, J.M. and Doonan, J.H. (2001) 'Microinjection reveals cell-to-cell movement of green fluorescent protein in cells of maize coleoptiles', *Planta* **212**, 692–695.

Yahalom, A., Lando, R., Katz, A. and Epel, B.L. (1998) 'A calcium-dependent protein kinase is associated with maize mesocotyl plasmodesmata', *Journal of Plant Physiology* **153**, 354–362.

Zhang, W.H. and Tyerman, S.D. (1991) 'Effect of low $O_2$ concentration and azide on hydraulic conductivity and osmotic volume of the cortical cells of wheat roots', *Australian Journal of Plant Physiology* **18**, 603–613.

Zhang, W.H. and Tyerman, S.D. (1997) 'Effect of low oxygen concentration of the electrical properties of cortical cells of wheat roots', *Journal of Plant Physiology* **150**, 567–572.

# Appendix

## 13.6 Solving the electric circuit model of a sister-cell pair

### *13.6.1 Model for cytoplasmic location of current-injecting electrode in a sister-cell pair*

The circuit model for cytoplasmic impalements, as shown in Figure 13.1a, may be solved algebraically to yield the resistances of the PM of each cell and of the junction between them:

*Definition of terms:*

$I$ amplitude of the injected current pulse
$\Delta V_1$ change in the membrane potential ($V_1$) in the injected cell (cell A)
$\Delta V_2$ change in the membrane potential ($V_2$) in the receiving cell (cell B)
$i_1$ current that passes to ground across the PM of cell A
$i_2$ current that passes through plasmodesmata to cell B and then to the ground
$R_{mA}$ PM resistance of cell A
$R_{mB}$ PM resistance of cell B
$R_j$ resistance of cell-to-cell junction
Surface area ratio = PM area cell A / PM surface area cell B

*Assumptions:*

(1) The resistivity ($\rho$) of the PM of cells A and B are the same:

$$\text{PM resistance} = \frac{\rho}{\text{membrane surface area}}$$

Thus,

$$\mathrm{R}_{mA} = \frac{\rho}{\text{Area } A} \qquad R_{mB} = \frac{\rho}{\text{Area } B} \quad \text{and} \quad \alpha = \frac{\text{Area } A}{\text{Area } B} = \frac{R_{mB}}{R_{mA}}$$

for simplicity, let $R_{mA} = R_m$ then $R_{mB} = \alpha R_m$

(2) All the current traversing the cell-to-cell junction travels through plasmodesmata i.e. $R_j = R_{plasmodesmata}$.

(3) Resistances of the extracellular medium and the cytoplasm of each cell are so small as to be negligible.

Application of Ohm's law ( $\Delta V = IR$) and Kirchhoff's junction and loop rules to the circuit model yields the following equations:

(i) $I = i_1 + i_2$ by Kirchhoff's junction rule
(ii) $\Delta V_1 = i_1 R_m$ by Ohm's law
(iii) $\Delta V_1 = i_2(R_j + \alpha R_m)$ by Ohm's law and Kirchhoff's loop rule
(iv) $\Delta V_2 = i_2 \alpha R_m$ by Ohm's law

*Finding the current passing through the cell–cell junction ($i_2$) in terms of $I$, $\Delta V_1$ and $\Delta V_2$*

rearranging (ii), we get (v) $i_1 = \dfrac{\Delta V_1}{R_m}$

rearranging (iv), we get (vi) $i_2 = \dfrac{\Delta V_2}{\alpha R_m}$

dividing (v) by (vi), we get (vii) $\dfrac{i_1}{i_2} = \dfrac{\Delta V_1}{R_m}\dfrac{\alpha R_m}{\Delta V_2} = \dfrac{\alpha \Delta V_1}{\Delta V_2}$

and rearranging (vii), we get (viii) $i_1 = \dfrac{\alpha\Delta V_1 i_2}{\Delta V_2}$

Substituting (viii) in (i)

$$I = \frac{\alpha\Delta V_1 i_2}{\Delta V_2} + i_2 = i_2\left[1 + \frac{\alpha\Delta V_1}{\Delta V_2}\right] = i_2\frac{\alpha\Delta V_1 + \Delta V_2}{\Delta V_2}$$

and rearranging, we get (ix) $i_2 = \dfrac{I\Delta V_2}{\alpha\Delta V_1 + \Delta V_2}$

*Solving for the PM resistance of cell A ($R_m$)*
Adding (ii) to (iv) we get (x) $\Delta V_1 + \Delta V_2 = R_m(i_1 + \alpha i_2)$

substituting (i) in (x) $\Delta V_1 + \Delta V_2 = R_m\,[I + (\alpha - 1)i_2]$

substituting (ix) in (x) $\Delta V_1 + \Delta V_2 = R_m I\left[1 + \dfrac{(\alpha - 1)\Delta V_2}{\alpha\Delta V_1 + \Delta V_2}\right]$

$$\Delta V_1 + \Delta V_2 = R_m I\frac{\alpha\Delta V_1 + \Delta V_2 + \alpha\Delta V_2 - \Delta V_2}{\alpha\Delta V_1 + \Delta V_2}$$

$$\Delta V_1 + \Delta V_2 = R_m \alpha I\frac{\Delta V_1 + \Delta V_2}{\alpha\Delta V_1 + \Delta V_2}$$

rearranging

$$\mathbf{R_m = \frac{\alpha\Delta V_1 + \Delta V_2}{\alpha I}} \quad (1)$$

*Solving for the electrical resistance of the cell–cell junction ($R_j$)*
Subtracting (iv) from (iii) we get (xi) $\Delta V_1 - \Delta V_2 = i_2 R_j$

rearranging (xi), we get (xii) $R_j = \dfrac{\Delta V_1 - \Delta V_2}{i_2}$

substituting (ix) in (xii)

$$\mathbf{R_j = \frac{(\Delta V_1 - \Delta V_2)(\alpha\Delta V_1 + \Delta V_2)}{I\Delta V_2}} \quad (2)$$

#### *13.6.1.1 Model for vacuolar location of electrode tip in a sister-cell pair*

When the tip of the current injecting electrode is in the vacuole, an extra unknown quantity, the resistance of the tonoplast ($R_t$), is present in the model (see Figure 13.1b). Only three quantities, $I$, $V_1$ and $V_2$ are measured in the two-cell system, and these quantities are used to solve for $i_2$, $R_m$ and $R_j$ in the cytoplasmic case. In the case of a vacuolar impalement, there are four unknown quantities and only three measurable ones; therefore, it is not possible to solve for $R_m$ and $R_j$ in this situation. The best that can be done is to measure the 'cell resistance' and the coupling ratio.
Applying Ohm's law and Kirchhoff's rules to the vacuolar case yields:

(i) $I = i_1 + i_2$
  (a) $\Delta V_1 = IR_t + i_1 R_m$;
  (b) $\Delta V_1 = IR_t + i_2(R_j + \alpha R_m)$;
  (c) $\Delta V_2 = i_2\,\alpha R_m$

It so happens that when (a) and (c) are substituted into Equation 2 (the solution for $R_m$ in cytoplasmic impalements), what is gained is not, in fact, $R_m$, but ($R_m + R_t$) for the current-injected cell. So in the vacuolar case,

$$R_{cell} = R_m + R_t = \frac{\alpha\Delta V_1 + \Delta V_2}{\alpha I} \quad (3)$$

The coupling ratio, which is the ratio of change in the membrane potential of cell B to that in cell A, is, in terms of the surface area ratio and the PM, tonoplast and junctional resistances:

$$\frac{\Delta V_2}{\Delta V_1} = \frac{\alpha R_m^2}{R_t R_j + (\alpha + 1)R_t R_m + R_m R_j + \alpha R_m^2}$$

It is helpful to the interpretation of the coupling ratio to look at the limiting cases;
when $R_j \rightarrow \infty$ (i.e. plasmodesmata are all closed)

$$\text{then} \quad \frac{\Delta V_2}{\Delta V_1} \rightarrow 0$$

when $R_j \rightarrow 0$ (i.e. electrodes in the same cell)

$$\text{then} \quad \frac{\Delta V_2}{\Delta V_1} \rightarrow \frac{\alpha R_m}{(\alpha + 1)R_t + \alpha R_m}$$

if $R_j \rightarrow 0$ and $R_t = 0$ (i.e. actually in the cytoplasm of the same cell)

$$\text{then} \quad \frac{\Delta V_2}{\Delta V_1} \rightarrow \frac{\alpha\, R_m}{\alpha\, R_m} = 1$$

## *13.6.2 Solution of the circuit model for a single-ended file*

The equations for the circuit model for cytoplasmic current injections into the tip cell of an 'infinite' file of cells (Figure 13.2) may be solved algebraically to yield the resistance of the plasmodesmata-containing junction between cells. The length of file required ($n$ cells) for the model to be valid was found by calculating the resistance between ground and the $n$th cell and ground and the ($n$-1)th cell for increasing file lengths. Interestingly, if junctional resistance ($R_j$) and membrane resistance ($R_m$) have values of 100 MΩ, then cell files of four cells or more are, for the purposes of the model, equivalent to an 'infinite' single-ended file [as the difference in resistance between ground and the $n$th cell and ground and the $(n - 1)$th cell is less than 1%].

*Definition of terms:*

| | |
|---|---|
| $I$ | current pulse injected |
| $\Delta V_1$ | change in the membrane potential ($V_1$) in the injected terminal cell (cell A) |
| $\Delta V_2$ | change in the membrane potential ($V_2$) in the neighbouring cell (cell B) |
| $i_1$ | current that passes through plasmodesmata to cell B |
| $i_2$ | current that passes to ground across the PM of cell A |
| $i_3$ | current that passes to ground across the PM of cell B |
| $i_4$ | current that passes through plasmodesmata to cell C |
| $R_m$ | PM resistance of every cell in the file |
| $R_j$ | resistance of every cell-to-cell junction |

*Assumptions:*

(1) All cells in the file have the same PM resistance.
(2) The file is single-ended and infinite in length.

Application of Ohm's law ($\Delta V = IR$) to the circuit model yields the following equations:

(i) $I = i_1 + i_2$; (ii) $\Delta V_1 = i_2 R_m$
(iii) $\Delta V_1 = i_1 R_j + i_3 R_m$; (iv) $\Delta V_2 = i_3 R_m$

by conservation of current, we get (v) $i_1 = i_3 + i_4$
from the assumption that the file is single-ended and 'infinite' in length,

(vi) $\dfrac{i_2}{i_1} = \dfrac{i_3}{i_4}$

*Finding $i_3$ in terms of $i_1$ and $i_2$*

use (v) in (vi) $\dfrac{i_2}{i_1} = \dfrac{i_3}{i_1 - i_3}$

$$i_3 = \frac{i_2[i_1 - i_3]}{i_1}$$

$$i_3\left(1 + \frac{i_2}{i_1}\right) = i_2$$

Finally, (vii) $i_3 = \dfrac{i_1 i_2}{i_1 + i_2}$

*Solving for $R_j$*

substitute (iv) in (iii) $\Delta V_1 = i_1 R_j + \Delta V_2$

(viii) $R_j = \dfrac{\Delta V_1 - \Delta V_2}{i_1}$

now, dividing (ii) by (iv), we get (ix) $\dfrac{\Delta V_1}{\Delta V_2} = \dfrac{i_2}{i_3}$

substituting (vii) in (ix), we get (x) $\dfrac{\Delta V_1}{\Delta V_2} = \dfrac{i_2[i_1 + i_2]}{i_1 i_2} = \dfrac{i_1 + i_2}{i_1}$

substituting (i) in (x) $\dfrac{\Delta V_1}{\Delta V_2} = \dfrac{I}{i_1}$

and (xi) $i_1 = \dfrac{I \Delta V_2}{\Delta V_1}$

substituting (xii) in (vii)

$$R_j = \frac{V_1[\Delta V_1 - \Delta V_2]}{I V_2}$$

# Index

2,3-butanedione monoxime, 85
5(6) carboxyfluorescein, 245

$\beta$-1, 3-glucanases, 15–16

ABC model, 118
acellular macroalgae, *see* algae, macro-, acellular
acid phosphatase, 11, 17
acid shock, 291
actin, 3–4, 9, 11, 13–14, 57, 60–61, 73, 85, 97, 107, 189, 199, 254, 271, 291–2
actin regulating proteins, 11
agroinfiltration, 174, 220, 248, 250
agro-inoculation, 250
algae, 18, 33–50
algae, brown, 34, 38, 42, 49; *see* also *Chara*, macroalgae, Chlorophyta
algae, green, 33–5, 38–9, 44, 49
algae, macro-, acellular, 64
algae, red, 33, 37–8, 44–6, 48–9
algae, unicellular, 36, 48
aluminium toxicity, 15
anaerobiosis, 288, 290
antioxidant defense, 170
*Antirrhinum*, 118–20
AP2/EREBP, 114
APETALA1, 116, 121–2, 126
APETALA2/ethylene responsive element binding protein, 114
apical meristem, *see* meristem, apical
apomorphic condition, 34
apoplasm, 42, 45, 49, 137–8, 154, 178
apoplasmic loading, 138, 140–43, 153
apoplasmic transport, 135, 154
aquaporin proteins, 136
*Arabidopsis*, 17–18, 20, 23, 37, 61, 91, 98–101, 103, 113–23, 125, 127, 143–4, 149, 152, 169, 178, 181, 215–17, 220, 222, 225, 230, 245, 250, 252–3, 257, 265, 281, 285, 287, 290
ARGONAUTE1 (AGO1), 106, 215
asynchronously dividing cells, 286
AtISE, 58
atomic force microscopy, *see* microscopy, atomic force
ATPase, 11, 14, 139, 155, 201, 203
autophagy, 146
autoradiography, 75, 138
azide, 253, 283, 288–9
*Azolla*, 264, 270, 282–3, 285

*Beet yellows virus* (BYV), 203, 219
bHLH (basic helix-loop-helix), 114
bi-arsenic, 254
biolistic bombardment, 95, 248–52
biotinylated proteins, 258
*brassinosteroid hormone receptor*, 177
brefeldin A (BFA), 202
brown algae, *see* algae, brown; *Chara*; macroalgae; Chlorophyta
bundle sheath, 16, 21, 75, 79–80, 82, 98, 136–7, 139–41, 145, 153
butyric acid, 287–8, 290–91

$C_3$ plants, 75
$C_4$ plants, 76–7, 79
$Ca^{2+}$-dependent protein kinase, 292
$Ca^{2+}$-loaded BAPTA buffer, 287
cadmium-induced protein, 226
*Caenorhabditis elegans*, 214, 216, 225–7
caged fluorescein, 141, 246
caged probes, 245–6, 252
calcium, 11, 14–15, 85, 107, 152, 269, 271–3, 285, 287–8
calcium channels, 271
calcium wave, 272
callose, 2, 4, 11, 14–16, 21, 73, 85, 102, 146, 192–3, 195, 226, 241, 282, 293

callus formation, 91
calreticulin, 11, 14, 107, 152, 202
cap membrane, 45–6
CAPRICE (CPC), 54, 62, 67, 107, 123, 125–6
carboxyfluorescein, 147, 149, 152–4, 245
cargo protein, 55–6
Casparian strips, 137
*Cauliflower mosaic virus*, 16, 195, 203, 250
cell-specific promoters, 250–51
central cavity, 3, 7, 19–20
central rod, 3–4, 6–8
centrin, 11, 14, 85, 107, 287, 291, 292
channels, calcium, *see* calcium channels
channels, voltage-dependent ion, *see* voltage-dependent ion channels
*Chara*, 40, 73, 264–5, 271, 286
*Chara corralina*, 17, 39, 286
chimeras, 22, 119–20, 145
Chlorophyta, 34–7, 39, 41, 47–8
*Cm*HSC70-1,2, 54, 58, 64
*Cm*NCAPP1, 58
*Cm*PP16, 54, 58, 64–5, 107, 170, 173, 200
*Cm*PP2, 54
*Cm*PP36, 54, 63–5, 167
collection phloem, *see* phloem, collection
collenchyma sheath, 271
Comoviruses, 201, 203
companion cell, 9, 22, 34, 76, 98, 125, 127, 136, 139–40, 145–7, 162, 164–5, 167–70, 172, 175, 178, 180–82, 250–51, 266–8, 273
confocal microscopy, *see* microscopy, confocal
connexin, 11, 17, 255
connexin-like proteins, 11, 17
*CONSTANS*, 127
contractile protein, 291
cortical endoplasmic reticulum, *see* endoplasmic reticulum, cortical
cotton fibre cells, 91
cotton fibres, 21
cotyledons, 101, 103, 154
coupling ratio, 264, 283, 285, 289–90, 299–300
*Cowpea mosaic virus* (CPMV), 201
*Cucumber mosaic virus (CMV)*, 15, 58, 169, 196, 219, 223, 251
current injection, 264, 266
cyanide, 283
cyanobacteria, 33, 37, 40
Cystatin, 54
cytochalasin, 9, 13, 189
cytochalasin D, 189
cytokinesis, 6, 62, 87, 93, 202
cytoplasmic annulus, 3, 73, 82–3, 280–82, 287, 289, 291–2
cytoplasmic pH, 290
cytoplasmic sleeve, 3–4, 7–9, 14, 23, 87, 145, 279–81
cytoplasmic streaming, 3, 9, 47, 50, 242
cytoskeletal proteins, 10, 189
cytoskeleton, 9–10, 13, 57, 59–62, 164, 193, 195, 199, 228
cytosolic esterase, 245

DEAD-box RNA Helicase, 105
DEFICIENS (DEF), 54, 107, 118–20
dephosphorylation, 152, 291–3
deplasmolysis, 15, 288–9, 291
depolarization, 264, 270–71
desmotubule, 2–4, 6–11, 14, 19, 20, 23, 40, 42, 45, 49, 78, 80–85, 87, 141, 146, 150, 244, 280–82, 285, 289, 291–2
diaminobenzidine, 256
Dicer, 173, 213–16, 218–19, 223
dicer-like enzymes, 213–15
diffusional transport, 138, 291–3
Diphtheria toxin, 257
dissociation constant ($pK_a$), 245
divalent cations, 287
DNA, 94–6, 114, 127, 171, 173–4, 181, 189, 201, 204, 213–14, 217, 219–20, 223, 225, 228, 230, 249–50, 258
DNA methylation, 213–14, 217, 225, 230
DnaJ proteins, 202–203
docking complex, 55, 57, 59
dormancy, 15, 91, 154
double-barrelled microelectrodes, *see* microelectrodes, double-barrelled
double-stranded (ds)-RNA, 212
Drosha, 214
*DsRed*, 247–8
dual voltage clamp system, 264
dye coupling, 16, 264, 266
dye loading, 245–6, 257

early flowering mutants, 99
electrical conductance, 263–5, 270, 272, 279–80, 283
electrical coupling, 15, 263–7, 269–70, 272, 274, 283, 287, 289–90
electrical discontinuity, 265, 267
electrical potential waves, 266, 269–70
electrical resistance, 263–4, 279–82, 290
electrical signaling, 263, 266, 269, 273–4, 290
electrical signals, 263, 270–71, 273, 282
electron microscopy, *see* microscopy, electron
electropotential waves, 269–71, 273–4
embryo-defective phenotypes, 104
embryogenesis, 61, 98, 100–104, 108
embryo-lethal lines, 104
endodermal cells, 105, 122–3, 125
endodermal fate, 122, 124
endodermis, 54, 122–5, 137, 140, 151–2, 251
endodermoid sheath, 77
endomembrane system, 7, 9, 18, 202, 280
endoplasmic reticulum, 2–3, 6–10, 14, 19–20, 22, 40, 54–5, 57–60, 62, 65, 78–85, 87, 97, 106, 140–41, 146–7, 152, 164, 170, 180, 189, 191–5, 198–202, 228, 244, 248, 254, 269, 280, 285, 291
endoplasmic reticulum, cortical, 2, 6, 9, 14, 22, 191–2, 200, 244
endosymbiosis, 38
enhanced GFP (eGFP), 252–3
ester loading, 245
ethanol-regulated elements, 258
evolution of plasmodesmata, *see* plasmodesmata, evolution of

FITC(F)-dextrans, 101
FITC-conjugated peptides, 243
FlAsH, 254–7
floral meristem, *see* meristem, floral
floral morphogenesis, 99
FLORICAULA (FLO), 107, 120–21
florigen, 127
flower development, 106
fluorescein, 9, 135, 141, 149, 245–6, 253–6
fluorescence recovery after photobleaching (FRAP), 147, 228, 253–4, 283
fluorescence resonance energy transfer (FRET), 17, 204
fluorescent probes, 9, 74, 169, 241–2, 245, 281, 289
fluorescent protein, green (GFP), 8, 10, 13, 18, 63–4, 66–7, 90, 94–8, 101–103, 107, 114, 116–18, 121–7, 129, 135, 146–7, 149–51, 153, 155, 169, 172, 174–6, 188, 191–2, 198, 200–202, 220, 222–4, 226–7, 244, 246–54, 280–81
fluorescent protein, kindling (KFP), 252–3
fluorescent protein, red, monomeric, 247–8
fluorescent proteins, red, 247
fluorescent reporter proteins, 246–7
fluorophores, 245–6, 254, 256, 258–9
FRET, *see* fluorescence resonance energy transfer
functional genomics, 23
FURA-2 dextran, 289

galactinol synthase, 137, 139–40, 153
gallinstan expansion 'femtosyringe' (GEF), 242
gametophyte, 34, 43, 45
gap junctions, 1, 37, 48, 241, 246
gating, 62, 73, 117, 198, 244, 266–7, 270, 274, 290
gene guns, 248
gene silencing signals, 90
genetic screen, 108
genetics, 104
GFP, *see* fluorescent protein, green (GFP)
GFP–cDNA fusion, 67
GFP–GFP fusion, 121
GFP–KN1 fusion, 117
GFP–LFY fusion, 121
GLABRA2 (GL2), 125–6, 251
GLOBOSA (GLO), 54, 107, 118–20
Glutaredoxin, 54
Golgi, 10, 20, 22, 59, 146, 192–3, 202
Golgi apparatus, 146
Golgi vesicles, 22
graft unions, 22
grafting, 105, 127–8, 146, 174, 177, 179, 181, 183, 219, 224
*Grapevine fanleaf virus* (GFLV), 195, 202
green fluorescent protein (GFP), *see* fluorescent protein, green (GFP)

guard cells, 21, 34, 91, 135, 174, 226–8, 244
gymnosperms, 140, 143, 155
gynoecium, 286

Hc-Pro, 219, 223, 225, 230
helicase, RNA, *see* RNA helicase
helicases, 191, 198, 201
heterocyst, 40
heterozygotes, 104
homeobox proteins, 115
homeotic genes, 119–20
Hordeiviruses, 190–91, 195
host-parasite contacts, 22
HPTS, 95, 98–103, 245
HSC70, 64–6
H-shaped intermediate, 19
HSP70, 54–5, 57, 59–62, 64–6, 202–203
hydrostatic pressure, 136
hypersensitive response, 192
hypoxia, 288

INCREASED SIZE EXCLUSION LIMIT 1, 57, 61–2, 105
inner central zone, 91, 100
inositol polyphosphates, 287
intercellular junctions, 282
intercellular resistance, 282, 292
intermediary-type companion cells, 140
ionic strength, 280
iontophoresis, 266, 281, 287–90
iontophoretic injection, 243–4, 265, 283, 287, 289–90

JA biosynthesis, *see* jasmonic acid biosynthesis
jasmonic acid biosynthesis, 179–81, 183
jasmonic acid (JA), 178–83, 272–3
juvenile plasmodesmata, *see* plasmodesmata, juvenile

Kaede protein, 252
kinase, 1, 17, 57, 178, 196, 204–205, 273, 292
kinase, protein, *see* protein kinase/s
kinase, protein, $Ca^{2+}$-dependent, *see* $Ca^{2+}$-dependent protein kinase
kinases, pumpkin, *see* pumpkin kinases
kindling fluorescent protein (KFP), *see* fluorescent protein, kindling
kinesin, 57, 60
KNAT1, 107, 118
KNOLLE, 10, 202
KNOTTED 1 (KN1), 58, 60–61, 63–5, 106–107, 113, 115–19, 121, 127–9, 171, 173, 198, 244, 248
Kranz mesophyll, *see* mesophyll, Kranz

laser ablation, 257
latrunculin B, 199
leaf-blade bundles, 77
leaf-dip, 281
leaf primordia, 98–9, 115–16, 222
LEAFY (LFY), 54, 67, 106–107, 116, 120–22, 125, 129
leaves, 2, 16, 20, 73–6, 84, 86, 94–5, 97–100, 105–106, 115–17, 127, 137–8, 142–4, 148–9, 152–6, 163, 165–6, 168, 174, 177, 219–20, 224, 226–7, 243, 245, 247, 249–50, 252, 263, 265, 270, 273
lipid-based signals, 181, 183
lipids, 9, 147, 280
loading pathway, 73, 143, 155
local silencing, 212, 220, 223–4
long-distance signals, 166, 170, 175–6, 180–82, 270, 272
long-distance transport, 42, 65, 105, 136, 138, 155–6, 162, 164–6, 170, 172–5, 182, 224
Lucifer Yellow CH, 243–4

macromolecular cargo, 292
macromolecular signals, 104–105, 107–108
macromolecular trafficking, 95, 162, 165–7, 169, 171, 175–6, 180, 182–3, 228, 272–3, 281, 291, 293
macromolecules, 39–40, 47, 49, 53, 62, 67, 90, 94–5, 101–6, 113, 162, 164–5, 167, 170–71, 182–3, 189, 204–205, 243, 259, 273, 279–81, 292–3
MADS box proteins, 118
*Maize dwarf mosaic virus*, 15
mannitol, 73, 150, 152, 288
mass flow, 45–6, 48, 138, 144, 272–3
mastoporan, 287–8

*Medicago truncculata*, 176
membrane depolarization, 269
membrane potential mapping, 265
membrane resistance, 264, 283, 289, 300
meristem, 1, 2, 15, 20, 54, 70, 91, 94–5, 98–100, 102, 106, 113, 115–16, 118–22, 127, 129, 148, 152, 155, 163, 173, 176, 227, 244–5, 286
meristem, apical, 1–2, 94–5, 98, 100, 115–16, 122, 148, 152, 163, 166, 244–5, 286
meristem, apical, shoot, 94–5, 99–100, 108, 115–17, 244–5, 286
meristem, floral, 120–21
meristem surveillance mechanism, 227
meristematic cells, 20, 90, 227, 251, 257
mesophyll, 5, 17–18, 20–21, 73, 75–9, 82–3, 86, 97, 115–19, 136–40, 142, 144–5, 148, 153, 167, 169, 174–5, 182, 226, 243, 263
mesophyll, Kranz, 76–7, 79, 82–3
mestome sheath, 75–8, 80–86
metabolic inhibition, 283, 287–90, 292
metaphloem, *see* phloem, meta-
microelectrode, 263, 265
microelectrodes, double-barrelled, 265
microinjection, 54, 63, 65, 74, 117, 128, 167, 169–70, 174, 183, 227, 229, 241–4, 246, 248–50, 252, 258–9
microRNAs, 106, 127–8, 175, 182, 213–16, 222, 230
microscopy, atomic force, 8
microscopy, confocal, 99, 101
microscopy, electron, 6, 8, 39–40, 117, 137, 150, 255–6, 279, 290
microtubule, 6–7, 13, 58, 60–61, 195–6
middle lamella, 7, 19–20, 22, 76–87
mid-torpedo stage, 101–104
*Mimosa*, 268, 270–71
molecular chaperones, 170
monocotyledons, 75
monomeric red fluorescent protein (mRFP), *see* fluorescent protein, red, monomeric
morphogenesis, 20, 90, 92–3, 98–9, 106, 286
morphogenic field, 286
mosaic analysis, 114
Mouse ears (Me), 105, 128, 173
MPB2C, 200
MPI7, 17
mRNA, 10, 47, 49, 54, 64, 90, 105–107, 113–14, 116–18, 121–2, 124, 126, 128–9, 172–3, 180, 215–17, 222–3, 225, 229, 244, 272
mRNA trafficking, 10, 117, 229
multicolour imaging, 258
Münch's mass-flow mechanism, 138
MYB, 114, 127
myosin, 4, 9, 11–14, 60, 73, 85, 107, 152, 291–2
myosin VIII, 12, 14, 152, 291–92

NAC-domain, 173
NCAP pathway, 168
NCAPP1, 57–8, 62, 105–106, 129, 170–71, 200
near-infrared (NIR), 257
Nepoviruses, 201, 203, 218
*Nicotiana clevelandii*, 17
*Nitella*, 40, 263
*Nitella translucens*, 263
nodule primordia, 176
non-cell autonomous proteins, 53–4, 57–8, 60–67, 136, 152, 155, 168, 170
nontargeted movement, 96–7, 121, 128–9
NtMPB2C, 58, 60, 62
nuclear localization signal, 64, 66, 97
nuclear targeting, 122
nucleic acid binding domain, 114
nucleic acids, 113, 136, 166, 170–72, 174, 212, 227–8
nucleoprotein complex, 172
nucleus, 38, 54–5, 97, 122, 125, 129, 136, 164, 201, 217, 228–9, 242, 249, 252

oat coleoptiles, 283
oil/water partition coefficient, 245
oligomerisation, 247–8
O*s*RBB13–1, 54, 64–5
oxygen deficiency, 289–90

patch-clamping, 284–5
pathogen-related proteins, 273
PCMBS, 141, 143
pectin, 12, 16–17, 196
pectin methylesterase, 16

permanent sinks, 148, 150–51, 154
peroxidase, 12, 17
*PHABULOSA*, 106
phalloidin, 13
*PHAVULOTA*, 106
phloem, 91, 98–9, 107, 125, 127–8, 130, 136–56, 162–83, 200, 202, 221, 224, 226–8, 245–6, 253, 257, 265–8, 271–4, 287
phloem, collection, 136, 165
phloem exudates, 138, 146
phloem lectin, 146, 150
phloem loading, 16, 21, 75, 86, 136–44, 153, 155, 165, 170, 175, 177, 246, 287
phloem, meta-, 148, 152–4
phloem-mobile dyes, 245
phloem parenchyma, 139, 141, 145, 150, 156, 168, 177–8, 180, 266, 271, 273
phloem, primary, 148
phloem protein, 54, 58, 146, 149–50, 166–7, 169–71, 182
phloem, proto-, 145, 148, 150, 152–3, 245, 251
phloem, proto-, root, 145
phloem, release, 148, 165, 175
phloem sap, 65, 128, 136, 147, 165–7, 170–71, 173–5, 183, 200, 227–8, 272
phloem sap proteins, 54
phloem, secondary, 148
phloem translocation, 165, 167–9, 173, 179–80, 182
phloem, transport, 54, 136–8, 147–8, 156, 165, 266–8
phloem, wound, 145, 148
phosphorylated movement protein, 194, 196–7
phosphorylation, 12, 14, 54, 57, 64, 67, 85, 195–8, 200, 204, 291–2
photoactivatable GFP (PA-GFP), 249, 251–3
photoactivatable proteins, 252
photo-oxidation, 256
phragmoplast, 2, 58
phragmoplastin, 8
pin-genes, 272
pit connections, 46
pit plugs, 33, 36–38, 43–6, 48–50
pits, 21, 81, 84
plasma membrane, 2–10, 12, 14, 19, 22–3, 43, 45–6, 56–7, 59, 74, 78, 84–5, 87, 100–101, 135–6, 138–9, 141–2, 146–7, 155–6, 164–5, 178, 189, 194–5, 201–203, 242, 263–4, 267–9, 280, 284
plasma membrane transporters, 135
plasmodesmal gating, 266, 274
plasmodesmal receptor, 57–8, 62, 129
plasmodesmata, branched, 8, 18–22, 76, 93, 108, 139–40, 146, 153, 164, 267–8, 271
plasmodesmata, evolution of, 33, 39, 43
plasmodesmata, juvenile, 87
plasmodesmata, primary, 2, 21, 93, 102, 144
plasmodesmata, secondary, 2, 6, 18–19, 21–2, 87, 93–4, 102, 108, 144–5, 281, 285
plasmodesmata, simple, 2, 7, 18–20, 76, 153
plasmodesmata, specialized, 22, 87, 136, 138, 142, 164, 167, 182, 244, 251, 285
plasmodesmata, truncated, 265
plasmodesmatal frequency, 2, 6, 142–3
plasmodesmograms, 152
plasmolysis, 9, 15, 85, 138, 246, 285, 287–9, 291
plastids, 38, 40, 49, 55, 164, 180
*Poa semilatent virus*, 193
pollen, 269, 274
pollen tube, 269
polymer-trap loading, 142
polyol-translocating plants, 143
pore-plasmodesma units (PPUs), 22, 144–6, 149, 164, 268, 272–3
porins, 46
post-transcriptional gene silencing, 53, 166, 173–5, 182–3, 212
*Potato mop top virus*, 193
potato spindle tuber viroid, 153
*Potato virus X*, 15, 191, 219
Potexviruses, 190, 195
P-proteins, 166–7, 170
pre-phloem transport, 136–9, 144, 155
pressure differentials, 244, 287–9
pressure injection, 242–4, 281, 289
pressure-flow, 136, 147, 165

pressure-valve mechanism, 141, 143, 147, 156
primary phloem, *see* phloem, primary
primary pit fields, 18, 76, 80, 83
primary plasmodesmata, *see* plasmodesmata, primary
primary wall, 19, 78
PRms, 12, 18
propidium iodide, 257
prosystemin, 177–80, 272–3
protein kinase CK2, 204–205
protein kinase/s, 12, 14, 17, 170, 178, 196, 198, 204–205, 273, 292–3
proteinase inhibitor, 272
protein–protein interaction domain, 114
proton-motive force, 267–8
proton pumps, 135, 139, 141, 151, 267–8
protophloem, *see* phloem, proto-
protoplast fusion, 22
pulvinus, 270–71
pumpkin kinases, 54
PYROPHOSPHATE-DEPENDENT PHOSPHOFRUCTOKINASE (PFP), 128, 173

quantum dots, 258

Rab11, 10
raffinose, 139–40
raffinose-family oligosaccharides, 139–40
rays, 155
ReAsH, 254–6
receptors, 12, 16, 18, 23, 53, 55–9, 62, 65–6, 113, 129, 177–81, 183, 194, 201, 226, 272
red algae, *see* algae, red
red fluorescent proteins, *see* fluorescent proteins, red
release phloem, *see* phloem, release
RGD-binding motifs, 202
rhizoid, 41
RISC complex, 106
RNA, 1, 18, 39, 53–4, 58–9, 65, 67, 105–107, 117, 128, 136, 153, 168, 170–75, 178, 182–3, 189–92, 195–7, 204, 212–23, 225–30, 242, 246
RNA binding proteins, 53, 171, 174, 182, 227, 229
RNA-dependent RNA polymerase 1 (RdRP1), 197
RNA helicase, 105, 217
RNA helicase, DEAD-box, *see* DEAD-box RNA helicase
RNA-induced silencing complex, 106, 173, 214–16, 221, 223–5
RNA silencing, 106, 197, 212–14, 217–20, 222–3, 225–6, 228–9
RNA surveillance, 173
RNA transport, 53–4, 65, 105, 172, 229
RNB1, 54
root, 2, 7, 13, 17, 20, 47, 54, 61–2, 91, 100, 105, 107, 110, 122–6, 135, 145, 148–52, 163, 165, 228, 245, 251, 256–7, 264, 282–3, 285, 287, 289–90
root hair cells, 91
root hairs, 61, 125–6, 135, 265, 285, 287, 290
root protophloem, *see* phloem, proto-, root
RPP13-1, 65, 107, 167
RTM1 protein, 12

salicylic acid, 180, 181–2
SCARECROW (SCR), 105, 122, 124–6, 251
sclerenchyma sheath, 271
SE–CC complex, 164–5, 167
secondarily modified, 87, 153
secondary phloem, *see* phloem, secondary
secondary plasmodesmata, *see* plasmodesmata, secondary
sensory cells, 269
shoot, 20, 41, 54, 63, 91–2, 94–5, 98–9, 102, 113, 115–16, 118, 122, 124, 127, 148, 150, 152, 155, 163, 165–6, 244–5, 286
shoot apex, 54, 63, 91–2, 98–9, 102, 116, 127, 150
shoot apical meristem, *see* meristem, apical, shoot
short interfering RNAs (siRNAs), 173–5, 213–27, 230
SHORT ROOT (SHR), 54, 67, 105–7, 122–5, 251
SID-1, 226–7
sieve element, 9, 18, 22, 76, 136, 139, 140, 144–6, 151, 155, 162, 168, 242, 244–5, 250–51, 266, 268, 273

sieve plates, 16, 164
sieve pores, 136, 145, 148, 272–4
signal, nuclear localization, *see* nuclear localization signal
signal, targeting, *see* targeting signal
signals, 53, 57, 63, 65–7, 74, 90, 98–100, 104–105, 107–108, 115, 199, 127, 143, 150, 162–3, 166, 170, 172–3, 175–8, 180–83, 195, 212, 222, 227, 229, 263, 270–73, 280–82
signals, electrical, *see* electrical signals
signals, gene silencing, *see* gene silencing signals
signals, lipid-based, *see* lipid-based signals
signals, long-distance, *see* long-distance signals
signals, macromolecular, *see* macromolecular signals
silencing, local, *see* local silencing
silencing, post-transcriptional gene, *see* post-transcriptional gene silencing
silencing, RNA, *see* RNA silencing
silencing, RNA, systemic, *see* systemic RNA silencing
silencing signals, gene, *see* gene silencing signals
silencing suppressor proteins, viral, *see* viral silencing suppressor proteins
silencing suppressor proteins, virus-encoded, *see* virus-encoded silencing suppressor proteins
silencing, systemic, *see* systemic silencing
silencing, transcriptional, *see* transcriptional silencing
silencing, virus-induced, RNA, *see* virus-induced RNA silencing
simple plasmodesmata, *see* plasmodesmata, simple
sink-source transition, 8, 20–21, 74, 93, 108, 145, 153–4, 165
size exclusion limit, 97, 117, 135, 168, 227, 229, 241, 279
*SNARE proteins*, 10, 201
SNAREs, 10
sodium azide, 253
solute loading, 135–6, 139
solute unloading pathways, 151
specialized plasmodesmata, *see* plasmodesmata, specialized
sphincters, 78, 80, 82–7
spokes, 4, 14, 292
sporophyte, 34, 43–4
stachyose, 139–40
STM, 107, 115, 117–18, 128
Stokes radius, 103, 149, 244, 247–8, 254
storage parenchyma, 154
storage sinks, 148, 151, 153–4
Strasburger cells, 140–41, 144, 155
suberin lamella, 76–84, 86
suberin lamellae, 75, 77–80, 82, 86–7
subliminal infections, 229
SUC2 promoter, 250–51
SUC2 transporter, 144
sucrose, 21, 75, 78, 137–41, 143–7, 149–52, 155–6, 170, 268
sucrose/$H^+$ symport, 138
*sucrose-export-deficient mutant (sxd1)*, 16, 21, 98
SUCROSE TRANSPORTER-1 (SUT1), 172
sugarcane, 77–9, 83–4, 154
sugar retrieval, 155
surveillance system, 197
SUT1, 107, 143, 146, 155, 172
SUT1 transporter, 143
SVR motif, 66
symbioses, 33, 48–9
symplasm, 20, 40, 135–6, 138, 142, 153, 163, 241–2, 268
symplasmic continuity, 5, 15, 21–2, 257
symplasmic domains, 20–21, 40, 90–92, 101, 103, 136, 163, 175–6, 247, 251, 265–7, 269, 274
symplasmic loading, 140, 142–3, 155, 175
synchronously dividing cells, 286
syntaxin, 10
systemic acquired resistance (SAR), 166, 176–7, 180–81, 222
systemic induced resistance, 272
systemic RNA silencing, 173, 213, 225–6, 228
systemic silencing, 136, 174, 212, 214, 219–20, 222–8, 230
systemin, 177–81, 183, 272–3
systemin receptor, 177–8

tannic acid, 4, 80–81, 84
targeted movement, 96–7, 106, 247, 279
targeting signal, 55–6, 59, 61, 63–5
temporary sinks, 148, 152
tetracycline, 258
tetra-cysteine motifs, 254
Thioredoxin h, 54, 63–5, 107, 170
tissue-specific promoters, 250–51, 257
TMV movement protein, 10, 57–8, 60, 97
*tobacco etch virus*, 12, 18, 219
*tobacco mosaic virus*, 10, 15–16, 57–8, 60, 96–7, 105, 170, 174, 181, 193, 195–200, 219, 227, 244, 247, 250, 252, 282, 292
*tobacco rattle virus*, 244
Tobamoviruses, 193, 195, 201, 203–205
tocopherol cyclase, 98
*Tomato bushy stunt virus*, 15, 196, 219, 223
*Tomato mosaic virus (*ToMV*)*, 198, 219
*Tomato yellow leaf curl virus*, 169, 219
Tospoviruses, 201–203
*Tradescantia*, 289
transcription factors, 10, 53–4, 66–7, 74, 113, 143, 198, 215, 244, 248, 251–2, 254
transcriptional silencing, 213–14, 217
transfer cells, 135, 139, 141–3, 154
transfer complex, 55, 57, 59–60
trans-membrane domain, 106
transport phloem, *see* phloem, transport
trichomes, 17–18, 244, 249
triple gene block, 190, 195
truncated plasmodesmata, *see* plasmodesmata, truncated
tuberisation, 154
tubule-forming viruses, 201, 203–204
tubulin, 13–14, 61
tunica, 91, 100
turgor pressure gradient, 165, 286
*Turnip vein clearing virus*, 16

ubiquitin, 12, 16, 20, 63, 170
unicellular algae, *see* algae, unicellular
unloading pathway, 74, 149
UV activation, 246
UV laser, 246

vacuole, 242–3, 265, 284, 299
vascular parenchyma, 16, 21, 75–6, 79–81, 83, 98, 136, 149
vesicle targeting, 10
vesicle trafficking proteins, 12
viral coat protein (CP), 191–5, 202–205
viral proteins, 113, 169, 189, 191, 195, 198, 203–205, 218, 247
viral replication complex (VRC), 194–5, 196–7, 200
viral silencing suppressor proteins, 223
viral vectors, 250
virus-encoded silencing suppressor proteins, 219
virus-induced RNA silencing, 218
virus movement, 17, 58, 65, 105, 188–92, 195–8, 200, 203–5, 229, 244, 248
voltage-dependent ion channels, 271
voltage gating, 290

wall collar, 3–4, 19
WD40 repeat, 18
WD40 repeat-containing protein, 12
wound phloem, *see* phloem, wound
wound potentials, 271
wound response, 145, 166, 177–82
WUSCHEL (WUS), 127

xenobiotics, 153, 245
xylem, 7, 21, 75, 136–7, 154, 162–3

yeast two-hybrid system, 17